T0281549

PRINCIPLES
OF COPULA
THEORY

PRINCIPLES OF COPULA THEORY

FABRIZIO DURANTE

Free University of Bozen-Bolzano
Bozen-Bolzano, Italy

CARLO SEMPI

Università del Salento
Lecce, Italy

CRC Press
Taylor & Francis Group
Boca Raton London New York

CRC Press is an imprint of the
Taylor & Francis Group, an **informa** business

A CHAPMAN & HALL BOOK

Chapman & Hall/CRC Press
Taylor & Francis Group
6000 Broken Sound Parkway NW, Suite 300
Boca Raton, FL 33487-2742

First issued in paperback 2021

© 2016 by Taylor & Francis Group, LLC
CRC Press is an imprint of Taylor & Francis Group, an Informa business

No claim to original U.S. Government works

Version Date: 20150601

ISBN 13: 978-1-03-209847-0 (pbk)
ISBN 13: 978-1-4398-8442-3 (hbk)

Contents

Preface

The official history of copulas begins in 1959 with Sklar [1959]; but, as is often the case in Mathematics, for groundbreaking results there are forerunners and precedents. These latter ones involve asking the right question, proving a partial result, possibly in a special case and so forth. In the case of copulas, Hoeffding [1940, 1941] introduced a concept very close to that of a copula. He constructed what we now call a copula on the square $[-1/2, 1/2] \times [-1/2, 1/2]$ rather than on the unit square and gave some of the properties, including the bounds. Perhaps his papers did not initially attract the attention they deserved because they were published in a journal without a wide circulation during the war (see [Fisher and Sen, 1994]).

Slightly later Fréchet [1951] studied the following problem: given the probability distribution functions (d.f.'s) $F_1, \ldots, F_d$ of d random variables $X_1, \ldots, X_d$ defined on the same probability space, what can be said about the set $\Gamma(F_1, \ldots, F_d)$ of d–dimensional d.f.'s whose marginals are $F_1, \ldots, F_d$? The problem is hence related to the determination of a set of multivariate d.f.'s $\Gamma(F_1, \ldots, F_d)$, now called the *Fréchet class*, given the partial information about the one-dimensional d.f.'s of $F_1, \ldots, F_d$. In no case is the Fréchet class of $F_1, \ldots, F_d$ empty, since one may always construct a multivariate d.f. in $\Gamma(F_1, \ldots, F_d)$ assuming the independence among $F_1, \ldots, F_d$, but, at the time, it was not clear which the other elements of $\Gamma(F_1, \ldots, F_d)$ were.

For Fréchet's work see also [Dall'Aglio, 1972].

Preliminary studies about this problem were conducted by Féron [1956], Fréchet [1956], Gumbel [1958], but see also [Dall'Aglio, 1991] and [Schweizer, 1991] for a historical overview. In 1959, Sklar obtained the most significant result in this respect, by introducing the notion and the name of a *copula*[1] and proving the theorem that now bears his name [Sklar, 1959]. In his own words [Sklar, 1996]:

> [...] In the meantime, Bert (Schweizer) and I had been making progress in our work on statistical metric spaces, to the extent that Menger suggested it would be worthwhile for us to communicate our results to Fréchet. We did: Fréchet was interested, and asked us to write an announcement for the Comptes Rendus [Schweizer and Sklar, 1958]. This began an exchange of letters with Fréchet, in the course of which he sent me several packets of reprints, mainly dealing with the work he and his colleagues were doing on distributions with given marginals. These reprints, among the later arrivals of which I particularly single out that of Dall'Aglio [1959], were important for much of

---

[1] As for the grammatical meaning of the word, the electronic version of the Oxford English Dictionary gives: *"Logic and Gram.* That part of a proposition which connects the subject and predicate; the present tense of the verb to be (with or without a negative) employed as a mere sign of predication".

our subsequent work. At the time, though, the most significant reprint for me was that of Féron [1956].

Féron, in studying three-dimensional distributions, had introduced auxiliary functions, defined on the unit cube, that connected such distributions with their one-dimensional margins. I saw that similar functions could be defined on the unit n-cube for all $n \geq 2$ and would similarly serve to link n-dimensional distributions to their one-dimensional margins. Having worked out the basic properties of these functions, I wrote about them to Fréchet, in English. He asked me to write a note about them in French. While writing this, I decided I needed a name for these functions. Knowing the word "copula" as a grammatical term for a word or expression that links a subject and predicate, I felt that this would make an appropriate name for a function that links a multidimensional distribution to its one-dimensional margins, and used it as such. Fréchet received my note, corrected one mathematical statement, made some minor corrections to my French, and had the note published by the Statistical Institute of the University of Paris as [Sklar, 1959].[2]

The proof of Sklar's theorem was not given in [Sklar, 1959], but a sketch of it was provided by Sklar [1973] and, finally, showed in detail by Schweizer and Sklar [1974]. It turned out that for a few years practitioners in the field had to reconstruct it relying on the handwritten notes by Sklar himself; this was the case, for instance, of the second author. It should also be mentioned that some "indirect" proofs of Sklar's theorem (without mentioning copulas) were later discovered by Moore and Spruill [1975] and by Deheuvels [1978].

For about 15 years, most results concerning copulas were obtained in the framework of the theory of Probabilistic metric spaces [Schweizer and Sklar, 1983]. The event that arose the interest of the statistical community in copulas occurred in the mid-seventies, when Schweizer, in his own words [Schweizer, 2007],

> quite by accident, reread a paper by A. Rényi, entitled *On measures of dependence* and realized that [he] could easily construct such measures by using copulas.

See [Rényi, 1959] for Rényi's paper. These results were presented to the statistical community in the paper by Schweizer and Wolff [1981] (see also [Schweizer and Wolff, 1976; Wolff, 1977]).

However, for several years, Chapter 6 of the fundamental book by Schweizer and Sklar [1983], devoted to the theory of Probabilistic metric spaces and published in 1983, was the main source of basic information on copulas. Again in Schweizer's words from (Schweizer [2007]),

> After the publication of these articles and of the book . . . the pace quickened as more . . . students and colleagues became involved. Moreover, since interest in questions of statistical dependence was increasing, others came to the subject from different directions. In 1986 the enticingly entitled article *The joy of copulas* by Genest and MacKay [1986b], attracted more attention.

In 1990, Dall'Aglio organised the first conference devoted to copulas, aptly called "Probability distributions with given marginals" [Dall'Aglio et al., 1991]. This turned out to be the first in a series of conferences that greatly helped the development of the

---

[2]Curiously, it should be noted that in that paper, the author "Abe Sklar" is named as "M. Sklar".

field, since each of them offered the chance of presenting one's results and to learn those of other researchers; here we mention the conferences held in Seattle in 1993 [Rüschendorf et al., 1996], in Prague in 1996 [Beneš and Štěpán, 1997], in Barcelona in 2000 [Cuadras et al., 2002], in Québec in 2004 [Genest, 2005a,b]), in Tartu in 2007 [Kollo, 2009] and in Maresias, Brazil (2010). The Québec conference had a much larger attendance than the previous ones as a consequence of the fairly recent interest in copulas from the part of the investigators in finance and risk management.

In fact, at end of the nineties, the notion of copulas became increasingly popular. Two books about copulas appeared that were to become the standard references for the following decade. In 1997 Joe published his book on multivariate models [Joe, 1997], with a great part devoted to copulas and families of copulas. In 1999 Nelsen published the first edition of his introduction to copulas [Nelsen, 1999], reprinted with some new results [Nelsen, 2006].

But, the main reason for this increased interest is to be found in the discovery of the notion of copulas by researchers in several applied fields, like finance. Here we should like briefly to describe this explosion by quoting Embrechts's comments [Embrechts, 2009]:

> As we have seen so far, the notion of copula is both natural as well as easy for look-ing at multivariate d.f.'s. But why do we witness such an incredible growth in papers published starting the end of the nineties (recall, the concept goes back to the fifties and even earlier, but not under that name). Here I can give three reasons: finance, finance, finance. In the eighties and nineties we experienced an explosive development of quan-titative risk management methodology within finance and insurance, a lot of which was driven by either new regulatory guidelines or the development of new products; see for instance Chapter 1 in [McNeil et al., 2005] for the full story on the former. Two pa-pers more than any others "put the fire to the fuse"': the [...] 1998 RiskLab report by Embrechts et al. [2002] and at around the same time, the Li credit portfolio model [Li, 2001].

The advent of copulas in finance, which is well documented in [Genest et al., 2009a], originated a wealth of different investigations: see, for example, the books by Schönbucker [2003], Cherubini et al. [2004], McNeil et al. [2005], Malevergne and Sornette [2006], Trivedi and Zimmer [2007] and the more recent contributions by Jaworski et al. [2010, 2013], Cherubini et al. [2011a] and Mai and Scherer [2012b, 2014]. At the same time, different fields like environmental sciences [Genest and Favre, 2007; Salvadori et al., 2007], biostatistics [Hougaard, 2000; Song, 2007], decision science [Clemen and Reilly, 1999], machine learning [Elidan, 2013], etc., discovered the importance of this concept for constructing more flexible multivari-ate models. Copulas are used in several commercial statistical software for handling multivariate data and various copula-based packages are included in the R environ-ment (http://www.r-project.org/). Nowadays, it is near to impossible to give a complete account of all the applications of copulas to the many fields where they have been used. As Schweizer wrote [Schweizer, 2007]:

> The "era of i.i.d." is over: and when dependence is taken seriously, copulas naturally come into play. It remains for the statistical community at large to recognize this fact.

And when every statistics text contains a section or chapter on copulas, the subject will have come of age.

However, a word of caution is in order here. Several criticisms have been recently raised about copulas and their applications, and several people started to speak about "copula craze" [Embrechts, 2009].

In the academic literature, a very interesting discussion was raised by Mikosch [2006a,b] (see also the comments by Embrechts [2006], Genest and Rémillard [2006], de Haan [2006], Joe [2006], Lindner [2006], Peng [2006] and Segers [2006]). Starting with a critical point of view to the copula approach that, in his words,

> [...] it promises to solve all problems of stochastic dependence but it falls short in achieving the goal,

Mikosch posed a series of questions that have stimulated in the last years several theoretical investigations about copula models, especially related to statistical inference (see, e.g., [Genest et al., 2009b; Omelka et al., 2009; Choroś et al., 2010; Fermanian, 2013; Bücher et al., 2014] and references therein) and stochastic processes (see, e.g., Kallsen and Tankov [2006]; Bielecki et al. [2010]; Fermanian and Wegkamp [2012]; Patton [2012, 2013]; Härdle et al. [2015]). Nowadays, a decade after Mikosch's comments, we may say that most of his questions have been taken seriously into account and have promoted influential theoretical and practical contributions to the field that "will also remain for the future".

Outside the academic world, instead, the discussion was largely influenced by Salmon [2009] (reprinted in [Salmon, 2012]), which was also considered by Jones [2009]. In the aftermath of the Global financial crisis of 2008–2009, Salmon reported several fallacies of the Li's model for credit portfolio [Li, 2001], corresponding to a Gaussian copula model, and called it "the formula that killed Wall Street", an expression that became suddenly popular. Here the main concern is about the use of Gaussian copulas to describe dependence of default times. As noted by Nassim Nicholas Taleb (as cited in [Salmon, 2009]),

> People got very excited about the Gaussian copula because of its mathematical elegance, but the thing never worked. Co-association between securities is not measurable using correlation.

Now, it should be stressed that many academics have pointed out the limitations of the mathematical tools used in the finance industry, including Li's formula (see, e.g., [Embrechts et al., 2002; Brigo et al., 2010]). In fact, as summarised by Donnelly and Embrechts [2010],

> One of the main disadvantages of the model is that it does not adequately model the occurrence of defaults in the underlying portfolio of corporate bonds. In times of crisis, corporate defaults occur in clusters, so that if one company defaults then it is likely that other companies also default within a short time period. Under the Gaussian copula model, company defaults become independent as their size of default increases.

However early, these warnings seem to have been often ignored. Li himself understood the fallacy of his model saying "Very few people understand the essence of the model" and "it's not the perfect model" [Whitehouse, 2005]. In our opinion, as often the case in applications, the problem was not about the model *per se*, but its abuse. For a sociological study about model culture and the Gaussian copula story, we refer to MacKenzie and Spears [2014b,a].

However, although both previous criticisms come from different (somehow opposite) perspectives, they were a quite natural reaction to a wide diffusion of theory and applications of copulas, but not always in a well motivated way. It seems that several people have wrongly interpreted copulas as the solution to "all problems of stochastic dependence". This is definitely not the case! Copulas are an indispensable tool for understanding several problems about stochastic dependence, but they cannot be considered the "panacea" for all stochastic models.

Why another book on copulas?

Given the past and recent literature about copulas, made by influential contributions by esteemed colleagues, we should like to justify our decision to write this book.

This book is an attempt to introduce the reader to the present state of the art on copulas. Our starting point is that *"There is nothing as practical as a good theory"*. Therefore, further attempts to use copulas in a variety of different applications will remain valid for years (and will not be just fashions) if they are grounded on a solid and formal mathematical background. In this respect, this book should be considered as an *Introduction to Copula Theory*, since it has the ambition to embed the concept of copula into other mathematical fields like probability, real analysis, measure theory, algebraic structures, etc. Whether we have succeeded (at least partially) in our goal is left to the reader's judgment.

In the book, we have tried

- to present the theory of general d-dimensional, $d \geq 2$, copulas;
- to unify various methods scattered through the literature in some common frameworks (e.g., shuffles of copulas);
- to make a deliberate effort to be as simple as possible in proving the results we present, but we are aware that a few proofs are not *proofs from THE BOOK*, according to Erdős's definition, see [Aigner and Ziegler, 1999];
- to find connexions with related functions (quasi-copulas, semi-copulas, triangular norms) that have been used in different domains;
- to give at least an idea of the importance of copulas in applied fields.

In writing this book we have heavily relied on the works of the authors who have written on copulas before us. To this end, two environments have been used in the text, namely, "Historical remarks" and "Further readings", in order to acknowledge the works done by several estimeed scientists during the past years. As the field of copulas is a very active one, the number of new results, of streamlined presentations, of new proofs of old results and the like is very high and the relevant literature is quite large. Therefore, we wish to offer our apologies to those authors who may feel slighted by our overlooking their contribution; we can only excuse ourselves by saying that the literature on copulas has grown to such an extent that it is impossible to quote every paper devoted to some aspect of their theory.

Most of the figures presented in this book were realised in R [R Core Team, 2014] by using the contributed package copula [Hofert et al., 2014]. The authors and maintainers of this software are gratefully acknowledged.

Acknowledgements

We owe a great debt of gratitude to many persons with whom we have had the good fortune to discuss, and from whom we have learned, new aspects of copulas. Many of these researchers have become friends and, in some cases, coauthors. To all of them we address our sincere thanks. In particular, we should like to acknowledge those colleagues and friends that have provided us several interesting suggestions and corrections on preliminary versions of this book.

We express our intellectual and personal indebtedness and the gratitude we feel toward Abe Sklar, the "inventor" of the very concept, and of the name of *copula*, and toward Berthold Schweizer, who together with him, has shown the mathematical world how useful copulas could be. Bert Schweizer did not live to see even the beginning of the writing of this book; we feel much poorer for not having been able to rely, as in so many instances in the past, on the advice he so generously gave us. To them we wish to dedicate this book, whose idea came to us during the meeting held in Lecce in 2009 in order to celebrate the first 50 years of copulas.

The first author would like also to thank the Free University of Bozen–Bolzano (Italy) and Johannes Kepler University Linz (Austria) for the financial support of the research activities during the years of preparation of this book. The second author acknowledges the support by the University of Salento (Italy).

We thank Rob Calver and the team at Chapman & Hall for all their help in the production of this book. We are also grateful to our anonymous referees who provided us with exemplary feedback, which has shaped this book for the better.

Fabrizio Durante
Faculty of Economics and Management
Free University of Bozen–Bolzano
E-mail: fabrizio.durante@unibz.it

Carlo Sempi
Department of Mathematics and Physics "Ennio De Giorgi"
University of Salento
E-mail: carlo.sempi@unisalento.it

List of symbols

Chapter 1

Copulas: basic definitions and properties

In this chapter we introduce the definitions and properties that constitute the essentials of copula theory. After introducing the relevant notations (Section 1.1), in Section 1.2 we recall the bare minimum that is necessary in order to understand the meaning and the use of random variables and distribution functions, and introduce those concepts that are appropriate when studying their connexion with copulas. The remaining sections are devoted to the study of basic definitions and properties of copulas from both a probabilistic and an analytic point of view.

1.1 Notations

First of all, the following notations, conventions and symbols will be consistently adopted throughout the text.

- $\mathbb{N}$ denotes the set of natural numbers: $\mathbb{N} = \{1, 2, \dots\}$, while $\mathbb{Z}_+ = \mathbb{N} \cup \{0\}$ is the set of positive integers. The symbol $\mathbb{R}$ denotes the set of real numbers, and $\mathbb{R}_+$ denotes the positive real line $[0, +\infty[$. The extended real line $[-\infty, +\infty]$ is denoted by $\overline{\mathbb{R}}$.

- A real number x is said to be positive (respectively, negative) if $x \geq 0$, while if $x > 0$ (respectively, $x < 0$) one says that it is strictly positive (respectively, negative).

- Let $d \in \mathbb{N}$, $d \geq 2$. The symbol $\mathbb{R}^d$ (respectively, $\overline{\mathbb{R}}^d$) denotes the Cartesian product of d copies of $\mathbb{R}$ (respectively, $\overline{\mathbb{R}}$). $\mathbf{0}$ and $\mathbf{1}$ are the vectors of $\mathbb{R}^d$ whose components are all equal to 0 and 1, respectively. A point in $\mathbb{R}^d$ is denoted by $\mathbf{x} = (x_1, \dots, x_d) \in \mathbb{R}^d$.

- The symbol $\mathbb{I}$ denotes the closed unit interval $[0, 1]$; thus, $\mathbb{I} := [0, 1]$. The Cartesian product of d copies of $\mathbb{I}$ is denoted by $\mathbb{I}^d$.

- If $\mathbf{x}, \mathbf{y} \in \mathbb{R}^d$ are such that $x_i \leq y_i$ for every $i \in \{1, \dots, d\}$, then one writes $\mathbf{x} \leq \mathbf{y}$. If $\mathbf{x} \leq \mathbf{y}$, a d-dimensional (closed) *rectangle* (also called a *d-box*) is the set

$$[\mathbf{x}, \mathbf{y}] := [x_1, y_1] \times \cdots \times [x_d, y_d] .$$

A left-open rectangle is given by the set

$$]\mathbf{x}, \mathbf{y}] :=]x_1, y_1] \times \cdots \times]x_d, y_d] .$$

1

Similar definitions can be given for right-closed or open rectangles.

- In order to shorten formulas, the notation

$$\mathbf{u}_j(t) := (u_1, \ldots, u_{j-1}, t, u_{j+1}, \ldots, u_d) \tag{1.1.1}$$

will be adopted for $t \in \mathbb{I}$ and for $(u_1, \ldots, u_{j-1}, u_{j+1}, \ldots, u_d)$ in $\mathbb{I}^{d-1}$.

- For every function F, $\mathrm{Dom}(F)$ and $\mathrm{Ran}(F)$, respectively, denote the domain and the image (range) of F.

- The *indicator function* of a set $A \subseteq \overline{\mathbb{R}}^d$ will be denoted by $\mathbf{1}_A$.

- For every real-valued function φ defined on a subset A of $\mathbb{R}$, one says that it is *increasing* (respectively, *strictly increasing*) if $\varphi(x) \leq \varphi(x')$ (respectively, $\varphi(x) < \varphi(x')$) for all x and x' in A such that $x < x'$.

- We denote by $\ell^+\varphi(t)$ and by $\ell^-\varphi(t)$, respectively, the right- and left-limits of the function $\varphi : \mathbb{R} \to \mathbb{R}$ at t, namely

$$\ell^+\varphi(t) := \lim_{\substack{s \to t \\ s > t}} \varphi(s) \qquad \text{and} \qquad \ell^-\varphi(t) := \lim_{\substack{s \to t \\ s < t}} \varphi(s) \,,$$

when these limits exist.

- The lattice notations $x \vee y := \max\{x, y\}$ and $x \wedge y := \min\{x, y\}$ will be adopted.

1.2   Preliminaries on random variables and distribution functions

For an introduction to basic concepts in Probability Theory we refer to the books by Dudley [1989], Billingsley [1979] and Bauer [1996]. Here we only recall some definitions that will be extensively used in the sequel.

We shall often have to consider random variables X_1, ..., X_d defined on the same probability space $(\Omega, \mathscr{F}, \mathbb{P})$. A d-dimensional *random vector* (r.v.) $\mathbf{X}$ is a measurable mapping from Ω into $\mathbb{R}^d$; in this case, the word "measurable" means that the counterimage

$$\mathbf{X}^{-1}(B) := \{\omega \in \Omega : \mathbf{X}(\omega) \in B\}$$

of every Borel set B in $\mathscr{B}(\mathbb{R}^d)$ belongs to $\mathscr{F}$. It can be proved that a random vector can be represented in the form $\mathbf{X} = (X_1, \ldots, X_d)$ where X_1, ..., X_d are (one-dimensional) random variables.

When d random variables X_1, ..., X_d are given, it is of interest to investigate the relationship between them. The introduction of copulas allows to answer this problem in an admirable and thorough way.

Given X a random variable on the probability space $(\Omega, \mathscr{F}, \mathbb{P})$, a probability measure $\mathbb{P}_X$ may be defined on the measurable space $(\mathbb{R}, \mathscr{B}(\mathbb{R}))$ by

$$\forall B \in \mathscr{B}(\mathbb{R}) \qquad \mathbb{P}_X(B) := \mathbb{P}\left(X^{-1}(B)\right) \,.$$

The probability measure $\mathbb{P}_X$ is called the *law* or *distribution* of X or, again, the *image probability of* $\mathbb{P}$ *under* X. A similar construction applies to a random vector

$\mathbf{X}$, the only difference, in the vector case, being that the image probability is defined on $(\mathbb{R}^d, \mathscr{B}(\mathbb{R}^d))$.

One of the crucial concepts in probability theory is that of independence, which we describe here in terms of r.v.'s. The random variables $X_1, \ldots, X_d$ defined on $(\Omega, \mathscr{F}, \mathbb{P})$ are said to be *independent*, if, for every choice of d Borel sets $B_1, \ldots, B_d$, one has

$$\mathbb{P}(X_1 \in B_1, \ldots, X_d \in B_d) := \mathbb{P}\left(\bigcap_{j=1}^{d} \{X_j \in B_j\}\right)$$

$$= \prod_{j=1}^{d} \mathbb{P}\left(X_j^{-1}(B_j)\right) = \prod_{j=1}^{d} \mathbb{P}\left(X_j \in B_j\right).$$

As known, the study of the law $\mathbb{P}_{\mathbf{X}}$ of a r.v. $\mathbf{X}$ is made easier by the knowledge of its distribution function. We shall start with the case $d = 1$, namely with the case where a single random variable X is given.

Definition 1.2.1. The *distribution function* F_X of a random variable X on the probability space $(\Omega, \mathscr{F}, \mathbb{P})$ is the function $F_X : \mathbb{R} \to [0, 1]$ defined by

$$F_X(t) := \mathbb{P}(X \leq t), \tag{1.2.1}$$

such that $\lim_{t \to -\infty} F(t) = 0$ and $\lim_{t \to +\infty} F(t) = 1$.

Very often we shall abbreviate the term "distribution function" and write simply d.f. The space of d.f.'s will be denoted by $\mathscr{D}$. ◇

The d.f. of a random variable can be characterised in terms of its analytical properties.

Theorem 1.2.2. Let $F : \mathbb{R} \to [0, 1]$. *The following statements are equivalent:*

- *there is a probability space $(\Omega, \mathscr{F}, \mathbb{P})$ and a random variable X on it such that F is the d.f. of X;*
- *F satisfies the following properties:*

  (a) *F is continuous on the right at every point of $\mathbb{R}$, i.e., for every $t \in \mathbb{R}$, $\ell^+ F(t) = F(t)$;*

  (b) *F is increasing, i.e., $F_X(t) \leq F_X(t')$ whenever $t < t'$;*

  (c) *F satisfies the following limits*

$$\lim_{t \to -\infty} F(t) = 0, \quad \lim_{t \to +\infty} F(t) = 1. \tag{1.2.2}$$

When appropriate, we shall slightly extend the domain of a d.f. to $\overline{\mathbb{R}}$, setting $F(-\infty) = 0$ and $F(+\infty) = 1$.

Remark 1.2.3. Distribution functions like those that have just been introduced are sometimes called *proper* in the literature. However, in the literature, one can also consider the more general set Δ of d.f.'s defined on $\overline{\mathbb{R}}$ such that they satisfy

$$F(-\infty) = 0 \qquad \text{and} \qquad F(+\infty) = 1 \,;$$

they are allowed to have discontinuities both at $-\infty$ and $+\infty$ with jumps given by $\lim_{t \to -\infty} F(t)$ and $1 - \lim_{t \to +\infty} F(t)$, respectively. Of course, $\mathscr{D}$ is strictly included in Δ. While, under certain respects, the restriction to $\mathscr{D}$ simplifies the theory, it has also the disadvantage of neglecting r.v.'s that may take the values $-\infty$ and/or $+\infty$ with probability different from zero. See, e.g., [Hammersley, 1974; Schweizer and Sklar, 1983], for possible uses of d.f.'s in Δ. However we shall restrict ourselves to proper distribution functions, or equivalently to real-valued r.v.'s, a choice that is usual in Probability Theory.   ∎

Given a random variable X with distribution function F, we use the notation $X \sim F$. In particular, when F is the uniform distribution on $\mathbb{I}$, i.e., $F(t) = t$ for every $t \in \mathbb{I}$, we write $X \sim \mathscr{U}(\mathbb{I})$. Under suitable conditions, it is possible to transform any r.v. X into another r.v. that is uniform on $\mathbb{I}$. Such a transformation requires the definition of a suitable inverse of a d.f. and it is given below.

Definition 1.2.4. For a d.f. $F : \mathbb{R} \to \mathbb{I}$, the *quasi-inverse* of F (also called *quantile function* or *percentile function*) is the function $F^{(-1)} : \mathbb{I} \to \mathbb{R}$ given, for every $t \in \,]0, 1]$, by

$$F^{(-1)}(t) := \inf\{x \in \mathbb{R} : F(x) \geq t\}, \qquad (1.2.3)$$

with $F^{(-1)}(0) := \inf\{x \in \mathbb{R} : F(x) > 0\}$.

Definition 1.2.4 appears throughout the literature; see, for instance, [Schweizer and Sklar, 1983; Klement et al., 1999; Embrechts and Hofert, 2013] and references therein. The quasi-inverse of a d.f. F coincides with the standard inverse when F is continuous and strictly increasing.

Theorem 1.2.5. *Let F be a d.f. and let $F^{(-1)}$ be its quasi-inverse. Then*

(a) $F^{(-1)}$ *is increasing. In particular, if F is continuous, then $F^{(-1)}$ is strictly increasing.*

(b) $F^{(-1)}$ *is left-continuous on $\mathbb{R}$.*

(c) *If $t \in \mathrm{Ran}(F)$, $F(F^{(-1)}(t)) = t$. In particular, if F is continuous, $F(F^{(-1)}(t)) = t$ for every $t \in \mathbb{I}$.*

(d) $F^{(-1)}(F(x)) \leq x$ *for every $x \in \mathbb{R}$. In particular, if F is strictly increasing, then $F^{(-1)}(F(x)) = x$ for every $x \in \mathbb{R}$.*

(e) *For every $x \in \mathbb{R}$ and $t \in \mathbb{I}$, $F(x) \geq t$ if, and only if, $x \geq F^{(-1)}(t)$.*

Proof. (a) Set $A_t := \{x \in \overline{\mathbb{R}} : F(x) \geq t\}$; then, if $t_1 < t_2$, one has $A_{t_2} \subseteq A_{t_1}$ so that $\inf A_{t_1} \leq \inf A_{t_2}$, namely $F^{(-1)}(t_2) \geq F^{(-1)}(t_1)$. For the second part, suppose that F is continuous at x and that $F^{(-1)}(t_1) = F^{(-1)}(t_2) = x$. By definition of $F^{(-1)}$, for every $\varepsilon > 0$ one has

$$F(x - \varepsilon) \leq t_1 < t_2 \leq F(x + \varepsilon),$$

that is $\ell^- F(x) < \ell^+ F(x)$, which is a contradiction. Thus $F^{(-1)}$ must be strictly increasing.

(b) Let $t_0 \in \,]0, 1]$. Let (t_n) be an arbitrary sequence monotonically increasing to t_0, $t_n \uparrow t_0$ as n tends to ∞. For all $n \in \mathbb{N}$, set $y_n = F^{(-1)}(t_n)$ and $y_0 = F^{(-1)}(t_0)$. Because of part (a), y_n is an increasing sequence bounded from above by y_0 that converges to a limit y. Suppose that $y < y_0$. By definition of $F^{(-1)}$, for all $n \in \mathbb{N}$ and $\varepsilon > 0$,

$$F(y_n - \varepsilon) < t_n \leq F(y_n + \varepsilon).$$

In particular, for $\varepsilon < (y_0 - y_n)/2$, $y_n + \varepsilon < y_0 - \varepsilon$ implies that

$$t_0 = \lim_{n \to +\infty} t_n \leq F(y_n + \varepsilon) \leq F(y_0 - \varepsilon) < t_0,$$

which is a contradiction. Thus, $y = y_0$ and $F^{(-1)}$ is left-continuous.

(c) Since $t \in \operatorname{Ran}(F)$ there is $y \in \mathbb{R}$ such that $F(y) = t$; y may not be unique, so set $\tilde{x} := \inf\{y \in \mathbb{R} : F(y) = t\}$. Then, by the right-continuity of F, one has $F(\tilde{x}) = t$ and

$$F(F^{(-1)}(t)) = F\left(\inf\{x \in \mathbb{R} : F(x) \geq t\}\right) = F(\tilde{x}) = t.$$

(d) For every $x \in \mathbb{R}$,

$$F^{(-1)}(F(x)) = \inf\{y \in \mathbb{R} : F(y) \geq F(x)\} \leq x.$$

For the second part, note that F being strictly increasing implies that there is no $z < x$ with $F(z) \geq F(x)$, thus $F^{(-1)}(F(x)) = x$.

(e) If $x \geq F^{(-1)}(t)$, then $F(x) \geq F(F^{(-1)}(t)) \geq t$, where the last inequality follows from the right-continuity of F. The converse implication follows by definition of $F^{(-1)}$. □

Thanks to these properties, the following classical results follow.

Theorem 1.2.6. *Let X be a random variable on $(\Omega, \mathscr{F}, \mathbb{P})$ whose d.f. is given by F.*

(a) *If F is continuous, then $F \circ X$ is uniformly distributed on $\mathbb{I}$.*

(b) *If U is a random variable that is uniformly distributed on $\mathbb{I}$, then $F^{(-1)} \circ U$ has distribution function equal to F.*

Proof. (a) For every $x \in \mathbb{R}$ one can decompose $\{\omega \in \Omega : F \circ X(\omega) \leq F(x)\}$ as (we omit ω in the following):

$$\{F \circ X < F(x)\}$$
$$= (\{F \circ X \leq F(x)\} \cap \{X \leq x\}) \bigcup (\{F \circ X \leq F(x)\} \cap \{X > x\}).$$

Since $\{X \leq x\} \subseteq \{F \circ X \leq F(x)\}$ and $\{F \circ X < F(x)\} \cap \{X > x\} = \emptyset$, it follows that

$$\{F \circ X \leq F(x)\} = \{X \leq x\} \bigcup \left(\{F \circ X = F(x)\} \cap \{X > x\}\right).$$

Since $\{F \circ X = F(x)\} \cap \{X > x\}$ has probability zero because it implies that X lies in the interior of an interval of constancy of F, one has $\mathbb{P}\left(F \circ X \leq F(x)\right) = \mathbb{P}(X \leq x)$. Since F is continuous, for every $t \in \,]0,1[$, there exists $x \in \mathbb{R}$ such that $F(x) = t$. Thus

$$\mathbb{P}(F \circ X \leq t) = \mathbb{P}(X \leq x) = F(x) = t,$$

which is the desired assertion.

(b) Because of Theorem 1.2.5 (e), for every $t \in \mathbb{I}$

$$\mathbb{P}\left(F^{(-1)} \circ U \leq t\right) = \mathbb{P}(U \leq F(t)) = F(t),$$

which concludes the proof. $\square$

The transformation of part (a) of Theorem 1.2.6 is known as the *probability integral transformation* or *distributional transform*. It allows to say that transforming a random variable by its continuous distribution function always leads to the same distribution, the standard uniform distribution. However, transforming a random vector componentwise in this way may lead to multivariate distributions different from the multivariate standard uniform distribution, and it is strictly related to the copula of the random vector, as will be seen.

Remark 1.2.7. The probability integral transformation represents a standard tool in Statistics, which is applied in a variety of situations. For example, in statistical inference, if a continuous r.v. X has d.f. F that depends on an unknown parameter θ, i.e., $F = F_\theta$, then $F_\theta(X)$ is distributed as a uniform random variable, independently of the value of θ, and hence provides a pivotal function which can be used for determining confidence intervals and/or sets. See, for instance, [Angus, 1994], which also contains the proof of Theorem 1.2.6 (a).   ∎

Remark 1.2.8. The transformation of part (b) of Theorem 1.2.6 is usually known as the *quantile function theorem*. It is of practical importance because it forms a basis for generating random points from an arbitrary (not necessarily continuous) distribution function F, once a procedure is known to generate a random number in $\mathbb{I}$. See, for instance, [Mai and Scherer, 2012b].   ∎

Concepts of d.f.'s may be defined similarly in higher dimensions.

Definition 1.2.9. The distribution function of a random vector $\mathbf{X} = (X_1, \ldots, X_d)$ on the probability space $(\Omega, \mathscr{F}, \mathbb{P})$ is defined by

$$F_{\mathbf{X}}(x_1, \ldots, x_d) := \mathbb{P}\left(X_1 \leq x_1, \ldots, X_d \leq x_d\right),$$

for all $x_1, \ldots, x_d$ in $\mathbb{R}$. The space of d-dimensional d.f.'s will be denoted by $\mathscr{D}^d$. $\diamond$

Two r.v.'s $\mathbf{X}$ and $\mathbf{Y}$ are said to be *equal in distribution* (or to have *the same law*) if the d.f. of $\mathbf{X}$ is equal to the d.f. of $\mathbf{Y}$. One writes $\mathbf{X} \stackrel{d}{=} \mathbf{Y}$.

The analogue of Theorem 1.2.2 in the d-dimensional case is slightly more complicated. Before stating it, two new concepts are needed.

Definition 1.2.10. Let A be a rectangle in $\overline{\mathbb{R}}^d$. For a function $H : A \to \mathbb{R}$, the H-*volume* V_H of $]\mathbf{a}, \mathbf{b}] \subseteq A$ is defined by

$$V_H \left(]\mathbf{a}, \mathbf{b}]\right) := \sum_{\mathbf{v} \in \mathrm{ver}(]\mathbf{a}, \mathbf{b}])} \mathrm{sign}(\mathbf{v}) \, H(\mathbf{v}), \tag{1.2.4}$$

where

$$\mathrm{sign}(\mathbf{v}) = \begin{cases} 1, & \text{if } v_j = a_j \text{ for an even number of indices,} \\ -1, & \text{if } v_j = a_j \text{ for an odd number of indices,} \end{cases}$$

and $\mathrm{ver}(]\mathbf{a}, \mathbf{b}]) = \{a_1, b_1\} \times \cdots \times \{a_d, b_d\}$ is the set of vertices of $]\mathbf{a}, \mathbf{b}]$. $\diamond$

Definition 1.2.11. Let A be a rectangle in $\overline{\mathbb{R}}^d$. A function $H : \overline{\mathbb{R}}^d \to \mathbb{R}$ is d-*increasing* if the H-volume V_H of every rectangle $]\mathbf{a}, \mathbf{b}]$ is positive, i.e., $V_H \left(]\mathbf{a}, \mathbf{b}]\right) \geq 0$. $\diamond$

The d-increasing property is sometimes called the Δ-*monotone property* or L-*superadditivity*, where L stands for lattice [Marshall and Olkin, 1979, p. 213]. For $d = 1$, a 1-increasing function is a function that is increasing in the classical sense.

If the domain of H is $\mathbb{R}^2$, then H is also said to be *supermodular*. In such a case, $V_H \left(]\mathbf{a}, \mathbf{b}]\right)$ is written explicitly as

$$V_H \left(]\mathbf{a}, \mathbf{b}]\right) = H(b_1, b_2) - H(a_1, b_2) - H(b_1, a_2) + H(a_1, a_2).$$

Remark 1.2.12. The H-volume of $]\mathbf{a}, \mathbf{b}]$ can be also computed by considering that $V_H(]\mathbf{a}, \mathbf{b}]) = \Delta^d_{a_d, b_d} \cdots \Delta^1_{a_1, b_1} C(\mathbf{t})$, where, for every $j \in \{1, \dots, d\}$, for $a_j \leq b_j$, and for every $\mathbf{x} \in \overline{\mathbb{R}}^d$, $\Delta^j_{a_j, b_j}$ is the finite difference operator given by

$$\Delta^j_{a_j, b_j} H(\mathbf{t}) := H \left(\mathbf{x}_j(b_j)\right) - H \left(\mathbf{x}_j(a_j)\right),$$

where $\mathbf{x}_j(b) := (x_1, \dots, x_{j-1}, b, x_{j+1}, \dots, x_d)$. ■

As in dimension 1, one can prove (see, e.g., Billingsley [1979], p. 260) the following

Theorem 1.2.13. *Let* $F : \mathbb{R}^d \to \mathbb{I}$. *The following statements are equivalent:*
- *there exists a r.v.* $\mathbf{X}$ *on a probability space* $(\Omega, \mathscr{F}, \mathbb{P})$ *such that* F *is the d.f. of* $\mathbf{X}$;
- F *satisfies the following properties:*

(a) *For every* $j \in \{1, \ldots, d\}$ *and for all* $x_1, \ldots, x_{j-1}, x_{j+1}, \ldots, x_d$ *in* $\mathbb{R}$, *the function*

$$t \mapsto F(x_1, \ldots, x_{j-1}, t, x_{j+1}, \ldots, x_d)$$

is right-continuous;

(b) *F is d-increasing;*

(c) *$F(\mathbf{x}) \to 0$, if at least one of the arguments of $\mathbf{x}$ tends to $-\infty$;*

(d) $\lim_{\min\{x_1, \ldots, x_d\} \to +\infty} F(x_1, \ldots, x_d) = 1$.

As an immediate consequence it follows that, if F is a d.f., then for every $j \in \{1, \ldots, d\}$ and for all $x_1, \ldots, x_d$ in $\mathbb{R}$, the functions

$$\mathbb{R} \ni t \mapsto F(x_1, \ldots, x_{j-1}, t, x_{j+1}, \ldots, x_d)$$

are increasing. We shall adopt the convention of writing $f(\pm\infty)$ instead of the more cumbersome $\lim_{t \to \pm\infty} f(t)$. Moreover, when necessary, we may extend the domain of F to $\overline{\mathbb{R}}^d$ by fixing the values at infinity as in the corresponding limits.

By using the inclusion-exclusion formula for probabilities, if the r.v. $\mathbf{X}$ is distributed according to F, then $V_F(]\mathbf{a}, \mathbf{b}]) = \mathbb{P}(\mathbf{X} \in]\mathbf{a}, \mathbf{b}])$ (see, for instance, [Billingsley, 1979]). Moreover, by convention, for a r.v. $\mathbf{X} \sim F$ we set $V_F([\mathbf{a}, \mathbf{b}]) = \mathbb{P}(\mathbf{X} \in [\mathbf{a}, \mathbf{b}])$ for all $\mathbf{a}$ and $\mathbf{b}$ in $\mathbb{R}^d$ with $\mathbf{a} \leq \mathbf{b}$.

Remark 1.2.14. Contrary to previous definitions in the copula literature (see, e.g., Schweizer and Sklar [1983]; Nelsen [2006]), we have defined the formula (1.2.4) for the volume of a d.f. F on a left-open box, instead of on a closed box, of $\overline{\mathbb{R}}^d$. This choice is usually adopted when measure-theoretic aspects are of primary interest (see, for instance, [Billingsley, 1979; Ash, 2000]). Moreover, it allows a more direct probabilistic interpretation of the F-volume of a box, especially when F is not continuous. In fact, since, by convention, we are assuming that d.f.'s are right-continuous in each coordinate, the probability that the r.v. $\mathbf{X}$ on $(\Omega, \mathscr{F}, \mathbb{P})$ with d.f. F takes values in $]\mathbf{a}, \mathbf{b}]$ is equal to $V_F(]\mathbf{a}, \mathbf{b}])$.

In the following, if $\mathbf{X}$ is a r.v. (on a suitable probability space) with $\mathbf{X} \sim F$, then $V_F([\mathbf{a}, \mathbf{b}]) := \mathbb{P}(\mathbf{X} \in [\mathbf{a}, \mathbf{b}])$. Obviously, if F is continuous, $V_F(]\mathbf{a}, \mathbf{b}]) = V_F(\mathbf{X} \in [\mathbf{a}, \mathbf{b}])$ for every $\mathbf{a} \leq \mathbf{b}$.   ■

If, for a r.v. $\mathbf{X} \sim F$, there exists a Borel set A such that $\mathbb{P}(\mathbf{X} \in A) = 1$, then $\mathbf{X}$ is said to be *concentrated* on A. In this case, the d.f. F is uniquely determined by the value it assumes on A and, as a consequence, one usually refrains from specifying the value of F outside A. Thus one briefly says that F is *a d.f. on A*.

A fundamental notion throughout this book will be that of *marginal distribution* of a given d.f. H.

Definition 1.2.15. Let H be a d-dimensional d.f. and $\sigma = (j_1, \ldots, j_m)$ a subvector of $(1, \ldots, d)$, $1 \leq m \leq d - 1$. We call *σ-marginal* of H the d.f. $H^\sigma : \mathbb{R}^m \to \mathbb{I}$ defined by setting $d - m$ arguments of H equal to $+\infty$, namely, for all $u_1, \ldots, u_m$ in $\mathbb{I}$,

$$H^\sigma(u_1, \ldots, u_m) = H(v_1, \ldots, v_d), \tag{1.2.5}$$

where $v_j = u_j$ if $j \in \{j_1, \ldots, j_m\}$, and one lets v_j tend to $+\infty$ otherwise. ◇

As is known, if the d-dimensional r.v. $\mathbf{X}$ has d.f. H, then the marginal H^σ is the joint d.f. of $(X_{j_1}, \ldots, X_{j_m})$. Explicitly, the j-th 1-marginal of H is the 1-dimensional d.f. of X_j, the j-th component of $\mathbf{X}$,

$$F_j(t) := H(+\infty, \ldots, +\infty, t, +\infty, \ldots, +\infty). \tag{1.2.6}$$

If the random variables $X_1, \ldots, X_d$ are independent and if F_j denotes the d.f. of X_j $(j = 1, \ldots, d)$, then the d.f. of the r.v. $\mathbf{X} = (X_1, \ldots, X_d)$ can be written as the product of the marginals

$$F_{\mathbf{X}}(x_1, \ldots, x_d) = \prod_{j=1}^{d} F_j(x_j).$$

The variation of a multivariate d.f. is somehow controlled by the variation of its univariate marginals, as the following result shows.

Lemma 1.2.16. *Let H be a d-dimensional d.f. and let $F_1, \ldots, F_d$ be its marginals. Then for every pair of points $(t_1, \ldots, t_d)$ and $(s_1, \ldots, s_d)$ in $\mathbb{R}^d$,*

$$|H(t_1, \ldots, t_d) - H(s_1, \ldots, s_d)| \leq \sum_{j=1}^{d} |F_j(t_j) - F_j(s_j)|. \tag{1.2.7}$$

Proof. Let H be the d.f. of the random vector $\mathbf{X}$ defined $(\Omega, \mathscr{F}, \mathbb{P})$. Then, for every $j \in \{1, \ldots, d\}$, for $t < t'$ and for every $\mathbf{x} = (x_1, \ldots, x_{j-1}, x_{j+1}, \ldots, x_d) \in \mathbb{R}^{d-1}$,

$$H(x_1, \ldots, x_{j-1}, t', x_{j+1}, \ldots, x_d) - H(x_1, \ldots, x_{j-1}, t, x_{j+1}, \ldots, x_d)$$
$$= \mathbb{P}(X_1 \leq x_1, \ldots, X_j \leq t', \ldots, X_d \leq x_d)$$
$$\quad - \mathbb{P}(X_1 \leq x_1, \ldots, X_j \leq t, \ldots, X_d \leq x_d)$$
$$= \mathbb{P}(X_1 \leq x_1, \ldots, t < X_j \leq t', \ldots, X_d \leq x_d) \leq F_j(t') - F_j(t).$$

Applying the previous inequality d times, one obtains (1.2.7). □

Remark 1.2.17. Given a random vector $\mathbf{X}$ on the probability space $(\Omega, \mathscr{F}, \mathbb{P})$, one can associate to it a *survival function* defined by

$$\overline{F}_{\mathbf{X}}(x_1, \ldots, x_d) := \mathbb{P}(X_1 > x_1, \ldots, X_d > x_d),$$

for all $x_1, \ldots, x_d$ in $\mathbb{R}$. When $d = 1$, $\overline{F}_X = 1 - F_X$. For $d \geq 2$, the survival function $\overline{F}$ of $\mathbf{X}$ may be written in terms of the d.f. of $\mathbf{X}$ (and its marginals) by using the inclusion-exclusion formula for probabilities (see, for instance, [Billingsley, 1979]).

In fact, one has

$$\overline{F}(\mathbf{x}) = \mathbb{P}(X_1 > x_1, \ldots, X_d > x_d)$$

$$= 1 - \mathbb{P}\left(\bigcup_{j=1}^{d}\{X_j \leq x_j\}\right)$$

$$= 1 - \sum_{j=1}^{d}\mathbb{P}\left(X_j \leq x_j\right) + \sum_{i<j}\mathbb{P}(X_i \leq x_i, X_j \leq x_j) + \ldots$$

$$+ (-1)^k \sum_{j(1)<j(2)<\cdots<j(k)} \mathbb{P}(X_{j(1)} \leq x_{j(1)}, \ldots, X_{j(k)} \leq x_{j(k)})$$

$$+ (-1)^d \mathbb{P}(X_1 \leq x_1, \ldots, X_d \leq x_d)$$

for every $\mathbf{x} \in \mathbb{R}^d$.   ∎

Finally, a r.v. $\mathbf{X} = (X_1, \ldots, X_d)$ is said to be *absolutely continuous* if there exists a positive and integrable function $f_{\mathbf{X}} : \mathbb{R}^d \to \mathbb{R}_+$ such that

$$\int_{\mathbb{R}^d} f_{\mathbf{X}}\, d\lambda_d = 1\,,$$

where λ_d is the d-dimensional Lebesgue measure; in this case the d.f. $F_{\mathbf{X}}$ of $\mathbf{X}$ can be represented in the form

$$F_{\mathbf{X}}(x_1, \ldots, x_d) = \int_{-\infty}^{x_1} dt_1 \ldots \int_{-\infty}^{x_d} f_{\mathbf{X}}(t_1, \ldots, t_d)\, dt_d\,.$$

In particular, a r.v. X that is uniform on $\mathbb{I}$ is absolutely continuous with density $f_X = \mathbf{1}_{\mathbb{I}}$.

1.3   Definition and first examples

The definition of copula relies on that of multivariate distribution function.

Definition 1.3.1. For every $d \geq 2$, a *d-dimensional copula* (a *d-copula*) is a d-dimensional d.f. concentrated on $\mathbb{I}^d$ whose univariate marginals are uniformly distributed on $\mathbb{I}$. The set of d-copulas ($d \geq 2$) is denoted by $\mathscr{C}_d$.   ◇

Since a copula C concentrates the probability distribution on $\mathbb{I}^d$, one usually refrains from specifying the values that C assumes on $\mathbb{R}^d \setminus \mathbb{I}^d$ and regards a copula as a function with domain $\mathbb{I}^d$ and range $\mathbb{I}$.

Historical remark 1.3.2. The definition of "copula" originated in the paper by Sklar [1959]. Later on, other authors discovered the same concept under different names. For instance, Kimeldorf and Sampson [1975] referred to it as "uniform representation"; Deheuvels [1978] and Galambos [1978] called it "dependence function".

In the literature, the name "copula" has two plurals, *copulae* (also written as copulæ) and *copulas*. Here we have adopted the latter version following the original convention in [Sklar, 1959]. ∎

As a consequence of Theorem 1.2.13, to each copula C there corresponds a random vector $\mathbf{U}$ defined on a suitable probability space such that the joint d.f. of $\mathbf{U}$ is given by C. Such a (quite natural) probabilistic characterisation allows introducing the following three fundamental examples of copulas.

Example 1.3.3 (The copula M_d). Let U be a random variable defined on the probability space $(\Omega, \mathscr{F}, \mathbb{P})$. Suppose that U is uniformly distributed on $\mathbb{I}$. Consider the r.v. $\mathbf{U} = (U, \dots, U)$. Then, for every $\mathbf{u} \in \mathbb{I}^d$,

$$\mathbb{P}(\mathbf{U} \le \mathbf{u}) = \mathbb{P}(U \le \min\{u_1, \dots, u_d\}) = \min\{u_1, \dots, u_d\}.$$

Thus the d.f. given, for every $\mathbf{u} \in \mathbb{I}^d$, by

$$M_d(u_1, u_2, \dots, u_d) := \min\{u_1, \dots, u_d\}$$

is a copula, which will be called the *comonotonicity copula* (a term that will be clarified later in Section 2.5). ∎

Example 1.3.4 (The copula Π_d). Let $U_1, \dots, U_d$ be independent r.v.'s defined on the probability space $(\Omega, \mathscr{F}, \mathbb{P})$. Suppose that each U_i is uniformly distributed on $\mathbb{I}$. Consider the r.v. $\mathbf{U} = (U_1, \dots, U_d)$. Then, for every $\mathbf{u} \in \mathbb{I}^d$,

$$\mathbb{P}(\mathbf{U} \le \mathbf{u}) = \mathbb{P}(U_1 \le u_1) \cdots \mathbb{P}(U_d \le u_d) = \prod_{j=1}^{d} u_j.$$

Thus the d.f. given, for every $\mathbf{u} \in \mathbb{I}^d$, by

$$\Pi_d(u_1, u_2, \dots, u_d) := \prod_{j=1}^{d} u_j$$

is a copula, which will be called the *independence copula*. ∎

Example 1.3.5 (The copula W_2). Let U be a random variable defined on the probability space $(\Omega, \mathscr{F}, \mathbb{P})$. Suppose that U is uniformly distributed on $\mathbb{I}$. Consider the r.v. $\mathbf{U} = (U, 1 - U)$. Then, for every $\mathbf{u} \in \mathbb{I}^2$,

$$\mathbb{P}(\mathbf{U} \le \mathbf{u}) = \mathbb{P}(U \le u_1, 1 - U \le u_2) = \max\{0, u_1 + u_2 - 1\}.$$

Thus the d.f. given, for every $\mathbf{u} \in \mathbb{I}^2$, by

$$W_2(u_1, u_2) := \max\{0, u_1 + u_2 - 1\}$$

is a copula, which will be called the *countermonotonicity copula* (a term that will be clarified later in Section 2.5). ∎

The previous examples have introduced fundamental dependence structures among the components of a random vector, namely linear dependence and independence. It will be made clear in the sequel that a variety of dependence properties may be described by copulas. Here it is enough to present two other examples that "interpolate", in some sense, the probabilistic behaviour of the previous examples.

Example 1.3.6. On the probability space $(\Omega, \mathscr{F}, \mathbb{P})$ let Z be a random variable with uniform distribution on $\mathbb{I}$. Let $\mathscr{J}$ be a non-empty proper subset of $\{1, \ldots, d\}$. Consider the r.v. $\mathbf{U} = (U_1, \ldots, U_d)$ where $U_j = Z$ for $j \in \mathscr{J}$, while $U_j = 1 - Z$, otherwise. Then, for every $\mathbf{u} \in \mathbb{I}^d$

$$\mathbb{P}\left(\mathbf{U} \le \mathbf{u}\right) = \mathbb{P}\left(Z \le \min_{j \in \mathscr{J}} u_j, 1 - Z \le \min_{j \in \mathscr{J}^c} u_j\right)$$
$$= \max\{0, \min_{j \in \mathscr{J}} u_j + \min_{j \in \mathscr{J}^c} u_j - 1\},$$

where $\mathscr{J}^c := \{1, \ldots, d\} \setminus \mathscr{J}$. In other words, the d.f. given, for every $\mathbf{u} \in \mathbb{I}^d$, by

$$C(u_1, u_2, \ldots, u_d) = \max\{0, \min_{j \in \mathscr{J}} u_j + \min_{j \in \mathscr{J}^c} u_j - 1\}$$

is a copula [Tiit, 1996]. An interesting feature of the r.v. $\mathbf{U}$ is that the pairwise Pearson correlation coefficient between any two components of $\mathbf{U}$ is always equal to either 1 or -1. ∎

Example 1.3.7. On the probability space $(\Omega, \mathscr{F}, \mathbb{P})$ consider a r.v. $\mathbf{V}$ whose components are independent and identically distributed on $\mathbb{I}$ with d.f. F, which is assumed to be strictly positive on $]0, 1]$ and such that $G(t) = t/F(t)$ is a d.f. on $]0, 1]$. On the same probability space consider a r.v. Z, independent of $\mathbf{V}$, with d.f. G. Consider now the r.v. $\mathbf{U} = (U_1, \ldots, U_d)$ where $U_j = \max\{V_j, Z\}$ for every $j \in \{1, \ldots, d\}$. Then, for every $\mathbf{u} \in \mathbb{I}^d$

$$\mathbb{P}\left(\mathbf{U} \le \mathbf{u}\right) = \mathbb{P}\left(V_1 \le u_1, \ldots, V_d \le u_d, Z \le \min\{u_1, \ldots, u_d\}\right)$$
$$= F(u_1) \cdots F(u_d) \, G\left(\min\{u_1, \ldots, u_d\}\right)$$
$$= u_{(1)} F(u_{(2)}) \cdots F(u_{(d)}),$$

where $u_{(1)}, \ldots, u_{(d)}$ correspond to the order statistics of $u_1, \ldots, u_d$. Thus, the d.f. given, for every $\mathbf{u} \in \mathbb{I}^d$, by

$$C(u_1, u_2, \ldots, u_d) = u_{(1)} \, F(u_{(2)}) \cdots F(u_{(d)}) \tag{1.3.1}$$

is a copula, which has been studied in [Durante et al., 2007c; Mai and Scherer, 2009]. In the special case $d = 2$ and $F(t) = t^{1-\alpha}$ for $\alpha \in \mathbb{I}$, one has

$$C_\alpha^{\mathbf{CA}}(u_1, u_2) := (M_2(u_1, u_2))^\alpha (\Pi_2(u_1, u_2))^{1-\alpha},$$

which corresponds to the exchangeable members of the Marshall-Olkin family [Marshall and Olkin, 1967b; Cuadras and Augé, 1981], also denoted as the *Cuadras-Augé family* of copulas (see Section 6.4). ∎

Given a copula C, several methods have been presented in the literature to transform C into another copula C' sharing (possibly) different properties. Below one example of such type is presented (see Dolati and Úbeda Flores [2009]; Kolesárová et al. [2013]).

Example 1.3.8. Let $(\Omega, \mathscr{F}, \mathbb{P})$ be a probability space. Let (X_1, Y_1) and (X_2, Y_2) be two independent vectors of uniform random variables in $\mathbb{I}$ with equal distribution function given by a copula C. Let $(X_{(1)}, X_{(2)})$ be the vector of the order statistics of (X_1, X_2) and, analogously, $(Y_{(1)}, Y_{(2)})$ be the vector of the order statistics of (Y_1, Y_2). Consider the random vector

$$(Z_1, Z_2) = \begin{cases} (X_{(1)}, Y_{(2)}), & \text{with probability } 1/2, \\ (X_{(2)}, Y_{(1)}), & \text{with probability } 1/2. \end{cases}$$

The distribution function D of (Z_1, Z_2) is given, for every $(u, v) \in \mathbb{I}^2$, by

$$\begin{aligned} D(u, v) &= \mathbb{P}\left(Z_1 \leq u, Z_2 \leq v\right) \\ &= \frac{1}{2} \mathbb{P}\left(X_{(1)} \leq u, Y_{(2)} \leq v\right) + \frac{1}{2} \mathbb{P}\left(X_{(2)} \leq u, Y_{(1)} \leq v\right) \\ &= \frac{1}{2}\left(\mathbb{P}(Y_{(2)} \leq v) - \mathbb{P}(X_{(1)} > u, Y_{(2)} \leq v)\right) \\ &\quad + \frac{1}{2}\left(\mathbb{P}(X_{(2)} \leq u) - \mathbb{P}(X_{(2)} \leq u, Y_{(1)} > v)\right) \\ &= \frac{1}{2}\left(v^2 - (v - C(u, v))^2 + u^2 - (u - C(u, v))^2\right) \\ &= C(u, v)(u + v - C(u, v)), \end{aligned}$$

which is easily proved to have uniform univariate margins on $\mathbb{I}$ and, hence, it is a copula. ∎

Just as when considering d.f.'s it is possible to speak of the marginals of a copula.

Definition 1.3.9. Let C be a d-dimensional copula and $\sigma = (j_1, \ldots, j_m)$ a subset of $(1, \ldots, d)$, $1 \leq m \leq d - 1$. The σ-*marginal* copula of C, $C^\sigma : \mathbb{I}^m \to \mathbb{I}$, is defined by setting $d - m$ arguments of C equal to 1, namely, for all $u_1, \ldots, u_m$ in $\mathbb{I}$,

$$C^\sigma(u_1, \ldots, u_m) = C(v_1, \ldots, v_d), \tag{1.3.2}$$

where $v_j = u_j$ if $j \in \{j_1, \ldots, j_m\}$, and $v_j = 1$ otherwise. ◇

For instance, the three 2-marginals of a copula $C \in \mathscr{C}_3$ are given, for u_1, u_2 and u_3 in $\mathbb{I}$, by

$$C(u_1, u_2, 1), \qquad C(u_1, 1, u_3) \qquad \text{and} \qquad C(1, u_2, u_3).$$

It is worth noticing that, if C is the d.f. of the r.v. $\mathbf{U}$, then the σ-marginal C^σ is the d.f. of the subvector $(U_{j_1}, \ldots, U_{j_m})$.

Example 1.3.10. Let C_1 and C_2 be, respectively, a d_1-dimensional and a d_2-dimensional copula. Then, for every $\mathbf{u} = (u_1, \ldots, u_{d_1}, u_{d_1+1}, \ldots, u_{d_1+d_2})$ in $\mathbb{I}^{d_1+d_2}$,

$$C(\mathbf{u}) = C_1(u_1, \ldots, u_{d_1}) \cdot C(u_{d_1+1}, \ldots, u_{d_1+d_2})$$

is a $(d_1 + d_2)$-dimensional copula. In fact, if $\mathbf{U}_1$ and $\mathbf{U}_2$ are independent r.v.'s on the same probability space distributed according to the copulas C_1 and C_2, respectively, then C is the d.f. of $(\mathbf{U}_1, \mathbf{U}_2)$.

Moreover, it can be easily checked that C_1 and C_2 are the marginals of C corresponding to the subvectors $\sigma_1 = (1, \ldots, d_1)$ and $\sigma_2 = (d_1 + 1, d_1 + 2, \ldots, d_1 + d_2)$, respectively. ∎

1.4 Characterisation in terms of properties of d.f.'s

Besides its probabilistic interpretation, copulas can be characterised in an equivalent way in terms of the analytical properties of the d.f.'s, as an immediate consequence of Theorem 1.2.13.

Theorem 1.4.1. *A function $C : \mathbb{I}^d \to \mathbb{I}$ is a d-copula if, and only if, the following conditions hold:*

(a) $C(u_1, \ldots, u_d) = 0$ if $u_j = 0$ for at least one index $j \in \{1, \ldots, d\}$;

(b) *when all the arguments of C are equal to 1, but possibly for the j-th one, then*

$$C(1, 1, \ldots, 1, u_j, 1, \ldots, 1) = u_j;$$

(c) *C is d-increasing.*

Properties (a) and (b) together are called the *boundary conditions* of a d-copula. Specifically, property (a) expresses the fact that C is *grounded* (or *anchored*), while by property (b) the univariate marginals of C are uniform on $\mathbb{I}$.

If $\mathbf{U}$ is a r.v. distributed according to a copula C, then $V_C(]\,\mathbf{a}, \mathbf{b}])$ expresses the probability that $\mathbf{U}$ takes values in $]\,\mathbf{a}, \mathbf{b}]$[1].

Remark 1.4.2. It is worth noticing that, since a copula C is a d.f., it is *increasing in each place*, i.e., for every $j \in \{1, \ldots, d\}$ and for all $u_1, \ldots, u_{j-1}, u_{j+1}, \ldots, u_d$ in $\mathbb{I}$, $t \mapsto C(\mathbf{u}_j(t))$ is increasing. This property can be also derived from properties (a)-(c) of Theorem 1.4.1. In fact, the d-increasing property of a copula C when applied to the d-box

$$]\,u_1, u_1] \times \cdots \times]\,u_{j-1}, u_{j-1}] \times]\,v, v'] \times]\,u_{j+1}, u_{j+1}] \times \cdots \times]\,u_d, u_d]$$

shows that $C(\mathbf{u}_j(v')) - C(\mathbf{u}_j(v)) \geq 0$, which proves the assertion. ∎

It is useful to re-write Theorem 1.4.1 explicitly in the case $d = 2$.

---

[1] It should be noticed that there is no difference between $V_C(]\,\mathbf{a}, \mathbf{b}])$ and $V_C([\,\mathbf{a}, \mathbf{b}])$, since, as will be seen shortly, every copula C is continuous

Corollary 1.4.3. *A function $C : \mathbb{I}^2 \to \mathbb{I}$ is a 2-copula if, and only if, it satisfies the following conditions:*

(a) $C(0, t) = C(t, 0) = 0$, *for every* $t \in \mathbb{I}$;

(b) $C(1, t) = C(t, 1) = t$, *for every* $t \in \mathbb{I}$;

(c) *for all* a_1, a_2, b_1, b_2 *in* $\mathbb{I}$, *with* $a_1 \leq b_1$ *and* $a_2 \leq b_2$,

$$V_C \left(\,]\,\mathbf{a}, \mathbf{b}]\right) = C(a_1, a_2) - C(a_1, b_2) - C(b_1, a_2) + C(b_1, b_2) \geq 0 . \quad (1.4.1)$$

Inequality (1.4.1) can be graphically represented as in Figure 1.1. Specifically, it implies that the sum of the values (multiplied by $+1$ or -1) assumed by a copula at the vertices of any rectangle is non-negative.

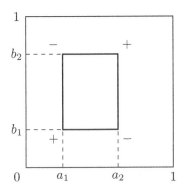

Figure 1.1: *Visualisation of the 2-increasing property.*

Two different strategies are usually adopted in order to prove that a function $C : \mathbb{I}^d \to \mathbb{I}$ is a copula: the first one aims at finding a suitable probabilistic model (i.e., a random vector) whose distribution function is concentrated on $\mathbb{I}^d$ and has uniform marginals; the second one consists of proving that properties (a), (b) and (c) of Theorem 1.4.1 are satisfied. However, this latter strategy is usually quite intricate in high dimensions. In order to simplify the calculations of the d-increasing property, the following results may be useful.

Lemma 1.4.4. *Let $F, G : \mathbb{I}^d \to \mathbb{I}$ be two functions. Let $]\,\mathbf{a}, \mathbf{b}]$ be a d-box in $\mathbb{I}^d$. Then:*

(a) $V_{F+G} \left(\,]\,\mathbf{a}, \mathbf{b}]\right) = V_F \left(\,]\,\mathbf{a}, \mathbf{b}]\right) + V_G \left(\,]\,\mathbf{a}, \mathbf{b}]\right)$;

(b) $V_{\alpha F} \left(\,]\,\mathbf{a}, \mathbf{b}]\right) = \alpha \, V_F \left(\,]\,\mathbf{a}, \mathbf{b}]\right)$ *for every* $\alpha > 0$;

(c) *if* $]\,\mathbf{a}, \mathbf{b}] = \cup_{j \in \mathscr{J}} B_j$ *where* $\mathscr{J}$ *has finite cardinality and all B_j's are all left-open d-boxes whose interiors are disjoint, then*

$$V_F \left(\,]\,\mathbf{a}, \mathbf{b}]\right) = \sum_{j \in \mathscr{J}} V_F(B_j) .$$

Proof. In view of the definition of H-volume of a function $H : \mathbb{I}^d \to \mathbb{R}$ (see Definition 1.2.10), statements (a) and (b) are simple and immediate consequences of the properties of the sum. For part (c), see [Billingsley, 1979, section 12]. □

As a direct application, one may prove the following result.

Theorem 1.4.5. *The set $\mathscr{C}_d$ is a convex set, i.e., for all $\alpha \in \mathbb{I}$ and C_0 and C_1 in $\mathscr{C}_d$ $C = \alpha\, C_0 + (1 - \alpha)\, C_1$ is in $\mathscr{C}_d$.*

Proof. Let C_0 and C_1 be in $\mathscr{C}_d$ and let α be in $\mathbb{I}$. Let C be the convex combination C_0 and C_1, given by $C = \alpha\, C_0 + (1 - \alpha)\, C_1$. It is trivial to prove that the univariate marginals of C are uniformly distributed on $\mathbb{I}$. Moreover, for every rectangle $]\,\mathbf{a}, \mathbf{b}] \subseteq \mathbb{I}^d$, direct application of Lemma 1.4.4 yields

$$V_C\left(]\,\mathbf{a}, \mathbf{b}]\right) = \alpha V_{C_0}\left(]\,\mathbf{a}, \mathbf{b}]\right) + (1 - \alpha)V_{C_1}\left(]\,\mathbf{a}, \mathbf{b}]\right) \geq 0,$$

which is the desired assertion. □

The previous result also admits a probabilistic proof, which is reported below.

Proof of Theorem 1.4.5. Let $\mathbf{U}_0$ and $\mathbf{U}_1$ be two d-dimensional random vectors on a probability space $(\Omega, \mathscr{F}, \mathbb{P})$ distributed according to the copulas C_0 and C_1, respectively. Let Z be a Bernoulli random variable such that $\mathbb{P}(Z = 0) = \alpha$ and $\mathbb{P}(Z = 1) = 1 - \alpha$. Suppose that $\mathbf{U}_0$, $\mathbf{U}_1$ and Z are independent. Now, consider the d-dimensional r.v. $\mathbf{U}^*$ defined by

$$\mathbf{U}^* = \sigma_0(Z)\, \mathbf{U}_0 + \sigma_1(Z)\, \mathbf{U}_1$$

where, for $j \in \{0, 1\}$, $\sigma_j(x) = 1$, if $x = j$, $\sigma_j(x) = 0$, otherwise. Then, for every $\mathbf{u} \in \mathbb{I}^d$

$$
\begin{aligned}
\mathbb{P}\left(\mathbf{U}^* \leq \mathbf{u}\right) &= \mathbb{P}\left(\mathbf{U}^* \leq \mathbf{u} \mid Z = 0\right) \mathbb{P}(Z = 0) + \mathbb{P}\left(\mathbf{U}^* \leq \mathbf{u} \mid Z = 1\right) \mathbb{P}(Z = 1) \\
&= \alpha\, \mathbb{P}\left(\mathbf{U}_0 \leq \mathbf{u}\right) + (1 - \alpha)\, \mathbb{P}\left(\mathbf{U}_1 \leq \mathbf{u}\right) \\
&= \alpha\, C_0\left(\mathbf{u}\right) + (1 - \alpha)\, C_1\left(\mathbf{u}\right).
\end{aligned}
$$

Thus, C is the d.f. of $\mathbf{U}^*$, and, since its univariate marginals are uniformly distributed on $\mathbb{I}$, it is a copula. □

Example 1.4.6. Consider the case $d = 2$ and let α and β be in $\mathbb{I}$ with $\alpha + \beta \leq 1$. Then, in view of the convexity of $\mathscr{C}_2$, $C_{\alpha,\beta} : \mathbb{I}^2 \to \mathbb{I}$ defined by

$$C_{\alpha,\beta}^{\mathbf{Fre}}(u_1, u_2) := \alpha\, M_2(u_1, u_2) + (1 - \alpha - \beta)\, \Pi_2(u_1, u_2) + \beta\, W_2(u_1, u_2)$$

is a copula. As the parameters α and β vary in $\mathbb{I}$ subject to the restriction $\alpha + \beta \leq 1$, the copula varies in a family of copulas known as the *Fréchet copulas*. These copulas were introduced by Fréchet [1958]. A subclass has been proposed by Mardia [1970]: a copula C_α belongs to *Mardia family* if it is defined by

$$C_\alpha(u,v) := \frac{1}{2}\,\alpha^2\,(1-\alpha)\,W_2(u,v) + (1-\alpha^2)\,\Pi_2(u,v) + \frac{1}{2}\,\alpha^2\,(1+\alpha)\,M_2(u,v),$$

for every $\alpha \in [-1,1]$. ∎

Theorem 1.4.5 can be further generalised in the following sense.

Corollary 1.4.7. *Let $\mathbb{P}_1$ be a probability measure on $\mathscr{B}(\mathbb{R})$ and let $(C_t)_{t\in\mathbb{R}}$ be a family of copulas in $\mathscr{C}_d$ such that $t \mapsto C_t(\mathbf{u})$ is Borel-measurable for every $\mathbf{u} \in \mathbb{I}^d$. Then, for every $\mathbf{u} \in \mathbb{I}^d$,*

$$C(\mathbf{u}) = \int_\mathbb{R} C_t(\mathbf{u})\,\mathrm{d}\mathbb{P}_1(t)$$

is a copula.

Proof. It follows by considering that, for every box $]\mathbf{a},\mathbf{b}] \subseteq \mathbb{I}^d$, one has

$$V_C\left(]\mathbf{a},\mathbf{b}]\right) = \int_\mathbb{R} V_{C_t}\left(]\mathbf{a},\mathbf{b}]\right)\,\mathrm{d}\mathbb{P}_1(t) \geq 0\,;$$

moreover, the boundary conditions for a copula are trivially satisfied by C. □

Example 1.4.8. Let C be in $\mathscr{C}_2$. For every $t \in \mathbb{I}$, consider the function

$$C_t(u,v) = \begin{cases} C(1+u-t,v) - C(1-t,v), & \text{if } u \leq t; \\ C(u-t,v) + v - C(1-t,v), & \text{if } u > t. \end{cases}$$

Then C_t is a copula. In fact, by direct application of Lemma 1.4.4(c), it can be proved that C_t is 2-increasing both in $[0,t] \times \mathbb{I}$ and $[t,1] \times \mathbb{I}$. Moreover, for every $s \in \mathbb{I}$, one has

$$C(s,0) = 0 = C(0,s), \qquad C(s,1) = s = C(1,s).$$

It is easy to prove using integration by substitution that for every copula C,

$$\int_\mathbb{I} C_t(u,v)\,\mathrm{d}t = uv\,,$$

i.e., the independence copula can be obtained as a mixture of copulas. ∎

Part (c) of Lemma 1.4.4 can be further refined in the case $d=2$ for some special kinds of rectangles. First, we define the sets

$$T_L := \{(u,v) \in \mathbb{I}^2 : u \geq v\}, \tag{1.4.2}$$

$$T_U := \{(u,v) \in \mathbb{I}^2 : u \leq v\}. \tag{1.4.3}$$

Lemma 1.4.9. *A function $A : \mathbb{I}^2 \to \mathbb{I}$ is 2-increasing if, and only if, the following conditions hold:*

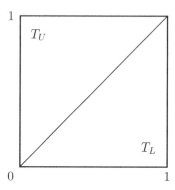

Figure 1.2: *Sets defined in* (1.4.2) *and* (1.4.3).

(a) $V_A(R) \geq 0$ *for every rectangle* $R \subseteq T_L$;

(b) $V_A(R) \geq 0$ *for every rectangle* $R \subseteq T_U$;

(c) $V_A(R) \geq 0$ *for every rectangle* $R =]s,t] \times]s,t] \subseteq \mathbb{I}^2$.

Proof. The desired assertion follows by noticing that the A-volume $V_A(R)$ of a rectangle $R \subseteq \mathbb{I}^2$ can be expressed as the sum $\sum_i V_A(R_i)$ of at most three terms, where the rectangles R_i may have a side in common and are one of the following three types: $R_i \subseteq T_L$; $R_i \subseteq T_U$; $R_i =]s,t] \times]s,t]$ for some s and t in $\mathbb{I}$ with $s < t$.   □

Example 1.4.10. Let C be the bivariate copula of eq. (1.3.1), i.e.,

$$C(u,v) = \min\{u,v\} \, F(\max\{u,v\}), \tag{1.4.4}$$

where F is increasing on $\mathbb{I}$, $F(1) = 1$ and $G(t) = t/F(t)$ is increasing for all $t \in \,]0,1]$. It can be easily checked that C has uniform margins. In order to prove that C is 2-increasing, in view of Lemma 1.4.9 and since $C(u,v) = C(v,u)$ for all $(u,v) \in \mathbb{I}^2$, it is enough to prove that $V_C(R) \geq 0$ for each rectangle R such that either $R \subseteq T_L$ or $R = \,]s,t]^2 \subseteq \mathbb{I}^2$. If $R = \,]s,s'] \times [t,t'] \subseteq T_L$, then $V_C(R) \geq 0$ because of monotonicity of F. Moreover, if $R = \,]s,t]^2$, then $V_C(R) = sF(s) + tF(t) - 2sF(t)$ is positive since G is increasing. It follows that C is a copula. Copulas of type (1.4.4) belong to a class of copulas originated from a seminal idea by Marshall [1996] and are called *exchangeable Marshall copulas* or *semilinear copulas*. For more details about copulas of this type, we refer to [Durante, 2006; Durante et al., 2008b; Jwaid et al., 2014; Durante and Okhrin, 2015].   ■

1.5 Continuity and absolutely continuity

Let denote by $\Xi(\mathbb{I}^d)$ the set of continuous real-valued functions on $\mathbb{I}^d$. The class of copulas is a subset of $\Xi(\mathbb{I}^d)$, as the following result shows.

Theorem 1.5.1. *A d-copula C satisfies the following condition:*

$$|C(u_1, \ldots, u_d) - C(v_1, \ldots, v_d)| \leq \sum_{j=1}^{d} |u_j - v_j| \qquad (1.5.1)$$

for all $\mathbf{u}$ and $\mathbf{v}$ in $\mathbb{I}^d$.

Proof. It follows from Theorem 1.2.16 by considering that each copula has uniform univariate marginals on $\mathbb{I}$. $\qquad\square$

Inequality (1.5.1) may be written in a different, but equivalent, way,

$$|C(\mathbf{u}) - C(\mathbf{v})| \leq \|\mathbf{u} - \mathbf{v}\|_1, \qquad (1.5.2)$$

where, for each $p \geq 1$, $\| \cdot \|_p$ denotes the $\ell^p(d)$-norm on $\mathbb{R}^d$,

$$\|\mathbf{u}\|_p := \left(\sum_{j=1}^{d} |u_j|^p \right)^{1/p}.$$

In the following, we refer to the inequality (1.5.1) or, equivalently, (1.5.2) as the Lipschitz condition with constant 1, or simply the 1-*Lipschitz condition*. One may more formally say that every copula $C \in \mathscr{C}_d$ is 1-Lipschitz continuous with respect to the $\ell_1(d)$-norm. In particular, every d-copula C is uniformly continuous on $\mathbb{I}^d$.

The condition (1.5.1) is actually a property of the one-dimensional sections of a copula, as the following result shows.

Lemma 1.5.2. *Let $F : \mathbb{I}^d \to \mathbb{I}$. The following statements are equivalent:*

(a) *F satisfies (1.5.1);*

(b) *F is 1-Lipschitz in each of its arguments, viz., for every $j \in \{1, \ldots, d\}$, for every point $(u_1, \ldots, u_{j-1}, u_{j+1}, \ldots, u_d)$ in $\mathbb{I}^{d-1}$, and for every all t and t' in $\mathbb{I}$,*

$$|F(\mathbf{u}_j(t')) - F(\mathbf{u}_j(t))| \leq |t' - t|. \qquad (1.5.3)$$

Proof. Only the implication (b) $\Longrightarrow$ (a) needs to be proved. Let $\mathbf{u}, \mathbf{v}$ be in $\mathbb{I}^d$. Then

$$
\begin{aligned}
|F(\mathbf{v}) - F(\mathbf{u})| &\leq |F(\mathbf{v}) - F(u_1, v_2, \ldots, v_d)| \\
&\quad + |F(u_1, v_2, \ldots, v_d) - F(u_1, u_2, v_3, \ldots, v_d)| \\
&\quad + |F(u_1, u_2, v_3, \ldots, v_d) - F(u_1, u_2, u_3, v_4, \ldots, v_d)| \\
&\quad + \cdots + |F(u_1, \ldots, u_{d-1}, v_d) - F(\mathbf{u})| \\
&\leq \sum_{j=1}^{d} |v_j - u_j|
\end{aligned}
$$

where the last inequality follows from condition (a). $\qquad\square$

Remark 1.5.3. It is possible to show that 1 is the best possible constant in the Lipschitz condition in (1.5.1); in other words, no constant $\alpha < 1$ exists such that

$$|C(v_1, \ldots, v_d) - C(u_1, \ldots, u_d)| \leq \alpha \|\mathbf{u} - \mathbf{v}\|_1 \,,$$

for every copula C. In fact, let $\mathbf{u} = (u, 1, \ldots, 1)$ and $\mathbf{v} = (v, 1, \ldots, 1)$ be such that $u < v$; then, for every copula $C \in \mathscr{C}_d$,

$$C(\mathbf{v}) - C(\mathbf{u}) = v - u = \|\mathbf{v} - \mathbf{u}\|_1 \,.$$

Moreover, since for a fixed $d \geq 2$, one has $\ell_1(d) \subseteq \ell_p(d)$, namely $\|\mathbf{x}\|_1 \geq \|\mathbf{x}\|_p$ for every real sequence $\mathbf{x} = (x_n)$, one may wonder whether there exists a number $p > 1$ such that

$$|C(\mathbf{v}) - C(\mathbf{u})| \leq \|\mathbf{v} - \mathbf{u}\|_p \,.$$

However, consider the copula C defined, for $\mathbf{u} \in \mathbb{I}^d$, by

$$C(\mathbf{u}) := W_2(u_1, u_2)\, u_3 \ldots u_d \,.$$

Take $\mathbf{u} = (u, u, 1, \ldots, 1)$ and $\mathbf{v} = (v, v, 1, \ldots, 1)$ in $\mathbb{I}^d$ with $1/2 < u < v$; then

$$\begin{aligned}
C(\mathbf{v}) - C(\mathbf{u}) &= W_2(v, v) - W_2(u, u) = 2\,(v - u) = \|\mathbf{v} - \mathbf{u}\|_1 \\
&> 2^{1/p}\,(v - u) = \|\mathbf{v} - \mathbf{u}\|_p \,.
\end{aligned}$$

In the bivariate case, De Baets et al. [2010] have proved that for every $p \in [1, +\infty]$ there exists a copula that is Lipschitz continuous with respect to the $\ell_p(2)$-norm. Specifically, for every $p \in [1, +\infty]$ such a copula equals M_2 for $p = +\infty$ and

$$C_p(u, v) = \max\left\{0, 1 - ((1 - u)^p + (1 - v)^p)^{1/p}\right\} \,,$$

for $p \in [1, +\infty[$.   ∎

A stronger notion of continuity can be introduced in $\mathscr{C}_d$.

Definition 1.5.4. A copula $C \in \mathscr{C}_d$ is *absolutely continuous* if it can be expressed in the form

$$C(\mathbf{u}) = \int_{[\mathbf{0}, \mathbf{u}]} c(\mathbf{t})\, d\mathbf{t}$$

for a suitable integrable function $c : \mathbb{I}^d \to \mathbb{R}_+$.   ◇

The function c is called the *density* of C. As an obvious example, the copula Π_d is absolutely continuous with density $c = 1$. Since a copula has uniform margins, the density of an absolutely continuous copula can be easily characterised by means of the following property: for every $t \in \mathbb{I}$ and for every $j \in \{1, \ldots, d\}$

$$\int_0^1 du_1 \ldots \int_0^1 du_{j-1} \int_0^t du_j \int_0^1 du_{j+1} \ldots \int_0^1 c(\mathbf{u})\, du_d = t \,. \tag{1.5.4}$$

1.6 The derivatives of a copula

Let C be a d-copula. For $j = 1, 2, \ldots, d$, we shall consistently adopt the notation

$$\partial_j C(u_1, \ldots, u_d) := \frac{\partial C(u_1, \ldots, u_d)}{\partial u_j}$$

for the j-th partial derivative of a copula C (where it exists).

Because of Lebesgue's theorem on the differentiation of a monotone function (see, e.g., [Tao, 2011]), and since the function $t \mapsto C(\mathbf{u}_j(t))$ is increasing for all $u_1,$ $\ldots, u_{j-1}, u_{j+1}, \ldots, u_d$ in $\mathbb{I}$, $\partial_j C(\mathbf{u}_j(t))$ exists for almost every $t \in \mathbb{I}$ and for all $u_1,$ $\ldots, u_{j-1}, u_{j+1}, \ldots, u_d$ in $\mathbb{I}$.

Moreover, as a consequence of Lipschitz condition (1.5.1), every copula C is absolutely continuous in each place separately, namely for all $u_1, \ldots, u_{j-1}, u_{j+1}, \ldots,$ u_d in $\mathbb{I}$, $t \mapsto C(\mathbf{u}_j(t))$ is absolutely continuous on $\mathbb{I}$. Thus (see, e.g., [Tao, 2011]) every d-copula C may be represented in the form

$$C(\mathbf{u}) = \int_0^{u_j} \partial_j C(u_1, \ldots, u_{j-1}, t, u_{j+1}, \ldots, u_d) \, dt \,. \tag{1.6.1}$$

By Vitali's theorem (see, e.g., [Vitali, 1905]), the derivatives $\partial_j C$ $(j = 1, \ldots, d)$ are Borel-measurable. This yields the first part of the following

Theorem 1.6.1. *The partial derivatives $\partial_j C$ $(j = 1, \ldots, d)$ of a d-copula C are Borel-measurable and satisfy*

$$0 \leq \partial_j C(\mathbf{u}_j(t)) \leq 1 \tag{1.6.2}$$

for all $u_1, \ldots, u_{j-1}, u_{j+1}, \ldots, u_d$ in $\mathbb{I}$ and for almost every $t \in \mathbb{I}$.

Proof. The inequalities (1.6.2) are an immediate consequence of the Lipschitz property (1.5.1). □

Remark 1.6.2. In the study of weak convergence of the empirical copula process (see, e.g., [Segers, 2012]) one usually requires that, for every $j \in \{1, \ldots, d\}$, the j-th first-order partial derivative exists and is continuous on the set $\{\mathbf{u} \in \mathbb{I}^d : 0 < u_j < 1\}$. Obviously, this condition is not satisfied by any copula; consider, for instance, the copulas M_d or W_2. ∎

Remark 1.6.3. Lipschitz condition (1.5.1) and Rademacher's theorem (see, e.g., [Tao, 2011]) imply that every d-copula C is totally differentiable at $x_0 \in \mathbb{I}^d$ for λ_d-almost every $x_0 \in \mathbb{I}^d$. In particular, it follows that $\partial_j C$ exists λ_d-almost everywhere for $j = 1, \ldots, d$. ∎

In the case $d = 2$, under some additional conditions, the derivatives can be used in order to provide a further characterisation of copulas.

Corollary 1.6.4. *Let $C : \mathbb{I}^2 \to \mathbb{I}$ be a copula such that the first partial derivatives of C exist on $\mathbb{I}^2$. Then, if u and v in $\mathbb{I}$ are such that $u < v$, then, for every t in $]0,1[$,*

$$\partial_2 C(v,t) \geq \partial_2 C(u,t) \qquad and \qquad \partial_1 C(t,v) \geq \partial_1 C(t,u), \tag{1.6.3}$$

or, equivalently, for every $u \in \mathbb{I}$, the functions

$$\mathbb{I} \ni t \mapsto \partial_2 C(t,u) \qquad and \qquad \mathbb{I} \ni t \mapsto \partial_1 C(u,t)$$

are increasing.

Proof. This follows immediately from Corollary 1.4.3 by showing that inequality (1.4.1) is equivalent to condition (1.6.3).   □

In order to formulate the previous result in a more general setting, one needs a generalised notion of derivative. This is the content of the next subsection, which is based on the results by Durante and Jaworski [2010]. For another characterisation based on derivatives, see also [Ghiselli Ricci, 2013b,a].

1.6.1   Characterisation of bivariate copulas via Dini derivatives

Before proceeding, the properties of the *Dini derivatives* of a real function f will be recalled.

Definition 1.6.5. Let $f : [a,b] \to \mathbb{R}$ be continuous and let x belong to $[a,b[$; the limits,

$$D^+ f(x) := \limsup_{\substack{h \to 0 \\ h > 0}} \frac{f(x+h) - f(x)}{h}, \tag{1.6.4}$$

$$D_+ f(x) := \liminf_{\substack{h \to 0 \\ h > 0}} \frac{f(x+h) - f(x)}{h}, \tag{1.6.5}$$

are called the *rightside upper* and *rightside lower Dini derivatives* of f at x, respectively.   ◇

These derivatives take values in $\mathbb{R}$. If f_1 and f_2 are continuous functions defined in $[a,b]$, it is easily shown that

$$D_+(f_1 + f_2)(x) \geq D_+ f_1(x) + D_+ f_2(x),$$
$$D^+(f_1 + f_2)(x) \leq D^+ f_1(x) + D^+ f_2(x),$$
$$D^+(f_1 - f_2)(x) \geq D^+ f_1(x) - D^+ f_2(x),$$

at every point $x \in [a,b[$ at which the Dini derivatives of f_1 and f_2 exist and are finite.

Obviously, if f is increasing in $[a,b]$, then both $D^+ f$ and $D_+ f$ are positive (≥ 0) for every $x \in [a,b[$. We shall need the converse of this latter statement [Lojasiewicz, 1988, Theorem 7.4.14].

Lemma 1.6.6. *If $f : [a, b] \to \mathbb{R}$ is continuous, if $D^+ f(x) \geq -\infty$ with the possible exception of an at most countable subset of $[a, b]$ and if $D^+ f(x) \geq 0$ a.e. in $[a, b]$, then f is increasing in $[a, b]$.*

Consider now a function $H : [x_1, x_2] \times [y_1, y_2] \to \mathbb{R}$ and its *horizontal section* at $y \in [y_1, y_2]$, $H_y : [x_1, x_2] \to \mathbb{R}$, defined by $H_y(x) := H(x, y)$.

The following lemma gives the core of the announced characterisation.

Lemma 1.6.7. *For a continuous function $H : \mathbb{I}^2 \to \mathbb{R}$ such that $D_+ H_0(x)$ and $D^+ H_1(x)$ are finite for every $x \in \mathbb{I} \setminus Z$, where Z is an at most countable subset of $\mathbb{I}$, the following statements are equivalent:*

(a) *H is 2-increasing;*

(b) *for all y and z in $\mathbb{I}$ with $z > y$, and for every $x \in \mathbb{I} \setminus Z$, $D^+ H_y(x)$ and $D_+ H_y(x)$ are finite and satisfy*

$$D_+ H_z(x) \geq D_+ H_y(x), \qquad D^+ H_z(x) \geq D^+ H_y(x) \, ;$$

(c) *for all y and z in $\mathbb{I}$ with $z > y$, and for every $x \in \mathbb{I} \setminus Z$, $D^+ H_y(x)$ is finite, and*

$$D^+ H_z(x) \geq D^+ H_y(x) \qquad \text{for almost every } x \in \mathbb{I}.$$

Proof. (a) $\Longrightarrow$ (b) The 2-increasing property for the rectangle $[x, x + h] \times [y, z]$ with $h > 0$ sufficiently small yields

$$H_y(x + h) - H_y(x) \leq H_z(x + h) - H_z(x),$$

and, as a consequence,

$$D^+ H_y(x) = \limsup_{\substack{h \to 0 \\ h > 0}} \frac{H_y(x + h) - H_y(x)}{h}$$

$$\leq \limsup_{\substack{h \to 0 \\ h > 0}} \frac{H_z(x + h) - H_z(x)}{h} = D^+ H_z(x) \, .$$

Replacing the limit superior by the limit inferior yields the other inequality. Since one has now

$$D_+ H_0(x) \leq D_+ H_y(x) \leq D^+ H_y(x) \leq D^+ H_1(x),$$

it is possible to conclude that $D_+ H_y(x)$ and $D^+ H_y(x)$ are finite for every $x \in \mathbb{I} \setminus Z$ and for every $y \in \mathbb{I}$.

The implication (b) $\Longrightarrow$ (c) is trivial.

(c) $\Longrightarrow$ (a) Take x_0, y and z in $\mathbb{I}$, with $y < z$. For $x \in \mathbb{I}$ with $x \geq x_0$ consider the function $f : \mathbb{I} \to \mathbb{R}$ defined via

$$f(x) := H_z(x) - H_y(x) - H_z(x_0) + H_y(x_0) \, .$$

Since $f(x_0) = 0$, in order to prove that H is 2-increasing, it suffices to show that f is increasing for arbitrary y and z in $\mathbb{I}$, with $y < z$. Now, for $x \in \mathbb{I} \setminus Z$, one has

$$D^+ f(x) = D^+ (H_z(x) - H_y(x)) \geq D^+ H_z(x) - D^+ H_y(x),$$

and, as a consequence, $D^+ f(x) \geq 0$ almost everywhere on $\mathbb{I}$; because of Lemma 1.6.6, it follows that f is increasing on $\mathbb{I}$.   □

The characterisation of 2-copulas is now an immediate consequence of the previous lemma.

Theorem 1.6.8. *For a function $C : \mathbb{I}^2 \to \mathbb{I}$ the following statements are equivalent:*

(a) *C is a copula;*

(b) *C satisfies the boundary conditions, is continuous and, for all y and z in $\mathbb{I}$ with $z > y$, and for every $x \in \mathbb{I} \setminus Z$, where Z is at most countable, $D^+ C_z(x)$ is finite and, for almost every $x \in \mathbb{I}$,*

$$D^+ C_z(x) \geq D^+ C_y(x).$$

In checking that a function $C : \mathbb{I}^2 \to \mathbb{I}$ is a copula, the above characterisation allows to prove (b) rather than directly the property of being 2-increasing.

An application of the above characterisation is provided by the following

Theorem 1.6.9. *Let $C \in \mathscr{C}_2$ be absolutely continuous with a continuous density c; if the functions f, $g : \mathbb{I} \to \mathbb{R}$ are continuous and satisfy*

$$f(0) = f(1) = g(0) = g(1) = 0,$$

then the following statements are equivalent:

(a) *the function $\widetilde{C} : \mathbb{I}^2 \to \mathbb{R}$ defined by $\widetilde{C}(u, v) := C(u, v) + f(u)\, g(v)$ is a bivariate copula;*

(b) *for all u and v in $\mathbb{I}$, the density $c(u, v)$ is greater than, or at most equal to,*

$$- \min\{D^+ f(u) D^+ g(v), D_+ f(u) D^+ g(v), D^+ f(u) D_+ g(v), D_+ f(u) D_+ g(v)\}.$$

Proof. First notice that it follows from

$$C(u, v) = \int_0^u \mathrm{d}s \int_0^v c(s, t)\, \mathrm{d}t$$

that the Dini derivative

$$D^+ C_v(u) = \partial_1 C(u, v) = \int_0^v c(u, t)\, \mathrm{d}t$$

is continuous as a function of either u or v. Moreover, for a fixed $u \in \mathbb{I}$, one has

$$D^+ \partial_1 C(u, v) = D_+ \partial_1 C(u, v) = \partial_2 \partial_1 C(u, v) = c(u, v).$$

For a fixed $u \in \mathbb{I}$, introduce the functions defined by

$$\Psi_u(v) := D^+ \widetilde{C}_v(u) \, \mathbf{1}_{\{v:g(v) \geq 0\}} + D_+ \widetilde{C}_v(u) \, \mathbf{1}_{\{v:g(v) < 0\}} \,,$$

$$\Phi_u(v) := D_+ \widetilde{C}_v(u) \, \mathbf{1}_{\{v:g(v) \geq 0\}} + D^+ \widetilde{C}_v(u) \, \mathbf{1}_{\{v:g(v) < 0\}} \,.$$

For a constant $a \in \mathbb{R}$, and for a function φ, one has

$$D^+ (a \, \varphi(t)) = \begin{cases} a \, D^+ \varphi(t) \,, & a \geq 0 \,, \\ a \, D_+ \varphi(t) \,, & a < 0 \,, \end{cases}$$

and

$$D_+ (a \, \varphi(t)) = \begin{cases} a \, D_+ \varphi(t) \,, & a \geq 0 \,, \\ a \, D^+ \varphi(t) \,, & a < 0 \,. \end{cases}$$

Thus

$$\begin{aligned} \Psi_u(v) &= \big(\partial_1 C(u,v) + g(v) \, D^+ f(u)\big) \, \mathbf{1}_{\{v:g(v) \geq 0\}} \\ &\quad + \big(\partial_1 C(u,v) + g(v) \, D^+ f(u)\big) \, \mathbf{1}_{\{v:g(v) < 0\}} \\ &= \partial_1 C(u,v) + g(v) \, D^+ f(u) \,, \end{aligned}$$

and, analogously,

$$\Phi_u(v) = \partial_1 C(u,v) + g(v) \, D_+ f(u) \,.$$

In the same way, one easily obtains

$$D^+ \Psi_u(v) \, \mathbf{1}_{\{D^+ f(u) \geq 0\}} + D^+ \Psi_u(v) \, \mathbf{1}_{\{D^+ f(u) < 0\}} = c(u,v) + D^+ f(u) \, D^+ g(v) \,,$$

$$D_+ \Psi_u(v) \, \mathbf{1}_{\{D^+ f(u) \geq 0\}} + D^+ \Psi_u(v) \, \mathbf{1}_{\{D^+ f(u) < 0\}} = c(u,v) + D^+ f(u) \, D_+ g(v) \,,$$

$$D^+ \Phi_u(v) \, \mathbf{1}_{\{D_+ f(u) \geq 0\}} + D_+ \Phi_u(v) \, \mathbf{1}_{\{D_+ f(u) < 0\}} = c(u,v) + D_+ f(u) \, D^+ g(v) \,,$$

$$D_+ \Phi_u(v) \, \mathbf{1}_{\{D_+ f(u) \geq 0\}} + D^+ \Phi_u(v) \, \mathbf{1}_{\{D_+ f(u) < 0\}} = c(u,v) + D_+ f(u) \, D_+ g(v) \,.$$

After these preliminaries the proof proceeds as follows.

(a) $\Longrightarrow$ (b) If $\widetilde{C}$ is a copula, then, by Theorem 1.6.8 and Lemma 1.6.7, both $v \mapsto D^+ \widetilde{C}_v(u)$ and $v \mapsto D_+ \widetilde{C}_v(u)$ are increasing, as are $v \mapsto \Psi_v(u)$ and $v \mapsto \Phi_v(u)$. Indeed both these latter functions are continuous and equal to $v \mapsto D^+ \widetilde{C}_v(u)$ and $v \mapsto D_+ \widetilde{C}_v(u)$, respectively, on intervals where the sign of g is constant. Therefore their Dini derivatives are positive and

$$c(u,v) + \min\{D^+ f(u) \, D^+ g(v), D_+ f(u) \, D^+ g(v),$$
$$D^+ f(u) \, D_+ g(v), D_+ f(u) \, D_+ g(v)\} \geq 0 \,.$$

(b) $\Longrightarrow$ (a) Conversely, the Dini derivatives of $v \mapsto \Psi_v(u)$ and $v \mapsto \Phi_v(u)$ are positive, so that both functions are increasing. As a consequence also $v \mapsto D^+ \widetilde{C}_v(u)$ is increasing; moreover it is finite, since

$$\widetilde{C}_0(u) = \widetilde{C}(u,0) = C(u,0) = 0 \quad \text{and} \quad \widetilde{C}_1(u) = \widetilde{C}(u,1) = C(u,1) = u \,,$$

which imply $D^+ \widetilde{C}_0(u) = 0$ and $D^+ \widetilde{C}(u,1) = 1$. Therefore, by virtue of Theorem 1.6.8, $\widetilde{C}$ is a 2-copula. $\qquad \square$

Example 1.6.10. Let $f, g : \mathbb{I} \to \mathbb{R}$ be differentiable functions satisfying the assumptions of Theorem 1.6.9. Consider the function $\widetilde{C} : \mathbb{I}^2 \to \mathbb{R}$ defined by $\widetilde{C}(u_1, u_2) := u_1 u_2 + f(u_1) g(u_2)$. Then, in view of Theorem 1.6.9, $\widetilde{C}$ is a copula if, and only if,

$$1 \geq - \min_{(u,v) \in \mathbb{I}^2} f'(u) g'(v).$$

In particular, for $\alpha \in [-1, 1]$ and $f(t) = g(t) = t(1 - t)$ it can be easily checked that the function

$$C_\alpha^{\mathbf{EFGM}}(u_1, u_2) = u_1 u_2 \left(1 + \alpha \left(1 - u_1\right) \left(1 - u_2\right)\right) \tag{1.6.6}$$

is a copula. These copulas are called *Eyraud-Farlie-Gumbel-Morgenstern* (EFGM), see Section 6.3. ∎

1.7 The space of copulas

Now, we consider the set of copulas as a *space*. In particular, in this section we focus on basic properties related to orderings and distances, as well as to transformations in the space $\mathscr{C}_d$ that are isomorphic to symmetries in $\mathbb{I}^d$.

1.7.1 Lattice properties

We start by introducing the standard partial order among real-valued functions in the space of copulas.

Definition 1.7.1. Let C, C' be in $\mathscr{C}_d$. C is less than C' in the *pointwise order*, and one writes $C \leq C'$, if, and only, if $C(\mathbf{u}) \leq C'(\mathbf{u})$ for every $\mathbf{u} \in \mathbb{I}^d$. ◇

When $C \leq C'$, one also says that C is *more positive lower orthant dependent* (more PLOD) (or *more concordant*) than C'. For this dependence notion, see also Definition 2.5.2.

The order on $\mathscr{C}_d$ introduced above is not total: in fact, not all copulas are comparable with respect to $\leq$ as the following example shows.

Example 1.7.2. Consider the 2-copula C given by:

$$C(u, v) = \begin{cases} \max\{0, u + v - \frac{1}{2}\}, & (u, v) \in [0, \frac{1}{2}]^2, \\ \min\{u, v\}, & \text{otherwise.} \end{cases}$$

Then

$$C\left(\frac{1}{4}, \frac{1}{4}\right) = 0 < \Pi_2\left(\frac{1}{4}, \frac{1}{4}\right) \quad \text{but} \quad C\left(\frac{3}{4}, \frac{3}{4}\right) = \frac{3}{4} > \Pi_2\left(\frac{1}{4}, \frac{1}{4}\right).$$

In other words, C and Π_2 are not comparable. ∎

The following result provides upper and lower bounds in $\mathscr{C}_d$ with respect to the given order. To this end, consider the function $W_d : \mathbb{I}^d \to \mathbb{I}$ defined by

$$W_d(\mathbf{u}) := \max\left\{0, \sum_{j=1}^{d} u_j - (d-1)\right\}. \tag{1.7.1}$$

Theorem 1.7.3. *For every d-copula C and for every point* $\mathbf{u} = (u_1, \ldots, u_d) \in \mathbb{I}^d$, *one has*

$$W_d(\mathbf{u}) \le C(\mathbf{u}) \le M_d(\mathbf{u}). \tag{1.7.2}$$

Proof. Let the copula C be the d.f. of a r.v. $\mathbf{U}$ that is defined on the probability space $(\Omega, \mathscr{F}, \mathbb{P})$ and which has uniform margins on $\mathbb{I}$. Then, for every index $j \in \{1, \ldots, d\}$ and for every $\mathbf{u} \in \mathbb{I}^d$, one has

$$\bigcap_{k=1}^{d} \{U_k \le u_k\} \subseteq \{U_j \le u_j\},$$

which implies that

$$C(\mathbf{u}) = \mathbb{P}\left(\bigcap_{k=1}^{d} \{U_k \le u_k\}\right) \le \min_{j \in \{1,\ldots,d\}} \mathbb{P}\left(U_j \le u_j\right) = M_d(\mathbf{u}).$$

Analogously, one has

$$C(\mathbf{u}) = \mathbb{P}\left(\bigcap_{j=1}^{d} \{U_j \le u_j\}\right) = 1 - \mathbb{P}\left(\bigcup_{j=1}^{d} \{U_j > u_j\}\right)$$

$$\ge 1 - \sum_{j=1}^{d} \mathbb{P}\left(U_j > u_j\right) = 1 - \sum_{j=1}^{d}(1 - u_j) = \sum_{j=1}^{d} u_j - (d-1).$$

Since C takes positive values, it follows that $C(\mathbf{u}) \ge W_d(\mathbf{u})$. $\qquad\square$

The functions W_d and M_d in (1.7 ?) are called the *lower* and *upper Hoeffding–Fréchet bounds*, respectively.

As shown in Example 1.3.3, M_d is a copula and, hence, it is the best-possible upper bound in $\mathscr{C}_d$; in other words, for every $\mathbf{u} \in \mathbb{I}^d$

$$M_d(\mathbf{u}) = \sup_{C \in \mathscr{C}_d} C(\mathbf{u}).$$

However, W_d is not a copula for $d > 2$. In fact, consider for instance the d-box

$$]\mathbf{1/2, 1}] =]1/2, 1] \times \cdots \times]1/2, 1].$$

The W_d-volume of this d-box is, for $d > 2$,

$$V_{W_d} \left(\,] \, 1/2, 1 \right]) = 1 - \frac{d}{2} < 0 \,.$$

It follows from Theorem 1.4.1 that W_d is not a copula for $d > 2$. However, the lower bound provided by inequality (1.7.2) is still the best possible as will be shown in Theorem 4.1.7.

Historical remark 1.7.4. The Hoeffding–Fréchet bounds appeared in the present form for the first time in an article by Fréchet [1951]. An earlier version, when the d.f.'s are continuous, had already been given by Hoeffding [1940], but with reference to the square $[-1/2, 1/2]^2$. Several proofs have since appeared, by Dall'Aglio [1972], by Wolff [1977], by Kimeldorf and Sampson [1978], by Mikusiński et al. [92]. For the upper bound see also [Dhaene et al., 2002a]. See also [Rüschendorf, 1981a] for the Hoeffding–Fréchet bounds in an abstract setting. The Hoeffding–Fréchet bounds can also be derived from the Bonferroni's inequalities [Bonferroni, 1936; Galambos, 1977].   ■

1.7.2   Metric properties

Consider the space $\left(\Xi(\mathbb{I}^d), d_\infty \right)$ of all continuous real-valued functions with domain $\mathbb{I}^d$. Moreover, define the distance d_∞ given, for all f_1 and f_2 in $\Xi(\mathbb{I}^d)$, by

$$d_\infty \left(f_1, f_2 \right) = \sup_{\mathbf{u} \in \mathbb{I}^d} \left| f_1 \left(\mathbf{u} \right) - f_2 \left(\mathbf{u} \right) \right| = \max_{\mathbf{u} \in \mathbb{I}^d} \left| f_1 \left(\mathbf{u} \right) - f_2 \left(\mathbf{u} \right) \right| \,.$$

As is well known, a sequence in $\Xi(\mathbb{I}^d)$ that converges with respect to d_∞ is said to converge *uniformly*. Moreover, the metric space $\left(\Xi(\mathbb{I}^d), d_\infty \right)$ is complete (see, e.g., Theorem 2.4.9 in [Dudley, 1989]). Properties of this space are inherited by the space of copulas.

Theorem 1.7.5. *If a sequence $(C_n)_{n \in \mathbb{N}}$ in $\mathscr{C}_d$ converges pointwise to C_0 as n goes to ∞, then C_0 is a copula.*

Proof. It is trivial to check that C_0 has univariate marginals that are uniformly distributed on $\mathbb{I}$. Moreover, one can note that, for all $\mathbf{a}$ and $\mathbf{b}$ in $\mathbb{I}^d$, the C_0-volume of $] \, \mathbf{a}, \mathbf{b}]$ can be expressed as the pointwise limit of the C_n-volumes of $] \, \mathbf{a}, \mathbf{b}]$ for n going to ∞, and hence

$$V_{C_0} \left( ] \, \mathbf{a}, \mathbf{b} \right]) = \lim_{n \to +\infty} V_{C_n} \left( ] \, \mathbf{a}, \mathbf{b} \right]) \geq 0.$$

As a consequence, C_0 is also a copula.   □

Theorem 1.7.6. *Let C_n $(n \in \mathbb{N})$ and C_0 be in $\mathscr{C}_d$. If the sequence $(C_n)_{n \in \mathbb{N}}$ converges to C_0 pointwise as n goes to ∞, then C_n converges to C_0 uniformly as $n \to \infty$.*

Proof. Let $\varepsilon > 0$. Because C_0 is uniformly continuous on $\mathbb{I}^d$, it follows that, for every $\mathbf{u} \in \mathbb{I}^d$, there is an open rectangular neighbourhood[2] $O_{\mathbf{u}}$ of $\mathbf{u}$ such that, for every $\mathbf{v} \in O_{\mathbf{u}}$,

$$|C_0(\mathbf{u}) - C_0(\mathbf{v})| < \varepsilon.$$

Moreover, $\mathbb{I}^d = \cup_{\mathbf{u} \in \mathbb{I}^d} O_{\mathbf{u}}$ and, since $\mathbb{I}^d$ is compact, it can be covered by a finite number of open neighbourhoods; thus there exist $k \in \mathbb{N}$ and k open sets $O_i =]\mathbf{a}_i, \mathbf{b}_i[\in \{O_{\mathbf{u}} : \mathbf{u} \in \mathbb{I}^d\}$ with $i = 1, \dots, k$ such that $\mathbb{I}^d = \cup_{i=1}^k]\mathbf{a}_i, \mathbf{b}_i[$. In view of the pointwise convergence of the sequence (C_n), there exists $\nu = \nu(\varepsilon) \in \mathbb{N}$ such that, for every $n \geq \nu$,

$$\max_{1 \leq i \leq k} (|C_n(\mathbf{a}_i) - C_0(\mathbf{a}_i)| \vee |C_n(\mathbf{b}_i) - C_0(\mathbf{b}_i)|) < \varepsilon.$$

Since copulas are increasing in each variable, one has, for every $\mathbf{u} \in]\mathbf{a}_i, \mathbf{b}_i[$,

$$C_n(\mathbf{u}) - C_0(\mathbf{u}) \leq C_n(\mathbf{b}_i) - C_0(\mathbf{b}_i) + C_0(\mathbf{b}_i) - C_0(\mathbf{u}) < 2\varepsilon,$$

and, analogously,

$$C_n(\mathbf{u}) - C_0(\mathbf{u}) \geq C_n(\mathbf{a}_i) - C_0(\mathbf{a}_i) + C_0(bfa_i) - C_0(\mathbf{u}) > -2\varepsilon.$$

Since the two previous inequalities are true for every $i = 1, \dots, k$, it follows that

$$\sup_{\mathbf{u} \in \mathbb{I}^d} |C_n(\mathbf{u}) - C_0(\mathbf{u})| \leq \max_{1 \leq i \leq k} \sup_{\mathbf{u} \in]\mathbf{a}_i, \mathbf{b}_i]} |C_n(\mathbf{u}) - C_0(\mathbf{u})| < 2\varepsilon,$$

which is equivalent to saying that C_n converges to C_0 uniformly. $\qquad\square$

An important property of $\mathscr{C}_d$ can now be proved.

Theorem 1.7.7. *The set $\mathscr{C}_d$ is a compact subset in $(\Xi(\mathbb{I}^d), d_\infty)$.*

Proof. Since $(\Xi(\mathbb{I}^d), d_\infty)$ is complete and $\mathscr{C}_d$ is closed in $\Xi(\mathbb{I}^d)$ (because of Theorem 1.7.5), also $\mathscr{C}_d$ is complete; see, e.g., [Dudley, 1989, Proposition 2.4.1]).

Moreover, $\mathscr{C}_d$ is uniformly bounded, since $\sup\{|C(\mathbf{u})| : \mathbf{u} \in \mathbb{I}^d, C \in \mathscr{C}_d\} \leq 1$ and equi-Lipschitz and, hence, equi-continuous, since every copula is 1-Lipschitz (see (1.5.1)). As a consequence of the Ascoli-Arzelà Theorem (see, e.g., [Dudley, 1989, 2.4.7]), $\mathscr{C}_d$ is totally bounded with respect to d_∞. Then, $\mathscr{C}_d$ is a complete and totally bounded metric space, and, as a consequence, it is compact (see, e.g., [Dudley, 1989, Theorem 2.3.1]). $\qquad\square$

Remark 1.7.8. To the best of our knowledge, the first results about convergence and compactness in $\mathscr{C}_d$ date back to Brown [1965, 1966] (see also Deheuvels [1978, 1979]). $\qquad\blacksquare$

---

[2]By the expression "rectangular neighbourhood" we mean a d-box containing $\mathbf{u}$ in its interior.

Since the family of copulas $\mathscr{C}_d$ is a compact (with respect to d_∞) and convex set, in view of the Krein-Milman Theorem [Dunford and Schwartz, 1958, Theorem V.8.4], it is the closure (with respect to d_∞) of convex hull of its extremal points, where the following definition holds.

Definition 1.7.9. A copula $C \in \mathscr{C}_d$ is said to be *extremal* if, given $A, B \in \mathscr{C}_d$, the equality $C = \alpha A + (1 - \alpha) B$ with $\alpha \in \,]0, 1[$ implies $C = A = B$. $\quad\diamond$

Examples of extremal copulas are M_d and W_2, since they are the sharp pointwise upper and lower bounds in $\mathscr{C}_d$ and $\mathscr{C}_2$, respectively. However, the complete characterisation of extremal copulas is still unknown.

Remark 1.7.10. Note that, as stressed by Beneš and Štěpán [1991], the most important reason for the interest in extremal points is that, in many situations, the search for a maximum, or a minimum, of a functional in $\mathscr{C}_d$ may be approached by considering the maximum, respectively the minimum, of the same functional in the smaller class of extremal copulas. $\quad\blacksquare$

1.7.3   Symmetry

A *symmetry* of $\mathbb{I}^d$ is a bijection ξ of $\mathbb{I}^d$ onto itself of the form

$$\xi(u_1, \ldots, u_d) = (v_1, \ldots, v_d),$$

where, for each i and for every permutation $(k_1, \ldots, k_d)$ of $(1, \ldots, d)$, either $v_i = u_{k_i}$ or $v_i = 1 - u_{k_i}$. The group of symmetries under the operation of composition is denoted by $\mathrm{Sym}(\mathbb{I}^d)$.

Symmetries can be used to transform a r.v. $\mathbf{U}$ with uniform marginals in $\mathbb{I}$ into another r.v. $\mathbf{U}'$, while preserving the marginal distributions. The usefulness of this property for copulas is summarised in the following result, whose proof is immediate by probabilistic arguments.

Theorem 1.7.11. *For a given copula* $C \in \mathscr{C}_d$, *let* $\mathbf{U}$ *be a random vector on* $(\Omega, \mathscr{F}, \mathbb{P})$ *whose d.f. is given by* C. *Let* ξ *be a symmetry in* $\mathbb{I}^d$ *and consider* $\mathbf{X} = \xi \circ \mathbf{U}$. *Then the d.f.* C^ξ *of* $\mathbf{X}$ *is a copula.*

Permutations and reflections are two elements of $\mathrm{Sym}(\mathbb{I}^d)$. A permutation $\tau : \mathbb{I}^d \to \mathbb{I}^d$ is defined by $\tau(x_1, \ldots, x_d) = (x_{k_1}, \ldots, x_{k_d})$ for a given one-to-one transformation

$$(1, \ldots, d) \mapsto (k_1, \cdots, k_d).$$

The *elementary reflections* $\sigma_1, \ldots, \sigma_d$ of $\mathbb{I}^d$ are defined via

$$\sigma_i(\mathbf{u}) = \mathbf{v}, \quad \text{where } v_j := \begin{cases} 1 - u_j, & \text{if } j = i, \\ u_j, & \text{if } j \neq i. \end{cases}$$

The composition $\sigma_1 \circ \sigma_2 \circ \cdots \circ \sigma_d$ will be denoted by $\sigma^{(d)}$; thus

$$\sigma^{(d)}(u_1, \cdots, u_d) = (1 - u_1, \cdots, 1 - u_d).$$

The sets of permutations and reflections are subgroups of $\text{Sym}(\mathbb{I}^d)$. Further, every symmetry ξ of $\mathbb{I}^d$ has a unique representation of the form $\sigma_{i_1} \cdots \sigma_{i_k} \tau$ (and another of the form $\tau' \sigma_{j_1} \cdots \sigma_{j_k}$) where τ is a permutation and $i_1 < \cdots < i_k$.

Thanks to previous concepts, two notions of symmetry in $\mathscr{C}_d$ can be defined.

Definition 1.7.12. A copula $C \in \mathscr{C}_d$ is said to be *exchangeable* if, for every permutation $\xi \in \text{Sym}(\mathbb{I}^d)$, one has $C = C^\xi$. $\diamond$

In the bivariate case, a copula C is exchangeable if $C = C^T$, where C^T denotes the copula obtained from C by permutating its arguments, also called *transpose* of C, i.e., $C^T(u, v) = C(v, u)$ for all $(u, v) \in \mathbb{I}^2$.

Remark 1.7.13. The property of being exchangeable is preserved by all the lower-dimensional marginals of a copula, in the sense that if a copula C is exchangeable, so are all its lower-dimensional marginals. However, the converse implication is not true, as shown, for instance, by the 3-copula $C(u_1, u_2, u_3) = M_2(u_1, u_2)\, u_3$. ∎

The following result may be useful in constructing an exchangeable copula from a given one and suitable auxiliary functions.

Theorem 1.7.14. *Let C be a d-copula and let F_1, ..., F_d be d.f.'s on $\mathbb{I}$ such that $\sum_{j=1}^{d} F_j(t) = d\,t$ for every $t \in \mathbb{I}$. Then the function $A : \mathbb{I}^d \to \mathbb{I}$ defined by*

$$A(\mathbf{u}) := \frac{1}{d!} \sum_{\sigma \in \mathscr{P}_d} C\left(F_1(u_{\sigma(1)}), \ldots, F_d(u_{\sigma(d)})\right), \tag{1.7.3}$$

where the sum is taken over all the permutations $\mathscr{P}_d$ of $\{1, \ldots, d\}$, is an exchangeable d-copula. Moreover, if C is absolutely continuous, so is D.

Proof. Since A is exchangeable by construction, it is enough to show that it is a copula. Now, each summand in (1.7.3) is a d.f., so that also A, which is a convex combination of d.f.'s, is a distribution function. Moreover

$$A(u_1, 1, \ldots, 1) = \frac{1}{d!} \sum_{j=1}^{d} (d-1)!\, F_j(u_1) = u_1. \tag{1.7.4}$$

This shows that A has uniform margins. Finally, let C be absolutely continuous with density c. For every $j \in \{1, \ldots, d\}$ and for $s < t$ one has, setting $G_j(t) := dt - F_j(t) = \sum_{i \neq j} F_i(t)$,

$$0 \leq F_j(t) - F_j(s) = (dt - G_j(t)) - (ds - G_j(s))$$
$$\leq d\,(t - s) - (G_j(t) - G_j(s)) \leq d\,(t - s),$$

which proves that F_j is a Lipschitz and, hence, absolutely continuous function; therefore is differentiable almost everywhere. An easy calculation now yields the density a of A

$$a(\mathbf{u}) = \frac{\partial^d A(\mathbf{u})}{\partial u_1 \ldots \partial u_d} = \frac{1}{d!} \sum_{\sigma \in \mathscr{P}_d} c\left(F_1(u_{\sigma(1)}), \ldots, F_d(u_{\sigma(d)})\right) \prod_{j=1}^{d} \frac{\partial F_j(u_{\sigma(j)})}{\partial u_{\sigma(j)}}$$

a.e. in $\mathbb{I}^d$.   □

Corollary 1.7.15. *Let C be a bivariate copula and let F be a d.f. on $\mathbb{I}$ such that $|F(t) - F(s)| \leq 2\,|t - s|$ for all s and t in $\mathbb{I}$. Then*

$$D(u, v) := \frac{1}{2}\,\{C(F(u), 2v - F(v)) + C(F(v), 2u - F(u))\}$$

is an exchangeable bivariate copula.

Example 1.7.16. Let (U, V) be a r.v. distributed according to $A \in \mathscr{C}_2$. Then $\delta(t) := \mathbb{P}(U \leq t, V \leq t) = A(t, t)$ is a d.f. on $\mathbb{I}$, which, in view of the 1-Lipschitz condition of a copula, satisfies $|\delta(t) - \delta(s)| \leq 2\,|t - s|$. Then by Corollary 1.7.15 it follows that, for every $C \in \mathscr{C}_2$,

$$D(u, v) = \frac{1}{2}\,\{C(\delta(u), 2v - \delta(v)) + C(\delta(v), 2u - \delta(v))\}$$

is a copula. In particular for $C = M_2$ one has

$$D(u, v) = \frac{1}{2}\,\{\min\{\delta(u), 2v - \delta(v)\} + \min\{\delta(v), 2u - \delta(u)\}\}$$
$$= \min\left\{u, v, \frac{\delta(u) + \delta(v)}{2}\right\},$$

which is a copula, introduced in [Fredricks and Nelsen, 1997; Nelsen and Fredricks, 1997], such that $D(t, t) = \delta(t)$. It will be called the *Fredricks-Nelsen copula*, and will be denoted by $C^{\mathbf{FN}}$ (see Section 2.6).   ■

A second form of symmetry is the *radial symmetry* (also called *reflection symmetry*) of a random vector.

Definition 1.7.17. A copula $C \in \mathscr{C}_d$ is said to be *radially symmetric* if $C = C^{\sigma^d}$ for the reflection $\sigma^{(d)} \in \mathrm{Sym}(\mathbb{I}^d)$.   ◇

Therefore, if C is radially symmetric, then $C = \widehat{C}$, where $\widehat{C}$ is the d.f. of the r.v. $(1 - U_1, \ldots, 1 - U_d)$, i.e.,

$$\widehat{C}(\mathbf{u}) = \mathbb{P}\,(1 - U_1 \leq u_1, \ldots, 1 - U_d \leq u_d) = \mathbb{P}(U_1 \geq 1 - u_1, \ldots, U_d \geq 1 - u_d)\,.$$

By using the inclusion-exclusion formula (see, e.g., [Billingsley, 1979]), one has

$$\widehat{C}(\mathbf{u}) = 1 + \sum_{k=1}^{d}(-1)^k \sum_{1 \leq i_1 < i_2 < \cdots < i_k \leq n} C_{i_1 i_2 \ldots i_k}(1 - u_{i_1}, 1 - u_{i_2}, \ldots, 1 - u_{i_k})\,,$$

$$\tag{1.7.5}$$

with $C_{i_1, \cdots, i_k}$ denoting the marginal of C related to $(i_1, \cdots, i_k)$.

Definition 1.7.18. Given a copula $C \in \mathscr{C}_d$, the copula $\widehat{C}$ of (1.7.5) is called the *survival copula associated with C*. ◇

The operation that associates to each copula its survival copula is involutory, i.e., $\widehat{(\widehat{C})} = C$, for every $C \in \mathscr{C}_d$.

Example 1.7.19. In dimensions 2 or 3, the survival copula can be easily computed. In fact, if $C \in \mathscr{C}_2$, then, by using (1.7.5), for every $(u_1, u_2) \in \mathbb{I}^2$, one has

$$\widehat{C}(u_1, u_2) = u_1 + u_2 - 1 + C(1 - u_1, 1 - u_2).$$

Moreover, the survival copula $\widehat{C}$ of the 3-copula C is given by

$$\widehat{C}(u_1, u_2, u_3) = 1 - C(u_1, u_2, 1) - C(u_1, 1, u_3) - C(1, u_2, u_3)$$
$$+ u_1 + u_2 + u_3 - C(u_1, u_2, u_3)$$

for every $(u_1, u_2, u_3) \in \mathbb{I}^3$. ∎

Remark 1.7.20. Let C be in $\mathscr{C}_d$ and let $\mathbf{U}$ be a r.v. on a suitable probability space $(\Omega, \mathscr{F}, \mathbb{P})$ having uniform margins and distributed according to C. Since C is (the restriction to $\mathbb{I}^d$ of) a d.f., it is possible to consider the associated *survival function*, denoted by $\overline{C}$, given, for every $\mathbf{u} \in \mathbb{I}^d$, by

$$\overline{C}(\mathbf{u}) := \mathbb{P}(U_1 > u_1, \ldots, U_d > u_d). \tag{1.7.6}$$

The survival function $\overline{C}$ is not a copula and, therefore, it should not be confused with the survival copula $\widehat{C}$ associated with C. Moreover, $\widehat{C}(\mathbf{1} - \mathbf{u}) = \overline{C}(\mathbf{u})$. ∎

Remark 1.7.21. If a copula C is absolutely continuous with density c, then the radial symmetry of C implies that

$$c(u_1, \ldots, u_d) = c(1 - u_1, \ldots, 1 - u_d)$$

for every $\mathbf{u} \in \mathbb{I}^d$. ∎

Finally, a third form of symmetry can be derived for a bivariate random vector.

Definition 1.7.22. A copula $C \in \mathscr{C}_2$ is said to be *jointly symmetric* if $C = C^{\sigma_1} = C^{\sigma_2}$. ◇

Obviously, every radially symmetric 2-copula is also jointly symmetric. An example of jointly symmetric copula is given by $(W_2 + M_2)/2$.

Further readings 1.7.23. For the study of symmetries of bivariate distribution functions and copulas, we refer the reader to [Nelsen, 1993]. Invariant 2-copulas under special symmetries are discussed by Klement et al. [2002]. Additional material about exchangeable copulas can be found in the survey by Genest and Nešlehová [2013]. ∎

1.7.4   Asymmetrisation of copulas

While most of the examples of copulas that are used in the literature are exchangeable, there is a statistical interest in having at one's disposal non-exchangeable copulas; these are better adapted to describe situations where the dependence structure is either not necessarily exchangeable or, even, inherently non-symmetric: it suffices to think of the case when one component of a r.v. influences the other one more than the other way round. Here we present a simple method to construct non-exchangeable multivariate copulas.

In this subsection, $\mathscr{G}$ denotes the set of functions $g : \mathbb{I} \to \mathbb{I}$ that are either identically equal to 1, or are continuous increasing bijections.

Theorem 1.7.24. *Let $C_1, \ldots, C_n$ be d-copulas, and, for all j and i in $\{1, \ldots, d\}$, let the functions g_{ji} be in $\mathscr{G}$. If $\prod_{j=1}^{n} g_{ji}(v) = v$ for all $v \in \mathbb{I}$ and $i = 1, \ldots, d$, then the function $\widetilde{C} : \mathbb{I}^d \to \mathbb{I}$ defined by*

$$\widetilde{C}(\mathbf{u}) := \prod_{j=1}^{n} C_j \left(g_{j1}(u_1), \ldots, g_{jd}(u_d) \right)$$

is also a d-copula.

Proof. Let

$$\mathbf{U_j} = \left(U_1^{(j)}, \ldots, U_d^{(j)} \right) \qquad (j = 1, \ldots, n)$$

be n independent random vectors on the same probability space $(\Omega, \mathscr{F}, \mathbb{P})$ whose d.f.'s are given, respectively, by the copulas $C_1, \ldots, C_n$. Since the inequality $v \leq g_{ji}(u)$ is equivalent to the other one $g_{ji}^{-1}(v) \leq u$, one has, for all u and v in $\mathbb{I}$,

$$\prod_{j=1}^{n} C_j \left(g_{j1}(u_1), \ldots, g_{jd}(u_d) \right)$$

$$= \mathbb{P} \left(U_1^{(j)} \leq g_{j1}(u_1), \ldots, U_d^{(j)} \leq g_{jd}(u_d) : j = 1, \ldots, n \right)$$

$$= \mathbb{P} \left(\max_{j=1,\ldots,n} \left\{ g_{j1}^{-1} \left(U_1^{(j)} \right) \leq u_1 \right\}, \ldots, \max_{j=1,\ldots,n} \left\{ g_{jd}^{-1} \left(U_d^{(j)} \right) \leq u_d \right\} \right),$$

so that $\widetilde{C}$ is the d.f. of the r.v.'s

$$\max \left\{ g_{ji}^{-1} \left(U_i^{(j)} \right)_{i=1,\ldots,d} : j \in J(i) \right\}$$

where $J(i)$ is the set of indices j in $\{1, \ldots, n\}$ for which g_{ji} is not identically equal to 1. Moreover, since $g_{ji}(1) = 1$, one has, for $i = 1, \ldots, d$,

$$\widetilde{C}(u_1, \ldots, u_{i-1}, 0, u_{i+1}, \ldots, u_d) = 0 \,,$$

$$\widetilde{C}(\underbrace{1, \ldots, 1}_{i-1}, v, 1, \ldots, 1) = \prod_{j=1}^{n} g_{ji}(v) = v \,;$$

hence $\widetilde{C}$ is a d-copula. $\square$

Example 1.7.25. Let C be any d-copula, $C \in \mathscr{C}_d$. In Theorem 1.7.24 take $n = 2$, $C_1 = C$, $C_2 = \Pi_d$, $g_{1i}(v) := v^{1-\theta_i}$ and $g_{2i}(v) = v^{\theta_i}$, where for every $i \in \{1, \ldots, d\}$, θ_i belongs to $]0, 1[$. Then the function $\widetilde{C} : \mathbb{I}^d \to \mathbb{I}$ defined by

$$\widetilde{C}(u_1, u_2, \ldots, u_d) := C\left(u_1^{1-\theta_1}, u_2^{1-\theta_2}, \ldots, u_d^{1-\theta_d}\right) \prod_{i=1}^{d} u^{\theta_i}$$

is a d-copula. Notice that, even when C is exchangeable, $\widetilde{C}$ is not exchangeable provided that $\theta_i \neq \theta_j$ for at least two indices i, j. This example is usually known as *Khoudraji's device* since it originated with Khoudraji [1995] and was also considered by Genest et al. [1998]. ∎

Further readings 1.7.26. The first example of asymmetrisation procedures for copulas originated with Khoudraji [1995]. Then, generalisations were provided by Liebscher [2008] and Durante [2009] among others. ∎

1.8   Graphical representations

As was said above, the study of copulas can be conducted from two perspectives, which are not necessarily disjoint: on the one hand, copulas are associated with r.v.'s having special marginals; on the other hand, copulas are functions on $\mathbb{I}^d$ satisfying particular assumptions.

When investigating a given copula, a preliminary assessment of its behaviour may be carried out via a graphical visualisation, at least in the two-dimensional case.

Definition 1.8.1. The *graph* of a copula $C \in \mathscr{C}_d$ is the set of all points $\mathbf{x}$ in $\mathbb{I}^{d+1}$ that can be expressed as $\mathbf{x} = (\mathbf{u}, C(\mathbf{u}))$ for $\mathbf{u} \in \mathbb{I}^d$. ◇

In view of the boundary conditions the graph of a copula takes fixed values on the boundaries on $\mathbb{I}^d$ and presents no jumps. Moreover, because of the Hoeffding–Fréchet bounds, the graph lies between the graphs of the surfaces $z = W_d(\mathbf{u})$, which bounds it from below, and $z = M_d(\mathbf{u})$, which bounds it from above. The three-dimensional graphs of the bivariate copula W_2, Π_2 and M_2 are shown in Figure 1.3.

Instead of the graph of the function, one may also consider the graphs related to the different levels assumed by the function, as stated below.

Definition 1.8.2. Let C belong to $\mathscr{C}_d$ and let t be in $\mathbb{I}$. The *t-level set*

$$L_C^t = \{\mathbf{u} \in \mathbb{I}^d : C(\mathbf{u}) = t\}$$

is the set of all points $\mathbf{u} \in \mathbb{I}^d$ such that $C(\mathbf{u}) = t$. ◇

Notice that, for every $t \in \mathbb{I}$, all the points of type $(t, 1, \ldots, 1)$, $(1, t, 1, \ldots, 1)$, $\ldots$, $(1, 1, \ldots, 1, t)$ belong to L_C^t. In the bivariate case, the level sets can be easily visualised as in Figure 1.4.

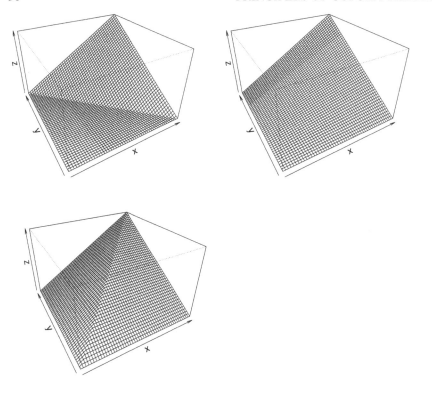

Figure 1.3 *3-d graphs of the basic copulas W_2 (upper right), Π_2 (upper left) and M_2 (lower left).*

The level set $L^t_{M_2}$ is the union of the two segments of the straight line joining the pairs of points (t, t) and $(t, 1)$, and (t, t) and $(1, t)$. Moreover, for $t > 0$, the level set $L^t_{W_2}$ coincides with the segment of endpoints $(t, 1)$ and $(1, t)$.

Remark 1.8.3. Let $C \in \mathscr{C}_d$. The 0-level set of C is also called the *zero set* of C. In view of the Hoeffding–Fréchet bounds, the zero set of C is included into the zero set of L_{W_d}. ∎

Together with the notion of level set, we may also consider the following

Definition 1.8.4. For a copula C in $\mathscr{C}_d$ $t \in \mathbb{I}$, the *t-upper level set* is the set of all points $\mathbf{u}$ such that $C(\mathbf{u}) \geq t$; in symbols,

$$L^{\geq t}_C = \left\{ \mathbf{u} \in \mathbb{I}^d : C(\mathbf{u}) \geq t \right\}.$$

Analogously,

$$L^{\leq t}_C = \left\{ \mathbf{u} \in \mathbb{I}^d : C(\mathbf{u}) \leq t \right\}$$

is the *t-lower level set*, namely the set of all points $\mathbf{u}$ such that $C(\mathbf{u}) \leq t$. ◇

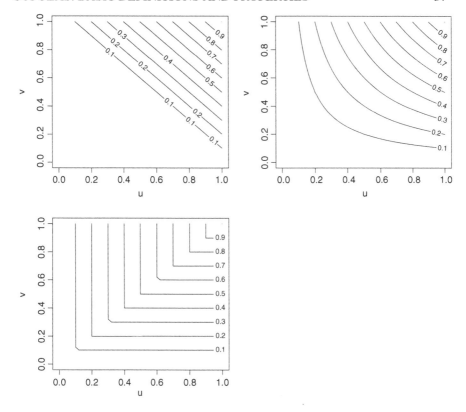

Figure 1.4 *Level plots of the basic copulas* W_2 *(upper right),* Π_2 *(upper left) and* M_2 *(lower left).*

As will be seen, the previous definitions are connected with notions of multivariate risks frequently used in hydrology (see, for instance, Salvadori et al. [2011, 2013]).

As a further tool, since a copula is the d.f. of a random vector **U**, we may also visualise its behaviour by random sampling points that are identically distributed as **U**. The resulting chart is usually called a *scatterplot*. For an example, see Figure 1.5.

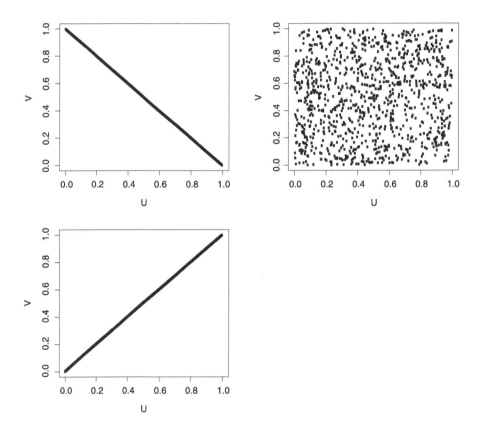

Figure 1.5 *Scatterplots of 1000 random points simulated from the basic copulas W_2 (upper right), Π_2 (upper left) and M_2 (lower left).*

Chapter 2

Copulas and stochastic dependence

In this chapter we consider the issue of modelling dependence among the components of a random vector using copulas. The main ingredient will be Sklar's theorem, which is the building block of the theory of copulas, since it allows to connect the probability law of any multivariate random vector to its marginal distributions through a copula. Because of this theorem copulas occupy a relevant position in Probability Theory and Statistics; the first part of this chapter is devoted to Sklar's theorem and presents several different proofs of it. Then we show how Sklar's theorem is used in order to express different dependence notions among random variables.

2.1 Construction of multivariate stochastic models via copulas

A *stochastic model* is a mathematical model expressed in the language of probability theory, according to the axioms introduced by Kolmogorov [1933]. In other words, the laws relating to the basic objects of the model are expressed in terms of random functions defined on a fixed probability space $(\Omega, \mathscr{F}, \mathbb{P})$.

In the following, we are mainly interested in *multivariate* (equivalently, *finite-dimensional*) *stochastic models* that are related to a system composed by several components whose behaviour can be described by a random vector $\mathbf{X} = (X_1, \ldots, X_d)$ defined on $(\Omega, \mathscr{F}, \mathbb{P})$ and taking values on $\mathbb{R}^d$ $(d \geq 2)$. Many real-world situations can be described by such models:

- In Portfolio Management, X_i's can represent (daily) returns of the assets constituting a portfolio of investments at a given reference time.

- In Credit Risk, X_i's can take values on the positive real line and have the interpretation of lifetimes (time-to-default) of firms.

- In Hydrology, X_i's may be random quantities related to the same environmental event. For instance, to each storm event one can associate the intensity and the duration of the raining event. Alternatively, X_i's may represent the same physical variable (e.g., storm variable) recorded at different sites and the main objective may be to interpolate precipitations across space.

Now, it is a general fact that in Probability, theorems are proved in the language of random variables over a probability space, while computations are usually carried out in the Borel space $(\mathbb{R}^d, \mathscr{B}(\mathbb{R}^d))$ endowed with the law of $\mathbf{X}$ (see [Loève, 1977, pp. 172-174] for a philosophical discussion about this aspect). Thus, in essence, for our purposes, a multivariate stochastic model can be conveniently represented by means of a d-dimensional d.f. $F = F_{\mathbf{X}}$ that describes the behaviour of the r.v. $\mathbf{X}$.

Every joint d.f. for a random vector implicitly contains the description of both the marginal behaviour, by which we mean the probabilistic knowledge of the single components of the random vector $\mathbf{X}$, and of their dependence structure; the copula approach provides a way of highlighting the description of the dependence structure.

In fact, suppose that the behaviour of the single components of a r.v. $\mathbf{X}$ is known in terms of their univariate d.f.'s $F_1, \ldots, F_d$; then a suitable multivariate model can be constructed by means of the following result whose proof is just a matter of straightforward verification.

Theorem 2.1.1. *Let $F_1, \ldots, F_d$ be univariate d.f.'s and let C be any d-copula. Then the function $H : \mathbb{R}^d \to \mathbb{I}$ defined, for every point $\mathbf{x} = (x_1, \ldots, x_d) \in \mathbb{R}^d$, by*

$$H(x_1, \ldots, x_d) = C\left(F_1(x_1), \ldots, F_d(x_d)\right), \tag{2.1.1}$$

is a d-dimensional d.f. with margins given by $F_1, \ldots, F_d$.

The previous result suggests a two-stage approach to multivariate model building. First, one may define the marginal d.f.'s with great flexibility (for instance, allowing each marginal to belong to a different family of distributions); then, a copula (any copula!) may be chosen in order to link the marginals in a common model. This approach seems quite promising since often the known families of multivariate d.f.'s (like Gaussian or elliptical distribution) require that their marginals are from the same family.

Definition 2.1.2. A *parametric copula-based stochastic model* is any model for a d-variate d.f. H that arises from (2.1.1) whenever, for $j = 1, \ldots, d$, $F_j = F_{\theta_j}$ and $C = C_\alpha$ for some parameters $\theta_1, \ldots, \theta_d$ and $\alpha = (\alpha_1, \ldots, \alpha_d)$, i.e.,

$$H_{\theta, \alpha} = C_\alpha\left(F_{\theta_1}, \ldots, F_{\theta_d}\right). \tag{2.1.2}$$

$\diamond$

Remark 2.1.3. The two-stage model building procedure just illustrated is also adopted in several statistical procedures as a criterion for the estimation of the parameters of the d.f. in (2.1.2). Specifically, first, the univariate marginals are estimated and hence, separately, the copula parameters are estimated by using the previous estimated parameters or by replacing the marginals by their empirical counterparts. See, for instance, the method of *inference function for margins* by Joe [1997, 2005] and the *maximum pseudo-likelihood estimation* by Genest et al. [1995]. For additional information and comparisons, see, also, [Kim et al., 2007; Choroś et al., 2010; Kojadinovic and Yan, 2010] and references therein.   ■

The separation that can be realised between dependence and marginal behaviour is potentially very useful, for instance, in risk management since it allows to test

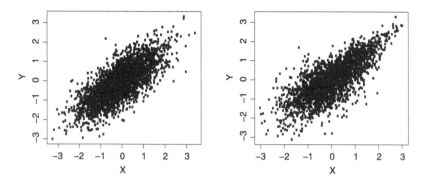

Figure 2.1 *Bivariate sample clouds of* 2500 *points from the d.f.* $F = C(F_1, F_2)$ *where* $F_1, F_2 \sim N(0,1)$, *the Spearman's* ρ *is equal to* 0.75, *and* C *is a Gaussian copula (left) or a Gumbel copula (right).*

several scenarios with different kinds of dependence between random riskFigures/s keeping the marginals fixed. As an illustration, we present in Figure 2.1 two random samples from two bivariate models with the same marginal distributions (standard Gaussian) but a different copula. Now, the shapes of the two sample clouds look quite different, especially in the upper right corner of the unit square, which corresponds to large realizations of both variables.

Moreover, for didactic purposes, Theorem 2.1.1 helps in understanding the dependence at a deeper level and allows to see potential pitfalls in the description of the relationships among r.v.'s, as shown in the sequel.

Example 2.1.4. Here we collect some examples that are usually presented in probability textbooks in order to understand dependence structures beyond the Gaussian paradigm. In this example Φ represents the d.f. of the standard Gaussian law $N(0,1)$.

- There exists a bivariate distribution that is not bivariate Gaussian but that has Gaussian marginal distributions: in fact, in (2.1.1) take $F_1 = F_2 = \Phi$ and $C = M_2$.

- Three pairwise independent Gaussian r.v.'s need not be related to a common trivariate Gaussian: in (2.1.1) take, for instance, $F_1 = F_2 = F_3 = \Phi$ and the copula

$$C(u_1, u_2, u_3) = u_1 u_2 u_3 \left(1 + (1 - u_1)(1 - u_2)(1 - u_3)\right).$$

- Random variables that are identically distributed need not be exchangeable: in (2.1.1) take any copula C that is not exchangeable. For instance, consider

$$C(u_1, u_2) = u_1^\alpha u_2^\beta M_2(u_1^{1-\alpha}, u_1^{1-\beta})$$

for $\alpha, \beta \in]0, 1[$ with $\alpha \neq \beta$.

- There exist random variables X, Y, Z such that X is uncorrelated with both Y and Z, but such that Y is correlated with Z: in (2.1.1) take

$$C(u_1, u_2, u_3) = u_1 \, M_2(u_2, u_3) \,.$$

Finally, it is worth noticing that Theorem 2.1.1 can be applied to continuous as well as discrete distribution functions. In other words, copulas allow generating suitable multivariate models for continuous as well as discrete data (this latter point will be discussed in greater detail later).

2.2   Sklar's theorem

We have seen in the previous section that a variety of multivariate models can be constructed with the help of copulas. Here, we prove that *any* multivariate model may be constructed by means of copulas. In other words, the representation of multivariate d.f.'s in terms of a copula and univariate margins is a general way for representing the law of multivariate vectors. This is the content of Sklar's theorem.

Theorem 2.2.1 (Sklar's theorem). *Let a random vector* $\mathbf{X} = (X_1, \ldots, X_d)$ *be given on a probability space* $(\Omega, \mathscr{F}, \mathbb{P})$, *let* $H(\mathbf{x}) := \mathbb{P}(X_1 \leq x_1, \ldots, X_d \leq x_d)$ *be the joint d.f. of* $\mathbf{X}$ *and let* $F_j(x_j) = \mathbb{P}(X_j \leq x_j)$ $(j = 1, \ldots, d)$ *be its marginals. Then there exists a d-copula* $C = C_\mathbf{X}$ *such that, for every point* $\mathbf{x} = (x_1, \ldots, x_d) \in \mathbb{R}^d$,

$$H(x_1, \ldots, x_d) = C\left(F_1(x_1), \ldots, F_d(x_d)\right). \tag{2.2.1}$$

If the marginals $F_1, \ldots, F_d$ *are continuous, then the copula C is uniquely defined.*

In essence, Sklar's theorem states that a multivariate d.f. may be expressed as a composition of a copula and its univariate marginals.

Remark 2.2.2. By direct calculation, if H is an absolutely continuous d.f. given by (2.2.1), then the density h of H is given, for almost all $\mathbf{x} \in \mathbb{R}^d$, by

$$h(\mathbf{x}) = c(F_1(x_1), \ldots, F_d(x_d)) \, f_1(x_1) \cdots f_d(x_d) \,,$$

where $F_1, \ldots, F_d$ are the univariate margins with densities $f_1, \ldots, f_d$, while c is the density of the copula C.   ∎

If all the marginals $F_1, \ldots, F_d$ of a d-dimensional d.f. are continuous the existence of the (unique) copula asserted by Sklar's theorem can be easily proved. We state this fact as a lemma.

Lemma 2.2.3. *Under the assumptions of Theorem 2.2.1, if* $F_1, \ldots, F_d$ *are continuous, then there exists a unique copula C associated with* $\mathbf{X}$ *that is the d.f. of the random vector* $(F_1 \circ X_1, \ldots, F_d \circ X_d)$. *It is determined, for every* $\mathbf{u} \in \mathbb{I}^d$, *via the formula*

$$C(\mathbf{u}) = H\left(F_1^{(-1)}(u_1), \ldots, F_d^{(-1)}(u_d)\right), \tag{2.2.2}$$

where, for $j \in \{1, \ldots, d\}$, $F_j^{(-1)}$ *is the quasi-inverse of* F_j.

Proof. First of all, notice that, in view of the univariate probability integral transformation (see Theorem 1.2.6), if F_j is continuous, then $F_j \circ X_j$ is uniform on $\mathbb{I}$ for each $j \in \{1, \ldots, d\}$. Thus, the d.f. of $(F_1 \circ X_1, \ldots, F_d \circ X_d)$ has uniform univariate marginals and, hence, it is a copula. Moreover, for every point $\mathbf{x} \in \mathbb{R}^d$, one has

$$
\begin{aligned}
H(\mathbf{x}) &= \mathbb{P}\left(X_1 \leq x_1, \ldots, X_d \leq x_d\right) \\
&= \mathbb{P}\left(F_1(X_1) \leq F_1(x_1), \ldots, F_d(X_d) \leq F_d(x_d)\right) \\
&= C\left(F_1(x_1), \ldots, F_d(x_d)\right),
\end{aligned}
$$

which is the assertion. □

Remark 2.2.4. As a consequence of Lemma 2.2.3 a copula represents a way of transforming the continuous r.v.'s $X_1, \ldots, X_d$ into other r.v.'s $F_1(X_1), \ldots, F_d(X_d)$ that are all uniformly distributed on $\mathbb{I}$; in other words, it is a way for providing a (uniform) representation for the dependence of a random vector. Of course, although it may be useful from a statistical point of view, there is no real, compelling mathematical reason for transforming the marginal d.f.'s of F to uniform d.f.'s on $\mathbb{I}$; Hoeffding [1940] used the interval $[-1/2, 1/2]$ (see [Embrechts, 2009]). In some instances, other ways for transforming r.v.'s to different margins may be more convenient (see [Klüppelberg and Resnick, 2008]). The interested reader may consult the paper by Mikosch [2006a] (and the related discussion) for possible inconveniences of transforming to uniform marginals, especially in multivariate extreme value theory. ∎

Remark 2.2.5. Lemma 2.2.3 also provides a way of visualising the dependence structure (i.e., the copula) starting with a random sample from an unknown distribution function H. In fact, suppose that $d = 2$. Consider a random sample (X_i, Y_i), $i = 1, \ldots, n$, from a random pair $(X, Y) \sim H$. In order to visualise the dependence arising from the data it is enough to estimate the marginal distributions F and G of X and Y, respectively, and then plot $(F(X_i), G(Y_i))$ for $i = 1, \ldots, n$. Notice that, since in several cases the marginal d.f.'s are not known, one may estimate them by means of the empirical distribution functions

$$
F_n(t) = \frac{1}{n} \sum_{i=1}^{n} \mathbf{1}_{\{X_i \leq t\}} \quad \text{and} \quad G_n(t) = \frac{1}{n} \sum_{i=1}^{n} \mathbf{1}_{\{Y_i \leq t\}}.
$$

Figure 2.2 illustrates the described transformation with an example. ∎

Remark 2.2.6. In view of the previous Lemma 2.2.3 it is not difficult to prove that if $\mathbf{U}$ is a r.v. distributed according to the copula C and $F_1, \ldots, F_d$ are univariate d.f.'s, then the random vector $(F_1^{-1}(U_1), \ldots, F_d^{-1}(U_d))$ is distributed according to $H = C(F_1, \ldots, F_d)$. This remark is particularly useful in simulations. In fact, in order to simulate a random sample from H it is enough to do the following:

1. Simulate $(u_1, \ldots, u_d)$ from the copula C,
2. Return $(F_1^{-1}(u_1), \ldots, F_d^{-1}(u_d))$.

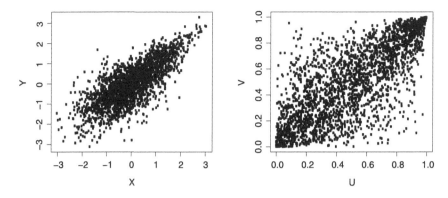

Figure 2.2 *Random sample of* 2500 *points from a r.v.* (X, Y) *(left) and the transformed sample* $(F(X), G(Y))$ *(right).*

Thus, as soon as the inverses of the univariate margins are available and a copula can be simulated, any multivariate d.f. can be simulated as well. ∎

In the literature a number of continuous multivariate d.f.'s H are known (see, for instance, [Hutchinson and Lai, 1990; Kotz et al., 2000; Lai and Balakrishnan, 2009]). In view of (2.2.2), one may find the copula associated with H simply by applying the inverse of the marginal d.f.'s to each component, as the following example shows.

Example 2.2.7 (Copula of the extreme order statistics). In the probability space $(\Omega, \mathscr{F}, \mathbb{P})$ let $X_1, \ldots, X_n$ be independent and identically distributed random variables with a common continuous d.f. F and let $X_{(j)}$ be the j-th order statistic so that $X_{(1)} \leq X_{(2)} \leq \cdots \leq X_{(n)}$. We wish to determine the copula of the vector $(X_{(1)}, X_{(n)})$. It is known (see, e.g., [David, 1981]) that the joint d.f. of the j-th and the k-th order statistics is

$$H_{j,k}(x, y) = \sum_{h=k}^{n} \sum_{i=j}^{h} \frac{n!}{i!(h-i)!(n-h)!} F^i(x) \left[F(y) - F(x)\right]^{h-i} \left[1 - F(y)\right]^{n-h} ,$$

for $x < y$, and $F_{(k)}(y)$, for $x \geq y$, where $F_{(k)}$ is the d.f. of the k-th order statistic, i.e.,

$$F_{(k)}(t) = \sum_{i=k}^{n} \binom{n}{i} F^i(t) \left[1 - F(t)\right]^{n-i} .$$

Hence, setting first $j = 1$ and then $k = n$, one has, for x and y in $\mathbb{R}$,

$$F_{(1)}(x) = 1 - \left[1 - F(x)\right]^n , \qquad F_{(n)}(y) = F^n(y) .$$

Now, the joint d.f. of $X_{(1)}$ and $X_{(n)}$ is given by

$$H_{1,n}(x,y) = \begin{cases} \sum_{i=1}^{n} \binom{n}{i} F^i(x) \left[F(y) - F(x) \right]^{n-i}, & x < y, \\ F_{(n)}(y), & x \geq y, \end{cases}$$

$$= \begin{cases} F^n(y) - \left[F(y) - F(x) \right]^n, & x < y, \\ F^n(y), & x \geq y. \end{cases}$$

By recourse to Lemma 2.2.3, one can now write the (unique) copula of $X_{(1)}$ and $X_{(n)}$

$$C_{1,n}(u,v) = H_{1,n} \left(F_{(1)}^{-1}(u), F_{(n)}^{-1}(v) \right)$$

$$= \begin{cases} v - \left[v^{1/n} + (1-u)^{1/n} - 1 \right]^n, & 1 - (1-u)^{1/n} < v^{1/n}, \\ v, & 1 - (1-u)^{1/n} \geq v^{1/n}. \end{cases}$$

The copula $C_{1,n}$ is related to a 2-copula from the Mardia-Takahasi-Clayton copulas family (see Example 6.5.17). In fact, by using the symmetry $\xi(u,v) = (1-u,v)$, one has

$$C_{1,n}^{\xi}(u,v) = \begin{cases} \left(v^{1/n} + u^{1/n} - 1 \right)^n, & v^{1/n} + u^{1/n} - 1 > 0, \\ 0, & \text{elsewhere}, \end{cases}$$

so that $C_{1,n}^{\xi}$ belongs to the Mardia-Takahasi-Clayton family with $\theta = -1/n$ (see Example 6.5.17).

Finally, it can be proved that

$$\lim_{n \to +\infty} C_{1,n}(u,v) = \Pi_2(u,v) \tag{2.2.3}$$

for every $(u,v) \in \mathbb{I}^2$. In other words, the first and last order statistics of n independent and identically distributed r.v.'s tend to be independent as $n \to \infty$. ■

Further readings 2.2.8. Example 2.2.7 is taken from [Schmitz, 2004]; but see also the much earlier contribution by Sibuya [1960]. The ease with which eq. (2.2.3) has been obtained ought to be compared with the standard proof of this fact, for which we refer, e.g., to [Arnold et al., 1992, p. 221]. ■

We wish to draw the reader's attention to the consequences of the lack of uniqueness of the connecting copula if the marginal distributions are not all continuous.

Lemma 2.2.9. *Under the assumptions of Theorem 2.2.1, a copula C associated with* $\mathbf{X}$ *is uniquely determined on* Ran $F_1 \times \cdots \times$ Ran F_d*, where* Ran $F_j = F_j(\overline{\mathbb{R}})$ *for every* $j = 1, \ldots, d$.

Proof. For every pair of points $\mathbf{t} = (t_1, \ldots, t_d)$ and $\mathbf{s} = (s_1, \ldots, s_d)$ in $\overline{\mathbb{R}}^d$, equation (1.2.7) holds; as a consequence, if $F(s_j) = F(t_j)$ for every $j = 1, \ldots, d$, then $H(\mathbf{s}) = H(\mathbf{t})$; in other words, the value $H(\mathbf{t})$ depends, for every $\mathbf{t} \in \overline{\mathbb{R}}^d$, only on the numbers $F_j(t_j)$ $(j = 1, \ldots, d)$. This is equivalent to saying that there exists a unique function C' with domain $\operatorname{Ran} F_1 \times \cdots \times \operatorname{Ran} F_d$ such that (2.2.1) holds. $\qquad\square$

Example 2.2.10. Let X and Y be Bernoulli random variables on the probability space $(\Omega, \mathscr{F}, \mathbb{P})$ with

$$\mathbb{P}(X = 0) = p, \qquad \text{and} \qquad \mathbb{P}(Y = 0) = q.$$

Let also

$$\mathbb{P}(X = 0, Y = 0) = r \in [W_2(p, q), M_2(p, q)].$$

Then one has

$$F_X(t) = \begin{cases} 0, & t < 0, \\ p, & t \in [0, 1[, \\ 1, & t \geq 1, \end{cases} \quad \text{and} \quad F_Y(t) = \begin{cases} 0, & t < 0, \\ q, & t \in [0, 1[, \\ 1, & t \geq 1. \end{cases}$$

Every copula C associated with the vector (X, Y) is uniquely defined on $\operatorname{Ran} F_X \times \operatorname{Ran} F_Y$; as a consequence, C satisfies the following constraints

$$\begin{aligned}
C(0, 0) &= 0, & C(0, q) &= 0, & C(0, 1) &= 0, \\
C(p, 0) &= 0, & C(p, q) &= r, & C(p, 1) &= p, \\
C(1, 0) &= 0, & C(1, q) &= q, & C(1, 1) &= 1.
\end{aligned}$$

In general there are several copulas satisfying these constraints. For instance, assume $p = q = r \in \,]0, 1[$. Then all possible copulas C associated with (X, Y) may be represented in the form

$$C(u, v) = \begin{cases} r\, C_1\left(\dfrac{u}{r}, \dfrac{v}{r}\right), & (u, v) \in [0, r]^2, \\[2mm] r + (1 - r)\, C_2\left(\dfrac{u - r}{1 - r}, \dfrac{v - r}{1 - r}\right), & (u, v) \in [r, 1]^2, \\[2mm] \min\{u, v\}, & \text{elsewhere,} \end{cases} \quad (2.2.4)$$

where C_1 and C_2 are arbitrary 2-copulas; see Section 3.8 on ordinal sums in order to check that C thus defined is indeed a copula. $\qquad\blacksquare$

Remark 2.2.11. As a consequence of Sklar's theorem, one will therefore be able to speak of *the* (unique!) copula $C = C_{\mathbf{X}}$ of the random vector $\mathbf{X}$ when all the components of $\mathbf{X}$ are continuous. Otherwise, all the copulas of $\mathbf{X}$ are uniquely defined on the set $\operatorname{Ran}(F_1) \times \cdots \times \operatorname{Ran}(F_k)$ and one needs to choose a suitable criterion (constraint) in order to single out a unique copula. A criterion often used is that of multilinear interpolation of Lemma 2.3.4 below. Other possible extensions are discussed in the literature (see, for instance, [Deheuvels, 2009; de Amo et al., 2012; Genest et al., 2014]). $\qquad\blacksquare$

Further readings 2.2.12. Example 2.2.10 is taken from [Genest and Nešlehová, 2007] and shows, in a perhaps extreme way, what the lack of uniqueness of the representing copula may lead to. The effects are actually much deeper; referring to topics that will be dealt with later in this book, we shall mention that

- measures of concordance may depend on the marginals;
- the probabilistic and analytical definitions of Kendall's τ and of Spearman's ρ may not coincide;
- as was seen above, the knowledge of the copula may not characterise the dependence between the random variables X and Y;
- perfect monotone dependence may not imply $|\tau| = |\rho| = 1$.

The reader is therefore strongly advised to consult [Genest and Nešlehová, 2007] for further details. ∎

2.2.1 Sklar's theorem for survival functions

Along with Sklar's theorem for d.f.'s, one can consider an analogous result for survival functions associated with a vector of r.v.'s. The proofs are similar to those of the original case and will therefore not be given.

Theorem 2.2.13. *Let $\mathbf{X} = (X_1, \ldots, X_d)$ be a vector of r.v.'s on a given probability space $(\Omega, \mathscr{F}, \mathbb{P})$. Let $\overline{H}$ be the d-dimensional survival function of $\mathbf{X}$, $\overline{H}(\mathbf{x}) = \mathbb{P}(\cap_{j=1}^d X_j > x_j)$ with survival marginals $\overline{F}_1, \ldots, \overline{F}_d$ given by*

$$\overline{F}_j(x_j) = \mathbb{P}(X_j > x_j) \quad (j = 1, \ldots, d).$$

Then there exists a d-copula C, called the survival copula of $\mathbf{X}$, such that, for every $\mathbf{x} \in \overline{\mathbb{R}}^d$,

$$\overline{H}(\mathbf{x}) = C\left(\overline{F}_1(x_1), \ldots, \overline{F}_d(x_d)\right). \tag{2.2.5}$$

The copula C is uniquely determined on $\operatorname{Ran} \overline{F}_1 \times \cdots \times \operatorname{Ran} \overline{F}_d$, and it is given by

$$C(\mathbf{u}) = \overline{H}\left(\overline{F}_1^{(-1)}(u_1), \ldots, \overline{F}_d^{(-1)}(u_d)\right),$$

for every $\mathbf{u} \in \operatorname{Ran} \overline{F}_1 \times \cdots \times \operatorname{Ran} \overline{F}_d$. Conversely, given a d-copula C and d 1-dimensional survival functions $\overline{F}_1, \ldots, \overline{F}_d$, the function $\overline{H} : \overline{\mathbb{R}}^d \to \mathbb{I}$ defined by (2.2.5) is a d-dimensional survival function with marginals $\overline{F}_1, \ldots, \overline{F}_d$ and copula C.

As noted in Remark 1.2.17 a joint survival function $\overline{H}$ can be expressed, for every

$\mathbf{x} \in \mathbb{R}^d$, in the form

$$\overline{H}(\mathbf{x}) = \mathbb{P}(X_1 > x_1, \ldots, X_d > x_d)$$

$$= 1 - \sum_{j=1}^{d} F_j(x_j) + \sum_{i<j} \mathbb{P}(X_i \leq x_i, X_j \leq x_j) + \ldots$$

$$+ (-1)^k \sum_{j(1)<j(2)<\cdots<j(k)} \mathbb{P}(X_{j(1)} \leq x_{j(1)}, \ldots, X_{j(k)} \leq x_{j(k)})$$

$$+ (-1)^d \mathbb{P}(X_1 \leq x_1, \ldots, X_d \leq x_d).$$

Notice also that if C denotes the d-copula that connects the joint d.f. H of $\mathbf{X}$ to its marginals $F_1,\ldots,F_d$, all the terms in the previous expression, starting from the third one, can be expressed in terms of these marginals and of the copula C.

Example 2.2.14. In the bivariate case, consider two r.v.'s X and Y, and let H be their joint d.f., F and G be their marginals and C be their copula. Then their joint survival function can be expressed in the following form:

$$\overline{H}(x,y) = \mathbb{P}(X > x, Y > y)$$
$$= 1 - \mathbb{P}(X \leq x) - \mathbb{P}(Y \leq y) + \mathbb{P}(X \leq x, Y \leq y)$$
$$= 1 - F(x) - G(y) + H(x,y) = \overline{F}(x) + \overline{G}(y) - 1 + C\left(F(x), G(y)\right)$$
$$= \overline{F}(x) + \overline{G}(y) - 1 + C\left(1 - \overline{F}(x), 1 - \overline{G}(y)\right).$$

Thus the copula associated with $\overline{H}$ is given by

$$\widehat{C}(u,v) := u + v - 1 + C(1 - u, 1 - v), \tag{2.2.6}$$

and one can express $\overline{H}$ as

$$\overline{H}(x,y) = \widehat{C}(\overline{F}(x), \overline{G}(y)),$$

where $\widehat{C}$ is the survival copula of (X, Y). ∎

2.3   Proofs of Sklar's theorem

Here we present different proofs of Sklar's theorem. Each of them starts from a specific perspective and uses different properties of a copula. Sklar's theorem was presented in [Sklar, 1959], the paper that began the history of copulas. This paper has recently been republished, [Sklar, 2010]. Interestingly, the original proof of this theorem appeared neither in [Sklar, 1959] nor in [Sklar, 1973], where the theorem was re-stated, but was presented by Schweizer and Sklar [1974] for the bivariate case. A proof for the general case was later given by Sklar [1996].

The first proof (Subsection 2.3.1) is related to Lemma 2.2.9. Basically, when one knows the values a copula C of a random vector assumes on a specific subset A of the domain $\mathbb{I}^d$, one tries to extend the definition of C outside A. This extension is usually performed by multilinear interpolation; it was presented in detail by Schweizer and Sklar [1974] for the bivariate case, and later by Carley and Taylor [2003] in the multivariate case.

The second proof of Subsection 2.3.2 is due to [Moore and Spruill, 1975, Lemma 3.2] and it is based on Lemma 2.2.3. Since Sklar's theorem follows by direct application of the probability integral transform to each margin, the main idea here is to generalise this transformation for (possibly) non-continuous margins. The connexion of this approach with Sklar's theorem has been also highlighted by Rüschendorf [1981b] (see also [Rüschendorf, 2005, 2009]).

The third proof of Subsection 2.3.3 was presented by Durante et al. [2012]. It shows the existence of a copula associated with a random vector, but it does not show either analytically or probabilistically its form. The main idea is based on the fact that, since Sklar's theorem is obvious for continuous d.f.'s, a suitable strategy of proof might consist of approximating any d.f. H via a sequence of continuous d.f.'s H_n and, hence, construct a (possible) copula of H by means of the sequence of copulas C_n associated with H_n. This proof exploits the compactness of the space of copulas (Theorem 1.7.7). Curiously, the idea of such a proof is also related to a similar reasoning presented by [Deheuvels, 1979] and attributed by this author to J. Geffroy.

Further readings 2.3.1. In view of the relevance of Sklar's theorem in statistics, the reader should not be surprised to learn that it has been rediscovered several times and, that, as a consequence, in the literature one encounters several proofs of this theorem. Further recent proofs include the papers by Burchard and Hajaiej [2006]; Deheuvels [2009]; de Amo et al. [2012]; Durante et al. [2013b]; Faugeras [2013]. A generalisation to probability spaces has been presented by Scarsini [1989].   ∎

2.3.1 Proof of Sklar's theorem by extension

In order to proceed we need a further concept.

Definition 2.3.2. Let $A_1, \ldots, A_d$ be subsets of the unit interval $\mathbb{I}$ containing both 0 and 1. Then a *subcopula* is a function $C' : A_1 \times \cdots \times A_d \to \mathbb{I}$ such that

(a) $C'(u_1, \ldots, u_d) = 0$ if $u_j = 0$ for at least one index $j \in \{1, \ldots, d\}$;

(b) $C'(1, \ldots, 1, t, 1, \ldots, 1) = t$ for every $t \in A_j$ ($j \in \{1, \ldots, d\}$);

(c) For every rectangle $]\mathbf{a}, \mathbf{b}]$ having its vertices in $A_1 \times \cdots \times A_d$, one has $V_{C'}(]\mathbf{a}, \mathbf{b}]) \geq 0$.   ◇

In particular, a copula is a subcopula defined on $\mathbb{I}^d$, namely such that $A_j = \mathbb{I}$ for every $j \in \{1, \ldots, d\}$.

The first proof of Sklar's theorem we present will be given in two lemmata, the first of which is a slightly different formulation of results already stated.

Lemma 2.3.3. *For every d-dimensional d.f. H with marginals $F_1, \ldots, F_d$ there* *exists a unique subcopula C' defined on* $\operatorname{Ran} F_1 \times \cdots \times \operatorname{Ran} F_d$ *such that, for all* $\mathbf{x} = (x_1, \ldots, x_d) \in \mathbb{R}^d$,

$$H(\mathbf{x}) = C'(F_1(x_1), \ldots, F_d(x_d)). \tag{2.3.1}$$

The subcopula C' is given by

$$C'(\mathbf{t}) = H\left(F_1^{(-1)}(t_1), \ldots, F_d^{(-1)}(t_d)\right), \tag{2.3.2}$$

for all $\mathbf{t} \in \times_{j=1}^d \operatorname{Ran} F_j$, where, for each $j \in \{1, \ldots, d\}$, $F_j^{(-1)}$ is quasi-inverse of F_j.

Proof. The uniqueness of C' follows from Lemma 2.2.9. That this function C' is a subcopula follows directly from the properties of H. For instance, property (c) in Definition 2.3.2 follows from the fact that H is d-increasing, while properties (a) and (b) in Definition 2.3.2 follow by direct calculation.

Next, one has $F_j(F_j^{(-1)})(t_j) = t_j$ for every $t_j \in \operatorname{Ran} F_j$ $(j = 1, \ldots, d)$, because of Theorem 1.2.5 (c), so that

$$H\left(F_1^{(-1)}(t_1), \ldots, F_d^{(-1)}(t_d)\right) = C'\left(F_1\left(F_1^{(-1)}\right)(t_1), \ldots, F_d\left(F_d^{(-1)}\right)(t_d)\right)$$
$$= C'(\mathbf{t}),$$

for all $\mathbf{t} \in \operatorname{Ran} F_1 \times \cdots \times \operatorname{Ran} F_d$, viz., (2.3.2). $\square$

The "hard" part of the proof of Theorem 2.2.1 is the extension from a subcopula to a copula, which is reformulated in the next result.

Lemma 2.3.4. *Under the notation of Theorem 2.2.1, for every subcopula C' defined on $A_1 \times \cdots \times A_d$, there exists a (generally non-unique) copula C that extends it, viz., such that, for all $\mathbf{t} = (t_1, \ldots, t_d)$ in $A_1 \times \cdots \times A_d$,*

$$C(\mathbf{t}) = C'(\mathbf{t}).$$

For historical reasons, we first present the proof in the 2-dimensional case, which is contained in Schweizer and Sklar [1974], then we prove the general case. In the latter case, we use a result that will be presented in detail in Section 4.1.1.

Proof of Lemma 2.3.4: the bivariate case. Since each subcopula is uniformly continuous on its domain (as a consequence of Lemma 1.2.16), it is possible to extend C' to a function C'' defined on the set $\overline{A}_1 \times \overline{A}_2$, where, for $j = 1, 2$, $\overline{A}_j$ denotes the closure of A_j. Obviously, also the function C'' is a subcopula. It is possible to extend the definition of C'' to $\mathbb{I}^2$ in such a way that the extension C is a copula. Let (a_1, a_2) be any point in $\mathbb{I}^2$. For $j = 1, 2$, define

$$a_j' := \max\{t_j \in \overline{A}_j : t_j \le a_j\} \quad \text{and} \quad a_j'' := \min\{t_j \in \overline{A}_j : a_j \le t_j\}.$$

Obviously, if a_j belongs to $\overline{A}_j$, then one has $a'_j = a''_j = a_j$. Next, define

$$\alpha_j := \begin{cases} \dfrac{a_j - a'_j}{a''_j - a'_j}, & \text{if } a''_j > a'_j, \\ 1, & \text{if } a''_j = a'_j. \end{cases} \qquad (j = 1, 2)$$

Consider now the rectangle $[a'_1, a''_1] \times [a'_2, a''_2]$ and its vertices $\mathbf{b} = (b_1, b_2)$, where, for every index j, either $b_j = a'_j$ or $b_j = a''_j$. Finally, introduce the function $C : \mathbb{I}^2 \to \mathbb{I}$ through

$$\begin{aligned} C(a_1, a_2) : = & (1 - \alpha_1)(1 - \alpha_2)\, C''(a'_1, a'_2) + (1 - \alpha_1)\, \alpha_2\, C''(a'_1, a''_2) \quad (2.3.3) \\ & + \alpha_1 (1 - \alpha_2)\, C''(a''_1, a'_2) + \alpha_1\, \alpha_2\, C''(a''_1, a''_2)\,. \end{aligned}$$

The function C defined above satisfies the boundary conditions for a copula (see (a) and (b) of Definition 1.3.1). Moreover, C is defined on $\mathbb{I}^2$ and clearly coincides with C' on $A_1 \times \cdots \times A_d$. Therefore, in order to show that it is indeed a 2-copula, one has to prove that it satisfies the 2-increasing property (see (c) of Definition 1.3.1). To this end, let (b_1, b_2) be another point in $\mathbb{I}^2$ and assume $a_1 \leq b_1$ and $a_2 \leq b_2$. For $j = 1, 2$, define b'_j, b''_j and β_j in a fashion similar to that of a'_j, a''_j and α_j, respectively.

Consider the rectangle $R = [a_1, a_2] \times [b_1, b_2]$; in evaluating its V_C-volume, four cases will have to be considered according to whether or not there is a point in $\overline{A}_1$ strictly between a_1 and b_1 and to whether there is a point in $\overline{A}_2$ strictly between a_2 and b_2. If no such point exists in either $\overline{A}_1$ or $\overline{A}_2$, namely if

$$\overline{A}_1 \cap \,]a_1, b_1[\,= \emptyset \qquad \text{and} \qquad \overline{A}_2 \cap \,]a_2, b_2[\,= \emptyset,$$

then $a'_j = b'_j$ and $a''_j = b''_j$ $(j = 1, 2)$. A tedious calculation based on (2.3.3) and on the analogous expressions for the remaining vertices leads to

$$\begin{aligned} V_C(R) &= C(b_1, b_2) - C(b_1, a_2) - C(a_1, b_2) + C(a_1, a_2) \\ &= (\beta_2 - \alpha_2)(\beta_1 - \alpha_1)\, V_{C''}\left([a'_1, a''_1] \times [b'_1, b''_1]\right). \end{aligned}$$

Hence $V_C(R) \geq 0$, since $b_1 \geq a_1$ and $b_2 \geq a_2$ imply $\beta_1 \geq \alpha_1$ and $\beta_2 \geq \alpha_2$.

The most involved case occurs when

$$\overline{A}_1 \cap \,]a_1, b_1[\,\neq \emptyset \qquad \text{and} \qquad \overline{A}_2 \cap \,]a_2, b_2[\,\neq \emptyset.$$

In this case one has $a_1 < a''_1 \leq b'_1 < b_1$ and $a_2 < a''_2 \leq b'_2 < b_2$. A long calculation now leads to the following expression

$$\begin{aligned} V_C(R) &= C(b_1, b_2) - C(b_1, a_2) - C(a_1, b_2) + C(a_1, a_2) \\ &= (1 - \alpha_1)\, \beta_2\, V_{C''}\left([a'_1, a''_1] \times [b'_2, b''_2]\right) + \beta_2\, V_{C''}\left([a''_1, b'_1] \times [b'_2, b''_2]\right) \\ &\quad + \beta_1\, \beta_2\, V_{C''}\left([b'_1, b''_1] \times [b'_2, b''_2]\right) + (1 - \alpha_1)\, V_{C''}\left([a'_1, a''_1] \times [a''_2, b'_2]\right) \\ &\quad + V_{C''}\left([a''_1, b'_1] \times [a''_2, b'_2]\right) + \beta_2\, V_{C''}\left([b'_1, b''_1] \times [a'_2, b'_2]\right) \\ &\quad + (1 - \alpha_1)(1 - \alpha_2)\, V_{C''}\left([a'_1, u''_1] \times [a'_2, a''_2]\right) \\ &\quad + (1 - \alpha_2)\, V_{C''}\left([a'_1, b'_1] \times [a'_2, a''_2]\right) \\ &\quad + \beta_1 (1 - \alpha_2)\, V_{C''}\left([b'_1, b''_1] \times [a'_2, a''_2]\right). \end{aligned}$$

Each term in the last sum is positive so that, also in this case, $V_C(R) \geq 0$. This shows that C is a copula and completes the proof.   □

Proof of Lemma 2.3.4: the general case. As in the bivariate case, without loss of generality we may assume that the domain of the subcopula C' is the Cartesian product $D_1 \times \cdots \times D_d$ of closed intervals of $\mathbb{I}$.

For each D_j one can find a sequence of finite sets $A_{j1} \subseteq A_{j2} \subseteq \cdots \subseteq D_j$ such that both 0 and 1 belong to A_{jn} and such that the closure of the union of the sequence (A_{jn}) is equal to D_j, i.e.

$$\overline{\bigcup_n A_{jn}} = D_j \,.$$

In particular, if D_j is a finite union, one can take $A_{jn} = D_j$ for every $n \in \mathbb{N}$.

Now define $S_n : A_{1n} \times \cdots \times A_{dn} \to \mathbb{I}$ via

$$S_n(u_1, \ldots, u_d) := C'(u_1, \ldots, u_d) \,.$$

For every $n \in \mathbb{N}$, S_n is a d-subcopula because it is the restriction of a d-subcopula and because both 0 and 1 belong to each of the factors of the domain of S_n. Thus a sequence (S_n) of d-subcopulas has been constructed such that, for every $n \in \mathbb{N}$,

- the domain of S_n is finite;
- S_n is a restriction of C';
- the domain of S_n is a subset of that of S_{n+1};
- the countable union of the domains of the S_n's is a countable, dense subset of the domain of C'.

For every $n \in \mathbb{N}$ let C_n be a copula that can be constructed by extending the domain of each S_n to $\mathbb{I}^d$; the existence of such a copula is guaranteed by Section 4.1.1. Since the class of copulas is compact with respect to uniform convergence (Theorem 1.7.7), there exists a subsequence $(C_{n(k)})_{k \in \mathbb{N}}$ of $(C_n)_{n \in \mathbb{N}}$ that converges to a copula C (Theorem 1.7.5).

Now, it follows that C agrees with C' at every point of

$$\bigcup_{n \in \mathbb{N}} \mathrm{dom}\, S_n \subseteq \overline{\bigcup_{n \in \mathbb{N}} \mathrm{dom}\, S_n} = \mathrm{dom}\, S \,,$$

which concludes the proof.   □

Given Lemma 2.3.4, the proof of Sklar's theorem may proceed as follows.

Proof of Sklar's theorem by extension arguments. Let H be any d-dimensional d.f. By Lemma 2.3.3 there exists a unique subcopula C' such that eq. (2.3.1) holds for every $\mathbf{x} \in \mathbb{R}^d$. Then by Lemma 2.3.4 it is possible to extend the subcopula C' to a copula C. Since, for every $x_j \in \mathbb{R}$ $(j = 1, \ldots, d)$, $F(x_j)$ belongs to Ran F_j, eq. (2.2.1) holds.   □

2.3.2 Proof of Sklar's theorem by distributional transform

Here we present a proof of Sklar's theorem that is based on probabilistic arguments, namely a generalised version of the probability integral transformation. Let X and V be independent random variables on the probability space $(\Omega, \mathscr{F}, \mathbb{P})$. Let F be the d.f. of X and let V have uniform distribution on $(0, 1)$. Then define the *modified d.f.* $F(\cdot, \lambda)$, for $c \in \mathbb{I}$ and for $x \in \mathbb{R}$, by

$$F(x, c) := \ell^- F(x) + c \left(F(x) - \ell^- F(x) \right) . \tag{2.3.4}$$

The *generalised distributional transform of* $X \sim F$ is defined by

$$U := F(X, V) . \tag{2.3.5}$$

Obviously, if F is continuous, then $F(x, c) = F(x)$ for every $x \in \mathbb{R}$. Moreover, in this case, the generalised distributional transform coincides with the probability integral transformation.

However, Theorem 1.2.6 can be also adapted to this general construction, as follows.

Theorem 2.3.5. *Let U be the distributional transform of $X \sim F$ as defined in (2.3.5). Then U has uniform distribution $(0, 1)$ and $X = F^{(-1)}(U)$ almost everywhere.*

Proof. For $0 < \alpha < 1$, denote by q_α the value

$$q_\alpha := \sup\{t : F(t) < \alpha\}.$$

Then $F(X, V) \leq \alpha$ if, and only if, (X, V) belongs to the set

$$\{(t, c) : \mathbb{P}(X < t) + c\,\mathbb{P}(X = t) \leq \alpha\} .$$

Assume $\beta := \mathbb{P}(X = q_\alpha) > 0$ and set $q := \mathbb{P}(X < q_\alpha)$; then

$$\{(t, c) : \mathbb{P}(X < t) + c\,\mathbb{P}(X = t) \leq \alpha\}$$
$$= \{X < q_\alpha\} \bigcup \{X = q_\alpha, q + V\,\beta \leq \alpha\} .$$

Thus one has

$$\mathbb{P}(U \leq \alpha) = \mathbb{P}(F(X, V) \leq \alpha) - q + \beta\,\mathbb{P}\left(V \leq \frac{\alpha - q}{\beta}\right) = q + \beta\,\frac{\alpha - q}{\beta} = \alpha .$$

If $\beta = 0$, then

$$\mathbb{P}(F(X, V) \leq \alpha) = \mathbb{P}(X < q_\alpha) = \mathbb{P}(X \leq q_\alpha) = \alpha .$$

By definition of U,
$$\ell^- F(X) \leq U \leq F(X). \tag{2.3.6}$$

For every $t \in\,]\ell^- F(x), F(x)]$, one has $F^{(-1)}(t) = x$. Therefore, by (2.3.6), one has $F^{(-1)}(U) = X$ a.e., which concludes the proof. $\qquad\square$

The distributional transform is now used in order to provide a different proof of Sklar's theorem.

Proof of Sklar's theorem. Let $\mathbf{X} = (X_1, \ldots, X_d)$ be a random vector on the probability space $(\Omega, \mathscr{F}, \mathbb{P})$ with joint d.f. F. Let V be independent of $\mathbf{X}$ and uniformly distributed on $(0, 1)$, $V \sim \mathscr{U}(0, 1)$, and, for $j = 1, \ldots, d$ consider the distributional transforms $U_j := F_j(X_j, V)$; then Theorem 2.3.5 yields $U_j \sim \mathscr{U}(0, 1)$, and $X_j = F_j^{(-1)}(U_j)$ a.e. $(j = 1, \ldots, d)$. Thus, defining C to be the d.f. of $U = (U_1, \ldots, U_d)$ one has

$$F(\mathbf{x}) = \mathbb{P}\left(\bigcap_{j=1}^{d} \{X_j \leq x_j\} \right) = \mathbb{P}\left(\bigcap_{j=1}^{d} \left\{ F_j^{(-1)}(U_j) \leq x_j \right\} \right)$$

$$= \mathbb{P}\left(\bigcap_{j=1}^{d} \{U_j \leq F_j(x_j)\} \right) = C\left(F_1(x_1), \ldots, F_d(x_d) \right),$$

i.e., C is a copula of the d.f. F. $\square$

From the previous proof, it is clear that the distributional transform is not unique when F has some jumps, as an effect of the randomisation given by the r.v. V.

2.3.3 Proof of Sklar's theorem by regularisation

As has already been said above, the hard part in the proof of Sklar's theorem consists in establishing the existence of the copula associated with an arbitrary, viz., possibly with some discontinuity, distribution function. Because of Lemma 2.2.3, from now on, one's attention may be concentrated on the case in which at least one of the marginals of the d.f. H is not continuous.

Let $B_r(\mathbf{a})$ denote the open ball in $\mathbb{R}^d$ of centre $\mathbf{a}$ and radius r and consider the function $\varphi : \mathbb{R}^d \to \mathbb{R}$ defined by

$$\varphi(\mathbf{x}) := k \, \exp\left(\frac{1}{|\mathbf{x}|^2 - 1} \right) \mathbf{1}_{B_1(\mathbf{0})}(\mathbf{x}),$$

where the constant k is such that the L^1 norm $\|\varphi\|_1$ of φ is equal to 1. Further, for $\varepsilon > 0$, define $\varphi_\varepsilon : \mathbb{R}^d \to \mathbb{R}$ by

$$\varphi_\varepsilon(\mathbf{x}) := \frac{1}{\varepsilon^d} \, \varphi\left(\frac{\mathbf{x}}{\varepsilon} \right).$$

It is known (see, e.g., [Brezis, 1983, Chapter 4]) that φ_ε belongs to $C^\infty(\mathbb{R}^d)$, and that its support is the closed ball $\overline{B}_\varepsilon(\mathbf{0})$. Functions like φ_ε are called *mollifiers*.

If a d-dimensional d.f. H is given, then the convolution

$$H_n(\mathbf{x}) := \int_{\mathbb{R}^d} H(\mathbf{x} - \mathbf{y}) \, \varphi_{1/n}(\mathbf{y}) \, d\mathbf{y} = \int_{\mathbb{R}^d} \varphi_{1/n}(\mathbf{x} - \mathbf{y}) \, H(\mathbf{y}) \, d\mathbf{y} \qquad (2.3.7)$$

is well defined for every $\mathbf{x} \in \mathbb{R}^d$ and finite; in fact, H_n is bounded below by 0 and above by 1.

Lemma 2.3.6. *For every d-dimensional d.f. H and for every $n \in \mathbb{N}$, the function H_n defined by (2.3.7) is a d-dimensional continuous d.f. on $\mathbb{R}^d$.*

Proof. First, notice that H_n satisfies the limiting conditions for a d.f. in $\mathscr{D}^d$. In fact, given any $\varepsilon > 0$, there exists $M > 0$ such that $H(\mathbf{x}) > 1 - \varepsilon$ for every $\mathbf{x}$ such that $x_j > M$ for every j. It follows that, for every $\mathbf{x}$ such that every $x_j > M + 1/n$

$$H_n(\mathbf{x}) = \int_{B_{1/n}(\mathbf{0})} H(\mathbf{x} - \mathbf{y})\, \varphi_{1/n}(\mathbf{y})\, d\mathbf{y} > 1 - \varepsilon,$$

which implies that $H_n(\mathbf{x}) \to 1$ when all the components of $\mathbf{x}$ tend to $+\infty$. With similar arguments, one can conclude that $H_n(\mathbf{x}) \to 0$, if at least one argument tends to $-\infty$.

Now let $]\,\mathbf{a}, \mathbf{b}] \subseteq \mathbb{I}^d$ and consider its H_n-volume, according to Definition 1.2.10,

$$V_{H_n}(]\,\mathbf{a}, \mathbf{b}]) = \sum_{\mathbf{v}} \operatorname{sign}(\mathbf{v})\, H_n(\mathbf{v}) = \int_{\mathbb{R}^d} \sum_{\mathbf{v}} \operatorname{sign}(\mathbf{v})\, H(\mathbf{v} - \mathbf{y})\, \varphi_{1/n}(\mathbf{y})\, d\mathbf{y}$$

$$= \int_{\mathbb{R}^d} V_H(]\,\mathbf{a} - \mathbf{y}, \mathbf{b} - \mathbf{y}])\, \varphi_{1/n}(\mathbf{y})\, d\mathbf{y},$$

which is positive since H is a distribution function. Thus H_n defined by equation (2.3.7) is a d.f. too.

Finally we show that H_n is continuous. Since the function $\varphi_{1/n}$ is continuous on the compact $\overline{B}_1(\mathbf{0})$, it is uniformly continuous on it. Thus, for every $\varepsilon > 0$, there exists $\delta > 0$ such that $|\varphi_{1/n}(\mathbf{v}) - \varphi_{1/n}(\mathbf{u})| < \varepsilon$ whenever $|\mathbf{v} - \mathbf{u}| < \delta$. It follows that, if $|\mathbf{x} - \mathbf{y}| < \delta$, then one has

$$|H_n(\mathbf{x}) - H_n(\mathbf{y})| = \left| \int_{\mathbb{R}^d} H(\mathbf{u}) \left[\varphi_{1/n}(\mathbf{x} - \mathbf{u}) - \varphi_{1/n}(\mathbf{y} - \mathbf{u}) \right] d\mathbf{u} \right|$$

$$\leq \int_{\mathbb{R}^d} H(\mathbf{u}) \left| \varphi_{1/n}(\mathbf{x} - \mathbf{u}) - \varphi_{1/n}(\mathbf{y} - \mathbf{u}) \right| d\mathbf{u} < 2\varepsilon\, \lambda_d\left(B_1(\mathbf{0}) \right),$$

which proves that H_n is continuous. This concludes the proof. $\square$

Analogously, one can prove the following result.

Lemma 2.3.7. *Every one-dimensional marginal $F_{n,j}$ $(j = 1, \ldots, d)$ of H_n is continuous.*

Lemma 2.3.8. *If H is continuous at $\mathbf{x} \in \mathbb{R}^d$, then $\lim_{n \to +\infty} H_n(\mathbf{x}) = H(\mathbf{x})$.*

Proof. Let H be continuous at $\mathbf{x} \in \mathbb{R}^d$ so that, for every $\varepsilon > 0$, there exists $\delta = \delta(\varepsilon) > 0$ such that for every $\mathbf{y} \in B_\delta(\mathbf{0})$ one has $|H(\mathbf{x}) - H(\mathbf{x} - \mathbf{y})| < \varepsilon$. Then, for every $n > 1/\delta$,

$$
\begin{aligned}
|H_n(\mathbf{x}) - H(\mathbf{x})| &= \left| \int_{\mathbb{R}^d} H(\mathbf{x} - \mathbf{y})\, \varphi_{1/n}(\mathbf{y})\, d\mathbf{y} - \int_{\mathbb{R}^d} H(\mathbf{x})\, \varphi_{1/n}(\mathbf{y})\, d\mathbf{y} \right| \\
&\leq \int_{\mathbb{R}^d} |H(\mathbf{x} - \mathbf{y}) - H(\mathbf{x})|\, \varphi_{1/n}(\mathbf{y})\, d\mathbf{y} \\
&\leq \int_{B_{1/n}(\mathbf{0})} |H(\mathbf{x} - \mathbf{y}) - H(\mathbf{x})|\, \varphi_{1/n}(\mathbf{y})\, d\mathbf{y} < \varepsilon \int_{B_{1/n}(\mathbf{0})} \varphi_{1/n}(\mathbf{y})\, d\mathbf{y} = \varepsilon,
\end{aligned}
$$

which proves the assertion. $\qquad\qquad\qquad\qquad\qquad\qquad\qquad\qquad\qquad\qquad\square$

In a similar way one proves

Lemma 2.3.9. *Let H be a d.f. with univariate marginals $F_1, \ldots, F_d$. Let $j \in \{1, \ldots, d\}$. If F_j is continuous at $x \in \mathbb{R}$, then $\lim_{n \to +\infty} F_{n,j}(x) = F_j(x)$.*

We are now ready for the final step.

Proof of Sklar's theorem. For any given d.f. H, construct, for every $n \in \mathbb{N}$, the continuous d.f. H_n of (2.3.7), whose marginals $F_{n,1}, \ldots, F_{n,d}$ are also continuous; therefore, Lemma 2.3.3 ensures that there exists a d-copula C_n such that

$$
H_n(\mathbf{x}) = C_n\left(F_{n,1}(x_1), \ldots, F_{n,d}(x_d) \right).
$$

Because of the compactness of $\mathscr{C}_d$, there exists a subsequence $(C_{n(k)})_{k \in \mathbb{N}} \subseteq (C_n)_{n \in \mathbb{N}}$ that converges to a copula C.

Let $\mathbf{x} = (x_1, \ldots, x_d) \in \mathbb{R}^d$. Suppose that H is continuous at $\mathbf{x}$ and, for every $j \in \{1, \ldots, d\}$, its j-th marginal F_j is continuous at x_j. Thus by Lemma 2.3.8, $H_n(\mathbf{x})$ converges to $H(\mathbf{x})$ as $n \to +\infty$; as a consequence, so does its subsequence, $(H_{n(k)}(\mathbf{x}))$. Moreover, in view of Lemma 2.3.9,

$$
\left(C_{n(k)}\left(F_{n(k),1}(x_1), \ldots, F_{n(k),d}(x_d) \right) \right)_{k \in \mathbb{N}}
$$

converges to $C\left(F_1(x_1), \ldots, F_d(x_d) \right)$. Therefore

$$
H(\mathbf{x}) = C\left(F_1(x_1), \ldots, F_d(x_d) \right).
$$

Otherwise, choose a sequence $(\mathbf{x}_j)_{j \in \mathbb{N}}$ with $\mathbf{x}_j > \mathbf{x}$, componentwise, of continuity points for H such that, as j goes to ∞, $H(\mathbf{x}_j)$ converges to $H(\mathbf{x})$ and $F_i(x_{j,i})$ converges to $F_i(x_i)$ for every $i \in \{1, \ldots, d\}$ (notice that d.f.'s are assumed to be right-continuous).

Thus, for every $\varepsilon > 0$, there exists $j_0 \in \mathbb{N}$ such that $|H(\mathbf{x}_j) - H(\mathbf{x})| < \varepsilon$ and $|F_i(x_{j,i}) - F_i(x_i)| < \varepsilon/(2d)$ for $i \in \{1, \ldots, d\}$ whenever $j \geq j_0$.

Since $\mathbf{x}_{j_0}$ is a continuity point for H, Lemma 2.3.8 implies that, for k large enough, say for $k \geq k_0 = k_0(\varepsilon, j_0)$, $|(H_{n(k)}(\mathbf{x}_{j_0})) - H(\mathbf{x}_{j_0})| < \varepsilon$. Therefore,

$$|H_{n(k)}(\mathbf{x}_{j_0}) - H(\mathbf{x})| \leq |(H_{n(k)}(\mathbf{x}_{j_0})) - H(\mathbf{x}_{j_0})| + |H(\mathbf{x}_{j_0}) - H(\mathbf{x})| < 2\varepsilon,$$

namely $(H_{n(k)}(\mathbf{x}_{j_0}))$ converges to $H(\mathbf{x})$.

On the other hand, the sequence

$$\big(C_{n(k)} \left(F_1(x_{j_0,1}), \ldots, F_d(x_{j_0,d}) \right) \big)_{k \in \mathbb{N}}$$

converges to $C\left(F_1(x_1), \ldots, F_d(x_d) \right)$. In fact,

$$\begin{aligned}
&|C_{n(k)} \left(F_1(x_{j_0,1}), \ldots, F_d(x_{j_0,d}) \right) - C\left(F_1(x_1), \ldots, F_d(x_d) \right)| \\
&\quad \leq |C_{n(k)} \left(F_1(x_{j_0,1}), \ldots, F_d(x_{j_0,d}) \right) - C_{n(k)} \left(F_1(x_1), \ldots, F_d(x_d) \right)| \\
&\qquad + |C_{n(k)} \left(F_1(x_1), \ldots, F_d(x_d) \right) - C\left(F_1(x_1), \ldots, F_d(x_d) \right)| \\
&\quad \leq \sum_{j=1}^{d} |F_j(x_{j_0,j}) - F_j(x_j)| \\
&\qquad + |C_{n(k)} \left(F_1(x_1), \ldots, F_d(x_d) \right) - C\left(F_1(x_1), \ldots, F_d(x_d) \right)| \\
&\quad < d\,\frac{\varepsilon}{2d} + \frac{\varepsilon}{2} = \varepsilon,
\end{aligned}$$

where we have used Lemma 1.2.16 and the fact that $(C_{n(k)})$ converges (uniformly) to C. Therefore, one has

$$H(\mathbf{x}) = C\left(F_1(x_1), F_2(x_2), \ldots, F_d(x_d) \right),$$

which concludes the proof. $\qquad\square$

2.4 Copulas and rank-invariant property

In this section, we consider a kind of invariance property that points to a class of transformations of random vectors that do not change their copula.

Theorem 2.4.1 (Rank-invariant property of the copula of a r.v.). *Let $X_1, \ldots, X_d$ be continuous r.v.'s defined on the probability space $(\Omega, \mathscr{F}, \mathbb{P})$ and consider d continuous and strictly increasing mappings $\varphi_j : \mathrm{Ran}\, X_j \to \mathbb{R}$ $(j = 1, \ldots, d)$. Then the random vectors $\mathbf{X} = (X_1, \ldots, X_d)$ and $\mathbf{Y} = (\varphi_1 \circ X_1, \ldots, \varphi_d \circ X_d)$ have the same copula, $C_{\mathbf{X}} = C_{\mathbf{Y}}$.*

Proof. Because the univariate marginal F_j of X_j is continuous and φ_j is strictly increasing for every $j \in \{1, \ldots, d\}$, for every given $u_j \in\,]0, 1[$ there exists $x_j \in \mathbb{R}$ such that $F_j(x_j) = u_j$. Moreover, for every $t \in \mathbb{R}$

$$\begin{aligned}
F_{Y_j}(t) = \mathbb{P}(Y_j \leq t) &= F_{\varphi_j \circ X_j}(t) = \mathbb{P}(\varphi_j \circ X_j \leq t) \\
&= \mathbb{P}\left(X_j \leq \varphi_j^{-1}(t) \right) = F_{X_j}\left(\varphi_j^{-1}(t) \right).
\end{aligned}$$

Then it follows from Sklar's theorem that, for every $\mathbf{u} \in \mathbb{I}^d$,

$$
\begin{aligned}
C_{\mathbf{Y}}(u_1, \ldots, u_d) &= C_{\mathbf{Y}}\left(F_1(x_1), \ldots, F_d(x_d)\right) \\
&= C_{\mathbf{Y}}\left(F_{Y_1}(\varphi_1(x_1)), \ldots, F_{Y_d}(\varphi_d(x_d))\right) \\
&= H_{\mathbf{Y}}\left(\varphi_1(x_1), \ldots, \varphi_d(x_d)\right) \\
&= \mathbb{P}\left(\varphi_1 \circ X_1 \leq \varphi_1(x_1), \ldots, \varphi_d \circ X_d \leq \varphi_d(x_d)\right) \\
&= \mathbb{P}\left(X \leq x_1, \ldots, X_d \leq x_d\right) = H_{\mathbf{X}}(x_1, \ldots, x_d) \\
&= C_{\mathbf{X}}\left(F_{X_1}(x_1), \ldots, F_{X_d}(x_d)\right) = C_{\mathbf{X}}(u_1, \ldots, u_d),
\end{aligned}
$$

which shows that $\mathbf{X}$ and $\mathbf{Y}$ have the same copula. $\square$

The result contained in the last theorem is called the *rank-invariant property* of copulas. This invariance is an essential property of the statistical description. Following Schweizer and Wolff [1981], one may state that

> [...] it is precisely the copula which captures those properties of the joint distribution which are invariant under almost surely strictly increasing transformations. Hence the study of rank statistics — insofar as it is the study of properties invariant under such transformations — may be characterised as the study of copulas and copula-invariant properties.

Remark 2.4.2. From a practical point of view, rank-invariance property has the meaning that if one studies d "physical" quantities X_1, ..., X_d, then a change of the unit of measure, which is performed through the strictly increasing functions φ_1, ..., φ_d, leaves unaltered the copula of the considered quantities.

Analogously, in a financial context, the copula between the returns of two stock prices at some future time point does not change if their logarithmic values are considered. Moreover, the conversion into other currencies by multiplication with the respective exchange rates has no effect on the dependence.   ∎

When monotonic, but not necessarily increasing, transformations are applied to each component of a random vector, the following result is useful in calculating the resulting copula.

Theorem 2.4.3. *Let X_1, ..., X_d be continuous r.v.'s defined on the probability space $(\Omega, \mathscr{F}, \mathbb{P})$ and consider d continuous and strictly monotone mappings $\varphi_j : \operatorname{Ran} X_j \to \mathbb{R}$ $(j = 1, \ldots, d)$; then the copula of the random vector $\mathbf{Y} := (\varphi_1 \circ X_1, \ldots, \varphi_d \circ X_d)$ is given by C^ξ, where ξ is the symmetry of $\mathbb{I}^d$ given by $\xi = \sigma_{i_1} \cdots \sigma_{i_k}$, and σ_{i_1}, ..., σ_{i_k} are reflections of $\mathbb{I}^d$ and the indices i_1, ..., i_k correspond to those mappings φ_{i_1}, ..., φ_{i_k} that are strictly decreasing. Here $\xi = id_{\mathbb{I}^d}$ if all φ_j's are increasing.*

Proof. Notice that the d.f. of Y_j is $F_{Y_j} = F_{X_j} \circ \varphi_j^{-1}$ when φ_j is strictly increasing and $F_{Y_j} = 1 - F_{X_j} \circ \varphi_j^{-1}$, when φ_j is strictly decreasing. Here use has been made of the continuity of the d.f.'s $F_1, \ldots, F_d$. Hence

$$F_{Y_j} \circ Y_j = \begin{cases} F_{X_j} \circ X_j, & \text{if } \varphi_j \text{ is strictly increasing,} \\ 1 - F_{X_j} \circ X_j, & \text{if } \varphi_j \text{ is strictly decreasing.} \end{cases}$$

Since C is the d.f. of $(F_{X_1} \circ X_1, \ldots, F_{X_d} \circ X_d)$ (see Theorem 2.2.1), from the very definition of the copula induced by a symmetry (see Theorem 1.7.11) it follows that the d.f. of $(F_{Y_1} \circ Y_1, \ldots, F_{Y_d} \circ Y_d)$ is the copula C^ς, which concludes the proof. $\square$

Explicitly one has the following in the bivariate case

Corollary 2.4.4. *Let X and Y be continuous random variables defined on the probability space $(\Omega, \mathscr{F}, \mathbb{P})$ and consider the continuous mappings $\varphi : \mathrm{Ran}\, X \to \mathbb{R}$ and $\psi : \mathrm{Ran}\, Y \to \mathbb{R}$.*

(a) *If both φ and ψ are strictly increasing, then, for all $(u, v) \in \mathbb{I}^2$,*

$$C_{\varphi(X), \psi(Y)}(u, v) = C_{XY}(u, v) ;$$

(b) *if φ is strictly increasing while ψ is strictly decreasing, then, for all $(u, v) \in \mathbb{I}^2$,*

$$C_{\varphi(X), \psi(Y)}(u, v) = C^{\sigma_2}(u, v) = u - C_{XY}(u, 1 - v) ;$$

(c) *if φ is strictly decreasing while ψ is strictly increasing, then, for all $(u, v) \in \mathbb{I}^2$,*

$$C_{\varphi(X), \psi(Y)}(u, v) = C^{\sigma_1}(u, v) = v - C_{XY}(1 - u, v) ;$$

(d) *if both φ and ψ are strictly decreasing, then, for all $(u, v) \in \mathbb{I}^2$,*

$$C_{\varphi(X), \psi(Y)}(u, v) = C^{\sigma_1 \sigma_2}(u, v) = u + v - 1 + C_{XY}(1 - u, 1 - v) .$$

Proof. Let F and G be the d.f.'s of X and Y, respectively and let $H_{\varphi(X), \psi(Y)}$ be the joint d.f. of $\varphi(X)$ and $\psi(Y)$.

Although the assertions are direct consequences of the results of this section, we prefer to give a direct proof of (b), those of (c) and (d) being completely analogous.

(b) Because F and G are continuous, one has $F(\varphi^{-1}(x)) = F_{\varphi(X)}(x)$, while

$$G(\psi^{-1}(y)) = \mathbb{P}(Y \leq \psi^{-1}(y)) = \mathbb{P}(\psi(Y) \geq y) = 1 - F_{\psi(Y)}(y).$$

Since φ is strictly increasing and ψ is strictly decreasing, for all given u and v in $]0, 1[$, there exist x and y in $\mathbb{R}$ such that $F(\varphi^{-1}(x)) = F_{\varphi(X)}(x) = u$ and $F_{\psi(Y)}(y) = v$, so that $G(\psi^{-1}(y)) = 1 - v$. Thus, because of the continuity of G,

$$\begin{aligned} C_{\varphi(X), \psi(Y)}(u, v) &= C_{\varphi(X), \psi(Y)}(F_{\varphi(X)}(x), F_{\psi(Y)}(y)) \\ &= \mathbb{P}(\varphi(X) \leq x, \psi(Y) \leq y) = \mathbb{P}(X \leq \varphi^{-1}(x), Y \geq \psi^{-1}(y)) \\ &= \mathbb{P}(X \leq \varphi^{-1}(x)) - \mathbb{P}(X \leq \varphi^{-1}(x), Y < \psi^{-1}(y)) \\ &= \mathbb{P}(\varphi(X) \leq x) - C_{XY}(F(\varphi^{-1}(x)), G(\psi^{-1}(y))) \\ &= F_{\varphi(X)}(x) - C_{XY}(F_{\varphi(X)}(x), 1 - F_{\psi(Y)}(y)) \\ &= u - C_{XY}(u, 1 - v), \end{aligned}$$

which concludes the proof. $\square$

Finally, we present two measures of association for r.v.'s that are based on the ranks of the observations and, thus, may be characterised in terms of copulas. For more details about these measures and the related proofs see, e.g., [Joe, 1997; Nelsen, 2006; Schmid et al., 2010].

Definition 2.4.5. If C is the copula of two continuous r.v.'s X and Y defined on the same probability space $(\Omega, \mathscr{F}, \mathbb{P})$, then the *Spearman's rho* of X and Y, which will be denoted indifferently by $\rho(X, Y)$ or by $\rho(C)$, is given by

$$\rho(X,Y) = \rho(C) = 12 \int_{\mathbb{I}^2} C(u,v) \, \mathrm{d}u \, \mathrm{d}v - 3. \qquad (2.4.1)$$

$\diamond$

The coefficient 12 represents a normalisation: in fact, thanks to it, ρ_C is in $[-1, 1]$. Notice also that

$$\int_{\mathbb{I}^2} \Pi_2(u,v) \, \mathrm{d}u \, \mathrm{d}v = \frac{1}{4},$$

so that Spearman's rho may be written in the form

$$\rho(X,Y) = 12 \int_{\mathbb{I}^2} \{C(u,v) - u\,v\} \, \mathrm{d}u \, \mathrm{d}v. \qquad (2.4.2)$$

Since the random variables $F \circ X$ and $G \circ Y$ are uniformly distributed on $(0, 1)$, their mean value is $\mathbb{E}(F \circ X) = \mathbb{E}(G \circ Y) = 1/2$, while their variance is $V(F \circ X) = V(G \circ Y) = 1/12$. Therefore one can also write:

$$\rho(X,Y) := \frac{\mathrm{Cov}(F \circ X, G \circ Y)}{\sqrt{V(F \circ X)} \sqrt{V(G \circ Y)}}, \qquad (2.4.3)$$

where $\mathrm{Cov}(F \circ X, G \circ Y)$ denotes the covariance of $F \circ X$ and $G \circ Y$.

Definition 2.4.6. If C is the copula of two continuous r.v.'s X and Y defined on the same probability space $(\Omega, \mathscr{F}, \mathbb{P})$, then the *Kendall's tau* of X and Y, which will be denoted indifferently by $\tau(X, Y)$ or by $\tau(C)$, is given by

$$\tau(X,Y) = \tau(C) = 4 \int_{\mathbb{I}^2} C(u,v) \, \mathrm{d}C(u,v) - 1. \qquad (2.4.4)$$

$\diamond$

In view of (4.1.9), the expression for τ may be written as

$$\tau(X,Y) = 1 - 4 \int_{\mathbb{I}^2} \partial_1 C(u,v) \, \partial_2 C(u,v) \, \mathrm{d}u \, \mathrm{d}v. \qquad (2.4.5)$$

Both Spearman's rho and Kendall's tau are measures of concordance, in the sense that they satisfy a series of axioms for functionals in the class of copulas that allow their interpretation in terms of "degree of association" between two continuous r.v.'s with a specified copula. These axioms have been introduced by Scarsini [1984a,b] (see also [Dolati and Úbeda Flores, 2006; Taylor, 2007] for the multidimensional case) and are reproduced here.

Definition 2.4.7. A *measure of concordance* is a mapping $\mu : \mathscr{C}_2 \to \mathbb{R}$ such that

(μ1) μ is defined for every copula $C \in \mathscr{C}_2$;

(μ2) for every $C \in \mathscr{C}_2$, $\mu(C) = \mu(C^T)$, where C^T is the transpose of C;

(μ3) $\mu(C) \leq \mu(C')$ whenever $C \leq C'$;

(μ4) $\mu(C) \in [-1, 1]$;

(μ5) $\mu(\Pi_2) = 0$;

(μ6) $\mu(C^{\sigma_1}) = \mu(C^{\sigma_2}) = -\mu(C)$ for the symmetries σ_1, σ_2 of $\mathbb{I}^2$;

(μ7) weak continuity: if $C_n \xrightarrow[n \to +\infty]{} C$ uniformly, then $\lim_{n \to +\infty} \mu(C_n) = \mu(C)$. $\diamond$

In order to prove that Spearman's rho and Kendall's tau satisfy these axioms, some preliminary results are needed.

Following Kruskal [1958], for all 2-copulas C_1 and C_2 one can introduce the *concordance function* $Q(C_1, C_2)$ defined by

$$Q(C_1, C_2) = 4 \int_{\mathbb{I}^2} C_2(u, v) \, \mathrm{d}C_1(u, v) - 1 \,. \tag{2.4.6}$$

This function satisfies the following properties (for the proofs, see [Nelsen, 2006, Section 5.1]):

(a) Q is symmetric: $Q(C_1, C_2) = Q(C_2, C_1)$ for all $C_1, C_2 \in \mathscr{C}_2$;

(b) Q is increasing in each place with respect to the concordance order: if $C_1 \leq C_1'$ and $C_2 \leq C_2'$, then $Q(C_1, C_2) \leq Q(C_1', C_2')$;

(c) Q is invariant under the replacement of copulas by their survival copulas:

$$Q(C_1, C_2) = Q(\hat{C}_1, \hat{C}_2)$$

for all $C_1, C_2 \in \mathscr{C}_2$.

Thanks to this definition, it can be easily proved that

$$\rho(C) = 3 \, Q(C, \Pi_2), \qquad \tau(C) = Q(C, C). \tag{2.4.7}$$

Moreover, we recall that a sequence of finite measures (ν_n) on $(\mathbb{R}^d, \mathscr{B}(\mathbb{R}^d))$ is said to *converge vaguely* to ν if

$$\lim_{n \to +\infty} \int f \, \mathrm{d}\nu_n = \int f \, \mathrm{d}\nu \,,$$

for every continuous function f with compact support.

Lemma 2.4.8. *Let Ω be a subset of $\mathbb{R}^d$, (f_n) a sequence of positive functions defined on Ω that converges pointwise to f and (ν_n) a sequence of probability measures that converges vaguely to the probability measure ν. If, for every $n \in \mathbb{N}$,*

$$\int_\Omega f_n \, \mathrm{d}\, \nu_n < +\infty \,,$$

then the following statements are equivalent:

(a) $\lim_{n \to +\infty} \int_\Omega f_n \, \mathrm{d}\, \nu_n = \int_\Omega f \, \mathrm{d}\, \nu < +\infty$;

(b) $\lim_{a \to +\infty} \sup_{n \in \mathbb{N}} \int_{\{f_n > a\}} f_n \, \mathrm{d}\, \nu_n = 0$.

Proof. See [Serfozo, 1982]. $\qquad\qquad\qquad\qquad\qquad\qquad\qquad\qquad\qquad\qquad\qquad$ $\square$

Thus, we are able to show the following result.

Theorem 2.4.9. *Spearman's rho and Kendall's tau are measures of concordance.*

Proof. In view of eq. (2.4.7) properties $(\mu 1)$ and $(\mu 2)$ of Definition 2.4.7 are obvious. Let $C_1 \le C_2$; then

$$\rho(C_1) = Q(C_1, \Pi_2) = Q(\Pi_2, C_1) = \int_{\mathbb{I}^2} C_1 \, \mathrm{d}\Pi_2 \le \int_{\mathbb{I}^2} C_2 \, \mathrm{d}\Pi_2 = \rho(C_2) \,.$$

The same argument applies to Kendall's tau. Thus, property $(\mu 3)$ of Definition 2.4.7 holds for the considered indices.

As for property $(\mu 4)$ one has

$$\begin{aligned}
\rho(C) &= 3\, Q(C, \Pi_2) \le 3\, Q(M_2, \Pi_2) = 1 = \rho(M_2) \,, \\
\rho(C) &\ge 3\, Q(W_2, \Pi_2) = -1 = \rho(W_2) \,, \\
\tau(C) &\le Q(M_2, M_2) = 1 = \tau(M_2) \,, \\
\tau(C) &\ge Q(W_2, W_2) = -1 = \tau(W_2) \,.
\end{aligned}$$

Property $(\mu 5)$ follows from eq. (2.4.7) and from the properties of the concordance function Q.

Recall (Corollary 2.4.4 (c)) that $C^{\sigma_1}(u,v) := C'(u,v) = v - C(1-u,v)$. Then,

$$\begin{aligned}
\rho(C') &= 3\, Q(C', \Pi_2) = 3\, Q(\Pi_2, C') = 12 \int_{\mathbb{I}^2} C'(u,v) \, \mathrm{d}\Pi_2(u,v) - 3 \\
&= 12 \int_0^1 \{v - C(1-u,v)\} \, \mathrm{d}u \, \mathrm{d}v - 3 \\
&= 12 \int_0^1 v \, \mathrm{d}v - 12 \int_0^1 \mathrm{d}v \int_0^1 C(1-u,v) \, \mathrm{d}u - 3 \\
&= 3 - 12 \int_0^1 \mathrm{d}v \int_0^1 C(s,v) \, \mathrm{d}s = 3 - 12 \int_{\mathbb{I}^2} C(s,v) \, \mathrm{d}s \, \mathrm{d}v = -\rho(C) \,.
\end{aligned}$$

For Kendall's tau one has $\tau(C') = Q(C', C')$; as above

$$\tau(C') = 4 \int_{\mathbb{I}^2} C'(u, v) \, \mathrm{d}C'(u, v) - 1 = 1 - 4 \int_{\mathbb{I}^2} \partial_1 C'(u, v) \, \partial_2 C'(u, v) \, \mathrm{d}u \, \mathrm{d}v.$$

Since

$$\partial_2 C'(u, v) = \partial_2 \{v - C(1 - u, v)\} = 1 - \partial_2 C(1 - u, v),$$

one has

$$\tau(C') = 1 - 4 \int_{\mathbb{I}^2} \partial_1 C(1 - u, v) \{1 - \partial_2 C(1 - u, v)\} \, \mathrm{d}u \, \mathrm{d}v$$

$$= 1 - 4 \int_0^1 \mathrm{d}v \int_0^1 \partial_1 C(1 - u, v) \, \mathrm{d}u$$

$$+ 4 \int_0^1 \mathrm{d}v \int_0^1 \partial_1 C(s, v) \, \partial_2 C(s, v) \, \mathrm{d}s$$

$$= 1 - 4 \int_0^1 \mathrm{d}v \int_0^1 \partial_1 C(s, v) \, \mathrm{d}s + 4 \int_0^1 \mathrm{d}v \int_0^1 \partial_1 C(s, v) \, \partial_2 C(s, v) \, \mathrm{d}s$$

$$= -1 + 4 \int_0^1 \mathrm{d}v \int_0^1 \partial_1 C(s, v) \, \partial_2 C(s, v) \, \mathrm{d}s = -\tau(C).$$

Since analogous calculations can be done for C^{σ_1}, it has thus been proved that the two measures considered satisfy (μ6) of Definition 2.4.7.

Only the convergence property (μ7) remains to be proved. This poses no problem for Spearman's rho, since for these latter quantities a recourse to Lebesgue dominated convergence yields the conclusion. In the case of Kendall's tau, one has to apply Lemma 2.4.8 choosing $\Omega = \mathbb{I}^2$, $f_n = C_n$ and ν_n equals to the probability measure induced by the copula C_n (see Section 3.1). $\square$

2.5 Characterisation of basic dependence structures via copulas

Now, consider the dependence structures induced by the three basic copulas.

Since d continuous random variables $X_1, \ldots, X_d$ defined on the same probability space $(\Omega, \mathscr{F}, \mathbb{P})$ are independent if, and only if, for every $\mathbf{x} \in \mathbb{R}^d$, $H(\mathbf{x}) = \prod_{j=1}^d F_j(x_j)$, the following characterisation of independence is immediate.

Theorem 2.5.1. *For d independent and continuous random variables $X_1, \ldots, X_d$ defined on the same probability space $(\Omega, \mathscr{F}, \mathbb{P})$ the following statements are equivalent:*

(a) *they are independent;*

(b) *their copula $C_{\mathbf{X}}$ equals Π_d, $C_{\mathbf{X}} = \Pi_d$.*

In view of the previous result (and the invariance of the copula of a vector under strictly increasing transformation), the copula Π_d is often used as the starting point in the definition of various notions of positive (respectively, negative) dependence of copulas. Here we just consider the following concepts.

Definition 2.5.2. A copula $C \in \mathscr{C}_d$ is said to be

(a) *positively lower orthant dependent* (PLOD) if $C \geq \Pi_d$;

(b) *positively upper orthant dependent* (PUOD) if $\overline{C} \geq \overline{\Pi}_d$, where $\overline{C}$ and $\overline{\Pi}_d$ are the survival functions associated with C and Π_d, respectively;

(c) *positively orthant dependent* (POD) when it is both PLOD and PUOD.

The corresponding concepts of negative dependence are defined analogously by reversing the previous inequalities. $\diamond$

While M_d and Π_d are POD, W_2 is NOD. Moreover, in the bivariate case, C is PLOD if, and only if, C is PUOD. In this case, one also says that C is positive quadrant dependent (PQD).

Remark 2.5.3. Notice that even if all the bivariate margins of a given copula C are PQD (respectively, NQD), C is not necessarily positively dependent (neither in the PLOD nor in the PUOD sense). Consider, for instance, the copula

$$C(u_1, u_2, u_3) = u_1 u_2 u_3 (1 + \alpha(1 - u_1)(1 - u_2)(1 - u_3))$$

for any $\alpha \in [-1, 1]$, with $\alpha \neq 0$. Then all the bivariate margins of C are equal to Π_2, but, if $\alpha \geq 0$, $C \geq \Pi_3$; otherwise $C \leq \Pi_3$. Such a copula belongs to the EFGM family. For additional results, see also Durante et al. [2014b].   ∎

The strongest notion of positive dependence is given by *comonotonicity*, a concept that will be introduced in the sequel.

Definition 2.5.4. A subset A of $\mathbb{R}^d$ will be said to be *comonotonic* if either $\mathbf{x} \leq \mathbf{y}$ or $\mathbf{y} \leq \mathbf{x}$ for all points $\mathbf{x}$ and $\mathbf{y}$ in A. $\diamond$

Notice that if $\mathbf{x}$ and $\mathbf{y}$ belong to a comonotonic set A, and if there is an index i for which $x_i < y_i$, then $x_j \leq y_j$ for every index $j \in \{1, \ldots, d\}$. A subset of a comonotonic set is also comonotonic.

For a set $A \subseteq \mathbb{R}^d$ define the (i, j)-projection of A by

$$A_{i,j} := \{(x_i, x_j) : \mathbf{x} \in A\}.$$

The following result is evident.

Lemma 2.5.5. *For a set $A \subseteq \mathbb{R}^d$ the following statements are equivalent:*

(a) *A is comonotonic;*

(b) *for $i \neq j$ all the projections $A_{i,j}$ are comonotonic.*

For a set A the comonotonicity of the $(j, j + 1)$-projections $A_{j,j+1}$ does not necessarily imply that of A. To this purpose consider the set

$$A = \{(x_1, 1, x_3) : x_1, x_3 \in \mathbb{I}\}.$$

Then A is not comonotonic, although so are both $A_{1,2}$ and $A_{2,3}$.

Definition 2.5.6. A random vector $\mathbf{X} = (X_1, \ldots, X_d)$ on a probability space $(\Omega, \mathscr{F}, \mathbb{P})$ is said to be $\mathbb{P}$-*comonotonic* if it has a comonotonic support, i.e., if there exists a comonotonic set A such that $\mathbb{P}(\mathbf{X} \in A) = 1$. $\qquad\qquad\qquad \diamond$

The comonotonicity of a random vector is characterised in the following result, which also provides the probabilistic interpretation of the copula M_d.

Theorem 2.5.7. *For a continuous random vector* $\mathbf{X} = (X_1, \ldots, X_d)$ *on the probability space* $(\Omega, \mathscr{F}, \mathbb{P})$, *the following statements are equivalent:*

(a) $\mathbf{X}$ *is* $\mathbb{P}$-*comonotonic;*

(b) M_d *is the copula of* $\mathbf{X}$;

(c) *there exist a probability space* $(\Omega', \mathscr{F}', \mu)$, *a random variable* Z *on this space and d increasing functions* $\varphi_j : \mathbb{R} \to \mathbb{R}$ $(j = 1, \ldots, d)$ *such that* $\mathbf{X}$ *has the same law as* $(\varphi_1 \circ Z, \ldots, \varphi_d \circ Z)$.

Proof. (a) $\Longrightarrow$ (b) Let A be a comonotonic support of $\mathbf{X}$, let $\mathbf{x}$ be in $\mathbb{R}^d$ and define

$$A_j(\mathbf{x}) := \{\mathbf{y} \in A : y_j \leq x_j\} \quad (j = 1, \ldots, d) \, .$$

Since $A(\mathbf{x})$ is comonotonic there is an index i such that $A_i \subseteq A_j$ for every index $j \neq i$; thus

$$A_i(\mathbf{x}) = \bigcap_{j=1}^{d} A_j(\mathbf{x}) \, .$$

Therefore $F_i(x_i) \leq F_j(x_j)$, and, if H is the d.f. of $\mathbf{X}$, one has

$$H(\mathbf{x}) = \mathbb{P}(\mathbf{X} \leq \mathbf{x}) = \mathbb{P}\left(\mathbf{X} \in \bigcap_{j=1}^{d} A_j(\mathbf{x}) \right) = \mathbb{P}(\mathbf{X} \in A_i(\mathbf{x}))$$

$$= \min\{F_1(x_1), \ldots, F_d(x_d)\} = M_d(F_1(x_1), \ldots, F_d(x_d)) \, .$$

(b) $\Longrightarrow$ (c) Let H be the d.f. of $\mathbf{X}$ and let U be a random variable on the standard probability space $(\mathbb{I}, \mathscr{B}(\mathbb{I}), \lambda)$ with uniform law on $\mathbb{I}$; then, if F_j is the d.f. of X_j, the continuity of each F_j ensures that one has, for every $\mathbf{x} = (x_1, \ldots, x_d)$,

$$H(\mathbf{x}) = M_d\left(F_1(x_1), \ldots, F_d(x_d)\right) = \lambda\left[U \leq M_d\left(F_1(x_1), \ldots, F_d(x_d)\right)\right]$$

$$= \lambda\left(\bigcap_{j=1}^{d} \{U \leq F_j(x_j)\} \right) = \lambda\left(F_1^{(-1)}(U) \leq x_1, \ldots, F_d^{(-1)}(U) \leq x_d \right) \, .$$

One can now take $Z = U$ and $\varphi_j = F_j^{(-1)}$ $(j = 1, \ldots, d)$.

(c) $\Longrightarrow$ (a) Assume that there exist a random variable Z on a probability space $(\Omega', \mathscr{F}', \mu)$ with $\mathbb{P}(Z \in A) = 1$ and d increasing functions $\varphi_1, \ldots, \varphi_d$ such that the random vectors $\mathbf{X}$ and $(\varphi_1 \circ Z, \ldots, \varphi_d \circ Z)$ have the same law. Then the set of values taken by $\mathbf{X}$ is

$$\{\varphi_1(z), \ldots, \varphi_d(z) : z \in A\} \, ,$$

which is obviously comonotonic; this implies that $\mathbf{X}$ is $\mathbb{P}$-comonotonic. $\qquad\qquad \square$

Corollary 2.5.8. *For a continuous random vector* $\mathbf{X} = (X_1, \ldots, X_d)$ *on the probability space* $(\Omega, \mathscr{F}, \mathbb{P})$, *condition* (c) *of Theorem 2.5.7 is equivalent to the following one*

(d) *there are* $d - 1$ *increasing functions* f_j $(j = 2, \ldots, d)$ *such that, for every index* $j \in \{2, \ldots, d\}$, X_j *and* $f_j \circ X_1$ *have the same law.*

Proof. (c) $\Longrightarrow$ (d) The random variable Z has the same law as $\varphi_1^{-1}(X_1)$, hence it suffices to set $f_j = \varphi_j \circ \varphi_1^{-1}$ for $j = 2, \ldots, d$.

The proof of the implication (d) $\Longrightarrow$ (c) is immediate. $\square$

Since the first component plays no special rôle, when M_d is the copula of the d-dimensional random vector $\mathbf{X} = (X_1, \ldots, X_d)$, each component has the same distribution as an increasing function of the remaining ones.

Notice that comonotonicity is a property of the copula of a random vector $\mathbf{X}$, regardless the specific choice of the marginals. Moreover, given d univariate d.f.'s $F_1, \ldots, F_d$ it is always possible to find a joint d.f. H, and thus a random vector $\mathbf{X}$ on a suitable probability space, such that $\mathbf{X}$ is comonotonic.

Comonotonicity of a random vector is equivalent to comonotonicity of all its bivariate subvectors, as the following result shows.

Corollary 2.5.9. *For a continuous random vector* $\mathbf{X}$ *on* $(\Omega, \mathscr{F}, \mathbb{P})$ *the following conditions are equivalent:*

(a) $\mathbf{X}$ *is* $\mathbb{P}$-*comonotonic;*

(b) (X_i, X_j) *are* $\mathbb{P}$-*comonotonic for all indices* i *and* j *in* $\{1, \ldots, d\}$, *with* $i \neq j$.

Proof. (a) $\Longrightarrow$ (b) If $\mathbf{X}$ is $\mathbb{P}$-comonotonic, then it is obvious that every pair (X_i, X_j) of components from $\mathbf{X} = (X_1, \ldots, X_d)$ is also $\mathbb{P}$-comonotonic.

(b) $\Longrightarrow$ (a) Conversely, let (X_i, X_j) be $\mathbb{P}$-comonotonic for all i and j in $\{1, \ldots, d\}$, with $i \neq j$, and consider the set

$$A = \left\{ \left(F_1^{(-1)}(\alpha), \ldots, F_d^{(-1)}(\alpha) \right) : \alpha \in \,]0,1[\right\}.$$

In view of part (c) of Theorem 2.5.7, A is the support of $\mathbf{X}$. The (i,j)-projection of A is given by

$$A_{ij} = \left\{ \left(F_i^{(-1)}(\alpha), F_j^{(-1)}(\alpha) \right) : \alpha \in \,]0,1[\right\},$$

which is comonotonic. Therefore, in view of Lemma 2.5.5, A is comonotonic, which is the desired assertion. $\square$

A stronger statement holds in the case $d = 2$. In fact, from condition (c) of Theorem 2.5.7 (and from Corollary 2.5.8 (d)), it follows that if (X, Y) is a continuous and comonotonic vector on the probability space $(\Omega, \mathscr{F}, \mathbb{P})$, the support of (X, Y) is a set of type $\{(x, \varphi(x)) : x \in \text{supp}(X)\}$. It follows that $X = \varphi(Y)$ almost certainly (with respect to $\mathbb{P}$). By virtue of Corollary 2.5.9, the following result easily follows.

Corollary 2.5.10. *Under the assumptions of Theorem 2.5.7,* **X** *is* $\mathbb{P}$-*comonotonic if, and only if, the following condition holds:*

(e) *for all indices* i *and* j *in* $\{1, \ldots, d\}$, X_j *is almost certainly an increasing function of* X_i.

Remark 2.5.11. As a consequence of Corollary 2.5.10, two continuous r.v.'s X and Y on the same probability space $(\Omega, \mathscr{F}, \mathbb{P})$ are equal a.e., $X = Y$ a.e., if, and only if, $F_X = F_Y$ and M_2 is their copula. In particular, (X, X) is comonotonic.

Moreover, if U and V are r.v.'s whose d.f. is given by M_2, then (U, V) is obviously comonotonic and $\mathbb{P}(U = V) = 1$. ∎

In the bivariate case it is possible to introduce the concept of a $\mathbb{P}$-*countermonotonicity* in a similar manner.

Definition 2.5.12. A subset A of $\mathbb{R}^2$ is said to be *countermonotonic* if for all (x_1, y_1) and (x_2, y_2) in A one has $(x_2 - x_1)(y_1 - y_2) < 0$.
A random pair (X, Y) on the probability space $(\Omega, \mathscr{F}, \mathbb{P})$ is said to be $\mathbb{P}$-*countermonotonic* if it has a countermonotonic support, i.e., there exists a countermonotonic set A such that $\mathbb{P}((X, Y) \in A) = 1$. ◇

The proof of the following result will not be provided since it can be obtained by the methods already encountered.

Theorem 2.5.13. *For a continuous random vector* (X, Y) *on the probability space* $(\Omega, \mathscr{F}, \mathbb{P})$, *the following statements are equivalent:*

(a) (X, Y) *is* $\mathbb{P}$-*countermonotonic;*

(b) W_2 *is the copula of* **X**;

(c) *there exist a probability space* $(\Omega', \mathscr{F}', \mu)$ *and a random variable* Z *on this space, an increasing function* φ *and a decreasing function* ψ *such that* (X, Y) *has the same law as* $(\varphi(Z), \psi(Z))$;

(d) *there exists a decreasing function* f *such that* $Y = f \circ X$ *a.e.*

Corollary 2.5.14. *For two continuous r.v.'s* X *and* Y *on the same probability space* $(\Omega, \mathscr{F}, \mathbb{P})$ *the following statements are equivalent:*

(a) $\mathbb{P}(X + Y = c) = 1$ *where* $c \in \mathbb{R}$;

(b) *for every* $t \in \mathbb{R}$, $F_X(t) + F_Y(c - t) = 1$ *and the copula of* X *and* Y *is* W_2.

Proof. (a) $\Longrightarrow$ (b) If $X + Y = c$, then X and Y are a strictly decreasing function of each other and, hence, their copula is W_2. Also, for every $t \in \mathbb{R}$,

$$F_X(t) = \mathbb{P}(X \leq t) = \mathbb{P}(c - Y \leq t) = \mathbb{P}(Y \geq c - t) = 1 - \mathbb{P}(Y < c - t)$$
$$= 1 - \ell^- F_Y(c - t) = 1 - F_Y(c - t),$$

since F_Y is continuous.

(b) $\Longrightarrow$ (a) Since W_2 is the connecting copula of X and Y, M_2 is the copula of X and $Z := c - Y$. Let F_Z denote the d.f. of Z; then, for every $t \in \mathbb{R}$,

$$F_Z(t) = \mathbb{P}(c - Y \leq t) = \mathbb{P}(Y \geq c - t) = 1 - \mathbb{P}(Y < c - t)$$
$$= 1 - \ell^- F_Y(c - t) = 1 - F_Y(c - t) = F_X(t).$$

Because of Theorem 2.5.13, $c - Y = X$ a.e., which concludes the proof. □

Further readings 2.5.15. The comonotonicity of random vectors has been extensively considered in the literature: see, for instance, [Denneberg, 1994; Dhaene et al., 2002b,a; Rüschendorf, 2013] and references therein. It has also been extended to comonotonic random vectors by Puccetti and Scarsini [2010] and Ekeland et al. [2012]. For an overview of comonotonicity and countermonotonicity, see also Puccetti and Wang [2014]. ∎

2.6　Copulas and order statistics

Sklar's theorem provides an elegant solution to the problem of determining all multivariate d.f.'s with a given set of univariate marginal distributions. However, in several practical cases, the Fréchet class may be further constrained by additional information available about the multivariate random vector $\mathbf{X}$ we aim at modelling. In this section, we present an illustration of this fact in the case when some of the order statistics of $\mathbf{X}$ are known. For simplicity's sake, the discussion will be mainly devoted to dimension $d = 2$ and to r.v.'s with uniform margins.

Let $U_1, \ldots, U_d$ be r.v.'s defined on the same probability space $(\Omega, \mathscr{F}, \mathbb{P})$, having uniform distribution on $(0, 1)$ and C as their distribution function; then, for every $t \in \mathbb{I}$

$$\mathbb{P}(\max\{U_1, U_2, \ldots, U_d\} \leq t) = \mathbb{P}\left(\bigcap_{j=1}^{d} \{U_j \leq t\} \right) = C(t, t, \ldots, t).$$

In other words, the largest order statistics of $\mathbf{U}$ is only determined by the so-called *diagonal section* δ_C of the copula C, given by $\delta_C(t) := C(t, t, \ldots, t)$.

Moreover, in the case $d = 2$, it can be also proved that

$$\mathbb{P}(\min\{U_1, U_2\} \leq t) = \mathbb{P}(U \leq t) + \mathbb{P}(V \leq t) - \mathbb{P}(\{U \leq t\} \cap \{V \leq t\})$$
$$= 2t - \delta_C(t).$$

Thus, determining a bivariate copula C with a prescribed diagonal section δ is equivalent to determining a random vector (U, V) such that $(U, V) \sim C$ and the marginal d.f.'s of the order statistics of (U, V) are known.

This poses the natural question of establishing suitable conditions that ensure that a d.f. $\delta : \mathbb{I} \to \mathbb{I}$ is the diagonal section of a copula, i.e. to determine whether one can construct a copula with a prescribed distribution of the maximal order statistics of the random vector $\mathbf{U}$ associated with C. Necessary and sufficient conditions are stated in the next theorem.

Theorem 2.6.1. *A function $\delta : \mathbb{I} \to \mathbb{I}$ is the diagonal section of a d-copula C if, and only if, it satisfies the following properties:*

(a) $\delta(0) = 0$ and $\delta(1) = 1$;

(b) $\delta(t) \leq t$ for every $t \in \mathbb{I}$;

(c) δ is increasing;

(d) δ is d-Lipschitz, viz., for all t and t' in $\mathbb{I}$, $|\delta(t') - \delta(t)| \leq d\,|t' - t|$.

Proof. In view of basic properties of a d-copula it is not hard to show that any diagonal of a copula satisfies conditions (a)–(d).

Conversely, given a function δ satisfying (a)–(d) it is enough to show the existence of a copula whose diagonal coincides with δ. To this end, consider the copula A of type (1.7.3) where $F_i(t) = \frac{dt - \delta(t)}{d-1}$ for $i \in \{1, \ldots, d-1\}$, $F_d(t) = \delta(t)$, $A = M_d$. Then A may be rewritten as

$$A(\mathbf{u}) = \frac{1}{d} \sum_{i=1}^{n} M_d\left(F_1(u_{\tau^i(1)}), \ldots, F_d(u_{\tau^i(d)})\right), \qquad (2.6.1)$$

where $\tau^i(k) = k + 1 \bmod d$. Now, it is easy to prove that $A(t, \ldots, t) = \delta(t)$, from which the assertion follows. $\qquad \square$

Remark 2.6.2. In the case $d = 2$, the copula of type (2.6.1) can be rewritten as

$$C_\delta^{\mathbf{FN}}(u, v) := \min\left\{u, v, \frac{\delta(u) + \delta(v)}{2}\right\}, \qquad (2.6.2)$$

which coincides with the Fredricks–Nelsen copula of Example 1.7.16. A special feature of this copula is that it determines not only the univariate distribution functions of order statistics, but also the dependence among them. In fact, if $(U, V) \sim C_\delta^{\mathbf{FN}}$, then it follows by the very construction of these copulas (see Example 1.7.16) that the copula of $\min\{U, V\}$ and $\max\{U, V\}$ is M_2. $\qquad \blacksquare$

A function that satisfies properties (a)–(d) of Theorem 2.6.1 will be called a *d-diagonal* or, if no possibility of confusion arises, simply a *diagonal*. In the case $d = 2$, the graph of a diagonal δ lies in the triangle of vertices $(0, 0)$, $(1/2, 0)$ and $(1, 1)$; moreover, its slope takes values in the interval $[0, 2]$ (see Figure 2.3),

For a fixed diagonal δ we denote by $\mathscr{C}_{d,\delta} \subseteq \mathscr{C}_d$ the set of all d-copulas with diagonal section equal to δ. We have seen that $C_{d,\delta}$ is non-empty. However, for a fixed δ, in general, $C_{d,\delta}$ may contain more than one single element, as the following examples show.

Example 2.6.3. The diagonal δ_{W_2} of the 2-dimensional lower Hoeffding–Fréchet bound is given by

$$\delta_{W_2}(t) = W_2(t, t) = \max\{2t - 1, 0\} = \begin{cases} 0, & t \leq 1/2, \\ 2t - 1, & t \geq 1/2. \end{cases}$$

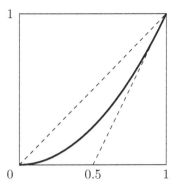

Figure 2.3: *Visualisation of a diagonal of a 2-copula.*

We shall now exhibit another 2-copula with the same diagonal as W_2.

On the unit square $\mathbb{I}^2$ consider the uniform law on the set

$$([0, 1/2] \times [1/2, 1]) \bigcup ([1/2, 1] \times [0, 1/2]) \; ;$$

this law has a density equal to 2 on each of the two squares and its d.f. is, of course, a 2-copula C_1. For $t \le 1/2$, $C_1(t, t) = 0$, while, for $t > 1/2$,

$$C_1(t, t) = 2 \int_0^{1/2} \mathrm{d}u \int_{1/2}^t \mathrm{d}v + 2 \int_{1/2}^t \mathrm{d}u \int_0^{1/2} \mathrm{d}v$$

$$= 2 \frac{1}{2} \left(t - \frac{1}{2} \right) + 2 \frac{1}{2} \left(t - \frac{1}{2} \right) = 2t - 1 \, .$$

Therefore C_1 and W_2 have the same diagonal, $\delta_{C_1} = \delta_{W_2}$, but $C_1 \ne W_2$. Indeed,

$$W_2 \left(\frac{1}{4}, \frac{3}{4} \right) = 0 \ne C_1 \left(\frac{1}{4}, \frac{3}{4} \right) = \frac{1}{8}.$$

Therefore, in this case, the knowledge of the diagonal does not suffice to single out a copula.   ■

Example 2.6.4. If a diagonal δ_C of a 2-copula C coincides with the identity function on $\mathbb{I}$, viz., $\delta_C = id_\mathbb{I}$, then $C = M_2$. In fact, assume, if possible, that there is a point $(u, v) \in \mathbb{I}^2$ for which $C(u, v) \ne M_2(u, v)$. In view of the upper Hoeffding–Fréchet bound, this latter relationship implies $C(u, v) < u \wedge v$. If $u \le v$, then

$$u = \delta_C(u) = C(u, u) \le C(u, v) < u \wedge v = u \, ,$$

which is a contradiction; similarly for $u > v$. Therefore $C(u, v) = M_2(u, v) = u \wedge v$ for every $(u, v) \in \mathbb{I}^2$.   ■

Given the richness of the class $\mathscr{C}_{d,\delta}$ it might also be convenient to determine (sharp) bounds for it, in order to have an idea of the range of possible dependencies that are covered by this subclass of copulas. Interestingly, again copulas of type (2.6.2) give a partial answer.

Lemma 2.6.5. *For every 2-diagonal δ and for every exchangeable copula $C \in \mathscr{C}_2$ one has $C \leq C_\delta^{\mathrm{FN}}$, where C_δ^{FN} is the Fredricks–Nelsen copula of Example 1.7.16.*

Proof. For all u and v one has $C(u,v) = C(v,u)$ and $C(v,v) - C(v,u) - C(u,v) + C(u,u) \geq 0$, so that

$$C(u,v) \leq \frac{\delta(u) + \delta(v)}{2}.$$

Since, on the other hand, $C(u,v) \leq \min\{u,v\}$, the last two inequalities yield the assertion. $\square$

In the bivariate case, a lower bound for $\mathscr{C}_\delta$ is given by the so-called *Bertino copula* introduced in the following eq. (2.6.5). This copula was considered by Fredricks and Nelsen [2003] by using a previous work by Bertino [1977].

For a given 2-diagonal δ we shall also consider the function $\hat{\delta} : \mathbb{I} \to \mathbb{I}$ defined by

$$\hat{\delta}(t) := t - \delta(t). \tag{2.6.3}$$

This function satisfies the following Lipschitz condition

$$\forall\, t, t' \in \mathbb{I} \qquad |\hat{\delta}(t') - \hat{\delta}(t)| \leq (d-1)\,|t' - t|. \tag{2.6.4}$$

Theorem 2.6.6. *For every 2-diagonal δ, the function $C_\delta^{\mathrm{Ber}} : \mathbb{I}^2 \to \mathbb{I}$ defined by*

$$C_\delta^{\mathrm{Ber}}(u,v) := \min\{u,v\} - \min\{\hat{\delta}(t) : t \in [u \wedge v, u \vee v]\} \tag{2.6.5}$$

$$= \begin{cases} u - \min_{t \in [u,v]}\{t - \delta(t)\}, & u \leq v, \\ v - \min_{t \in [v,u]}\{t - \delta(t)\}, & v < u, \end{cases}$$

is a symmetric 2-copula having the diagonal equal to δ, i.e., $C_\delta^{\mathrm{Ber}} \in \mathscr{C}_\delta$.

Proof. The equality $C_\delta^{\mathrm{Ber}}(t,t) = \delta(t)$ is immediate. Since $\hat{\delta}(0) = \hat{\delta}(1) = 0$, the boundary conditions of a 2-copula are satisfied, i.e., for every $t \in \mathbb{I}$,

$$C_\delta^{\mathrm{Ber}}(0,t) = C_\delta^{\mathrm{Ber}}(t,0) = 0 \qquad \text{and} \qquad C_\delta^{\mathrm{Ber}}(1,t) = C_\delta^{\mathrm{Ber}}(t,1) = t.$$

In order to show that C_δ^{Ber} is 2-increasing, we shall take advantage of Lemma 1.4.9.

If $R = [u, v] \times [u, v]$, set $\hat{\delta}(t_0) := \min\{\hat{\delta}(t) : t \in [u, v]\}$; then, in view of symmetry and of the Lipschitz properties satisfied by δ and $\hat{\delta}$,

$$V_{C_\delta^{\mathbf{Ber}}}(R) = C_\delta^{\mathbf{Ber}}(v, v) - C_\delta^{\mathbf{Ber}}(u, v) - C_\delta^{\mathbf{Ber}}(v, u) + C_\delta^{\mathbf{Ber}}(u, u)$$

$$= \delta(u) + \delta(v) - 2\,C_\delta^{\mathbf{Ber}}(u, v) = \delta(u) + \delta(v) - 2u + 2\hat{\delta}(t_0)$$

$$= \delta(v) - \delta(u) - 2\{\delta(t_0) - \delta(u)\} + 2\,(t_0 - u)$$

$$\geq \delta(v) - \delta(u) - 4\,(t_0 - u) + 2\,(t_0 - u) = \delta(v) - \delta(u) - 2\,(t_0 - u)$$

$$\geq \hat{\delta}(u) + u - 2t_0 + \delta(t_0) = \hat{\delta}(u) - \hat{\delta}(t_0) - (t_0 - u) \geq 0\,.$$

Let the rectangle $R = [u_1, u_2] \times [v_1, v_2]$ be entirely contained in the upper triangle T_U, namely, $u_1 \leq u_2 \leq v_1 \leq v_2$ and set $\hat{\delta}(t_{ij}) := \min\{\hat{\delta}(t) : t \in [u_i, v_j]\}$; then

$$V_{C_\delta^{\mathbf{Ber}}}(R) = C_\delta^{\mathbf{Ber}}(u_2, v_2) - C_\delta^{\mathbf{Ber}}(u_2, v_1) - C_\delta^{\mathbf{Ber}}(u_1, v_2) + C_\delta^{\mathbf{Ber}}(u_1, v_2)$$

$$= \hat{\delta}(t_{21}) - \hat{\delta}(t_{11}) - \hat{\delta}(t_{12}) + \hat{\delta}(t_{22})\,.$$

Since $[u_1, v_1] \cup [u_2, v_2] = [u_1, v_2]$, the point t_{12} at which $\hat{\delta}$ attains its minimum in the interval $[u_1, v_2]$ necessarily coincides with either t_{11} or t_{22}.

If $t_{12} = t_{11}$, then $V_{C_\delta^{\mathbf{Ber}}}(R) = \hat{\delta}(t_{21}) - \hat{\delta}(t_{22}) \geq 0$, while if $t_{12} = t_{22}$, then $V_{C_\delta^{\mathbf{Ber}}}(R) = \hat{\delta}(t_{21}) - \hat{\delta}(t_{11}) \geq 0$.

The remaining case, when the rectangle $R = [u_1, u_2] \times [v_1, v_2]$ is entirely contained in the lower triangle T_L, i.e., when $v_1 \leq v_2 \leq u_1 \leq u_2$, is dealt with in exactly the same manner as the previous one. $\square$

Example 2.6.7. Given a 2-diagonal δ, consider the function given, for all $(u, v) \in [0, 1]^2$, by

$$C_\delta(u, v) := \max\{0, \delta(u \vee v) - |u - v|\}. \tag{2.6.6}$$

As shown by Durante et al. [2006a], this function is a copula provided that there exists $a \in [0, 1/2]$ such that $\delta(x) = 0$ on $[0, a]$ and the function $t \mapsto (\delta(t) - t)$ is increasing on $[a, 1]$. Under these conditions, it is easily proved that C_δ is also a Bertino copula that satisfies the functional equation

$$C_\delta(u, v) + |u - v| = \delta(u \vee v),$$

whenever $C(u, v) > 0$. In particular, the family $\{C_\alpha : \alpha \in [0, 1]\}$ defined by

$$C_\alpha(u, v) := \begin{cases} \max\{0, u + v - \alpha\} & (u, v) \in [0, \alpha]^2\,; \\ \min\{u, v\} & \text{otherwise} \end{cases} \tag{2.6.7}$$

is formed by copulas of type (2.6.6).   ∎

Bertino copulas are a lower bound in the class of copulas with a given diagonal section, as the following result shows.

Theorem 2.6.8. *For every diagonal $\delta \in \mathscr{D}_2$ and for every copula $C \in \mathscr{C}_\delta$ one has*

$$C_\delta^{\mathbf{Ber}} \leq C . \tag{2.6.8}$$

Proof. Let C be in $\mathscr{C}_\delta$ and take u and v in $\mathbb{I}$ with $u \leq v$. For every $t \in [u, v]$, one has $C(t, v) \geq C(t, t) = \delta(t)$. Then $C(t, v) - C(u, v) \leq t - u$ by Lipschtiz condition, whence

$$C(u, v) \geq C(t, v) + u - t \geq u - t + \delta(t) = u - \hat{\delta}(t) ,$$

from which one has $C(u, v) \geq u - \min_{t \in [u,v]} \hat{\delta}(t) = C_\delta^{\mathbf{Ber}}(u, v)$.

A similar argument leads to the inequality $C(u, v) \geq C_\delta^{\mathbf{Ber}}(u, v)$ when $u > v$. This proves the assertion. □

Further readings 2.6.9. Extension of the results presented here may be found in [Jaworski and Rychlik, 2008; Jaworski, 2009]. Basically they are also contained in a series of papers by Rychlik [1993, 1994] about order statistics of random variables. Further results about copulas and order statistics have also been presented in [Navarro and Spizzichino, 2010].

The construction of a copula in $\mathscr{C}_d$ with a given diagonal section as in Theorem 2.6.1 is due to Jaworski [2009] (see also [Cuculescu and Theodorescu, 2001]).

The copula of type (2.6.2) was introduced by Fredricks and Nelsen [1997] (see also Nelsen and Fredricks [1997]). It is also related to the construction hairpin copulas (see [Seethoff and Shiflett, 1978; Sherwood and Taylor, 1988; Durante et al., 2014a]). ∎

2.6.1 Diagonal sections of copulas and tail dependence

The main relevance in applications of the diagonal section of a copula is its role in determining the so-called *tail dependence coefficients* (TDCs) of a bivariate copula. These coefficients were suggested by Sibuya [1960], but their copula-based representation is made explicit by Joe [1993].

Definition 2.6.10. Let X and Y be continuous r.v.'s with d.f.'s F_X and F_Y, respectively. The *upper tail dependence coefficient* λ_U (UTDC) of (X, Y) is defined by

$$\lambda_U = \lim_{\substack{t \to 1 \\ t < 1}} \mathbb{P}\left(Y > F_Y^{(-1)}(t) \mid X > F_X^{(-1)}(t) \right) ; \tag{2.6.9}$$

and the *lower tail dependence coefficient* λ_L (LTDC) of (X, Y) is defined by

$$\lambda_L = \lim_{\substack{t \to 0 \\ t > 0}} \mathbb{P}\left(Y \leq F_Y^{(-1)}(t) \mid X \leq F_X^{(-1)}(t) \right) , \tag{2.6.10}$$

provided that the above limits exist.

Therefore, the upper tail dependence coefficient indicates the asymptotic limit of the probability that one random variable exceeds a high quantile, given that the other variable exceeds a high quantile. Similar interpretation holds for the LTDC.

The application of probability integral transformations to equations (2.6.9) and (2.6.10) yields the following equivalent expressions:

$$\lambda_L = \lim_{\substack{t \to 0 \\ t > 0}} \frac{\delta_C(t)}{t} \quad \text{and} \quad \lambda_U = \lim_{\substack{t \to 1 \\ t < 1}} \frac{1 - 2t + \delta_C(t)}{1 - t}, \qquad (2.6.11)$$

where C is the copula of (X, Y).

The existence of the limits in (2.6.11) is not guaranteed; see, for instance, Example 3.8.7. Another example based on copulas of type (2.6.2) is given in [Kortschak and Albrecher, 2009].

It is interesting to note whether the existence of non-zero tail dependence coefficients may be related to properties of the derivatives of the corresponding copula, as illustrated below (see [Segers, 2012]).

Example 2.6.11. Let C be a 2-copula with first-order partial derivatives $\partial_1 C$ and $\partial_2 C$ and lower tail dependence coefficient $c > 0$. Now, on the one hand, for every $u \in \mathbb{I}$, $\partial_1 C(u, 0) = 0$. On the other hand, $\partial_1 C(0, v) = \lim_{u \to 0} C(u, v)/u \geq c > 0$ for every $v \in]0, 1]$. It follows that $\partial_1 C$ cannot be continuous at the point $(0, 0)$; similarly for $\partial_2 C$.

Analogously, for copulas with a positive upper tail dependence coefficient, the first-order partial derivatives cannot be continuous at the point $(1, 1)$. ∎

Now, while tail dependence coefficients give an asymptotic approximation of the behaviour of the copula in the tail of the distribution, it might be also of interest to consider the case when the tail behaviour is considered at some (finite) points near the corners of the unit square. An auxiliary function that may serve to clarify this point is defined below.

Definition 2.6.12. Let C be a copula. The *tail concentration function* associated to C is the function $q_C :]0, 1[\to \mathbb{I}$ given by

$$q_C(t) = \frac{\delta_C(t)}{t} \cdot \mathbf{1}_{]0, 0.5]} + \frac{1 - 2t + \delta_C(t)}{1 - t} \cdot \mathbf{1}_{]0.5, 1[}.$$

The tail concentration function of M_2 coincides with the constant 1, while it is equal to the constant 0 for W_2. Notice that $q_C(0.5) = (1 + \beta_C)/2$, where $\beta_C = 4C(0.5, 0.5) - 1$ is the Blomqvist's measure of association associated with C (see, e.g., [Blomqvist, 1950; Nelsen, 2006]).

Further readings 2.6.13. Alternative ways of defining tail dependence coefficients and relative multivariate extensions have been presented in the literature. See, for instance, [Li, 2013; Joe, 2014; Bernard and Czado, 2015] and references therein.

The tail concentration function has been defined, for instance, in [Patton, 2012] (see also Manner and Segers [2011]; Durante et al. [2015a]). ∎

Chapter 3

Copulas and measures

In this chapter, we present copulas from the point of view of measure theory. Moreover, we show the usefulness of this approach in studying special dependence structures (e.g., complete dependence). Finally, we present how a measure-theoretic approach provides useful tools to generate flexible constructions of copula models.

3.1 Copulas and d-fold stochastic measures

We start with a basic definition.

Definition 3.1.1. A measure μ on $(\mathbb{I}^d, \mathscr{B}(\mathbb{I}^d))$ will said to be *d-fold stochastic* if, for every $A \in \mathscr{B}(\mathbb{I})$ and for every $j \in \{1, \ldots, d\}$,

$$\mu(\underbrace{\mathbb{I} \times \cdots \times \mathbb{I}}_{j-1} \times A \times \mathbb{I} \times \cdots \times \mathbb{I}) = \lambda(A), \qquad (3.1.1)$$

where λ denotes the (restriction to $\mathscr{B}(\mathbb{I})$ of the) Lebesgue measure.　　$\diamondsuit$

By definition all d-fold stochastic measures are probability measures.

In the case $d = 2$, μ is usually referred to as a *doubly stochastic* measure or a *bistochastic measure*.

The defining equation (3.1.1) may be written in the equivalent form

$$\mu\left(\pi_j^{-1}(A)\right) = \lambda(A), \qquad (3.1.2)$$

where $\pi_j : \mathbb{I}^d \to \mathbb{I}$ is the j-th canonical projection, $\pi_j(u_1, \ldots, u_d) = u_j$ ($j = 1, \ldots, d$). In other words, μ is a d-fold stochastic measure on $\mathbb{I}^d$ if, and only if, the image measure of μ under any projection equals the Lebesgue measure on the Borel sets of $\mathbb{I}$. We recall that the *image measure* of μ under φ (also called *push-forward* of μ by φ) is the set function $\mu \circ \varphi^{-1}$ defined, for every $A \in \mathscr{F}_1$,

$$(\mu \circ \varphi^{-1})(A) := \mu\left(\varphi^{-1}A\right).$$

Other notations for image measures also include $\varphi_* \mu$ or $\varphi_\# \mu$.

Copulas and d-fold stochastic measures are connected through the following

Theorem 3.1.2. *Every copula* $C \in \mathscr{C}_d$ *induces a d-fold stochastic measure* μ_C *on the measurable space* $(\mathbb{I}^d, \mathscr{B}(\mathbb{I}^d))$ *defined on the rectangles* $R =]\mathbf{a}, \mathbf{b}]$ *contained in* $\mathbb{I}^d$, *by*

$$\mu_C(R) := V_C\left(]\mathbf{a}, \mathbf{b}]\right). \tag{3.1.3}$$

Conversely, to every d-fold stochastic measure μ *on* $(\mathbb{I}^d, \mathscr{B}(\mathbb{I}^d))$ *there corresponds a unique copula* C_μ *in* $\mathscr{C}_d$ *defined by*

$$C_\mu(\mathbf{u}) := \mu\left(]\mathbf{0}, \mathbf{u}]\right). \tag{3.1.4}$$

Proof. The d-increasing property (Theorem 1.4.1 (c)) implies that $\mu_C(R) \geq 0$ for every rectangle R. Once μ_C has been defined on the rectangles R contained in $\mathbb{I}^d$, the usual procedures of Measure Theory allow constructing a unique measure on $(\mathbb{I}^d, \mathscr{B}(\mathbb{I}^d))$. Notice, in fact, that the set of rectangles $R =]\mathbf{a}, \mathbf{b}]$ contained in $\mathbb{I}^d$ is a semiring $\mathscr{S}$, that this semiring generates the ring $\mathscr{R}(\mathscr{S})$ of the finite unions of disjoint rectangles, that μ_C has a unique extension to $\mathscr{R}(\mathscr{S})$ and, finally that, by Carathéodory's theorem, this latter measure admits a unique extension to the σ-field $\mathscr{B}(\mathbb{I}^d)$. (For the technical measure-theoretic details, see, e.g., [Kingman and Taylor, 1966, Chapters 3 and 4] or [Rao, 1987].) We shall again denote this unique extension by μ_C. That μ_C is a d-fold stochastic measure is a consequence of the observation that eq. (3.1.3) implies, for a and b in $\mathbb{I}$, with $a < b$, and for every $j \in \{1, \dots, d\}$,

$$\mu(\underbrace{\mathbb{I} \times \cdots \times \mathbb{I}}_{j-1} \times \,]a, b] \times \mathbb{I} \times \cdots \times \mathbb{I}) = b - a \,.$$

Now the same measure theoretic results we have just quoted, applied this time to the semiring of intervals contained in the unit interval, yield eq. (3.1.1), so that μ_C is indeed a d-fold stochastic measure.

Conversely, let μ be a d-fold stochastic measure and define C_μ by eq. (3.1.4). The boundary conditions for a copula are obviously satisfied. As for the d-increasing property, notice that for every rectangle $]\mathbf{a}, \mathbf{b}]$ in $\mathbb{I}^d$,

$$V_C(]\mathbf{a}, \mathbf{b}]) = \mu\left(]\mathbf{a}, \mathbf{b}]\right) \geq 0 \,,$$

which ends the proof. $\square$

One of the first measure-theoretic notions related to copulas is that of *support*. We recall that the support of a measure μ on the Borel sets $\mathscr{B}(E)$ of a topological space E is the complement of the union of all open subsets of E with μ-measure zero. This justifies the introduction of the following natural definition.

Definition 3.1.3. The *support* of a copula $C \in \mathscr{C}_d$ is the support of the d-fold stochastic measure μ_C it induces on $(\mathbb{I}^d, \mathscr{B}(\mathbb{I}^d))$. $\diamond$

Turning to the fundamental examples of copulas, one sees that Π_d has support equal to $\mathbb{I}^d$, while the supports of M_d and W_2 are given, respectively, by the sets $\{\mathbf{u} \in \mathbb{I}^d : u_1 = \cdots = u_d\}$ and $\{(u_1, u_2) \in \mathbb{I}^2 : u_1 + u_2 = 1\}$.

Remark 3.1.4. In the literature, integrals of the form

$$\int_{\mathbb{I}^d} \varphi(\mathbf{u}) \, dC(\mathbf{u}) \tag{3.1.5}$$

often appear; here $\mathbf{u}$ belongs to $\mathbb{I}^d$, $\varphi : \mathbb{I}^d \to \mathbb{R}$ is integrable and C is a d-copula. The above expression simply represents the integral with respect to the d-fold stochastic measure μ_C induced by $C \in \mathscr{C}_d$ on the Borel sets of $\mathbb{I}^d$ according to Theorem 3.1.2. Thus,

$$\int_{\mathbb{I}^d} \varphi(\mathbf{u}) \, dC(\mathbf{u}) = \int_{\mathbb{I}^d} \varphi \, d\mu_C \,,$$

which clarifies the meaning of the integral in eq. (3.1.5). Notice, however, that in the calculation of these integrals one can also take advantage of the probabilistic interpretation of a copula, as the following example shows. ∎

Example 3.1.5. Let $C_1, C_2 \in \mathscr{C}_2$. Suppose that we are interested in the evaluation of the following integral

$$\int_{\mathbb{I}^2} C_1 \, dC_2 \,.$$

Given a probability space $(\Omega, \mathscr{F}, \mathbb{P})$ and r.v.'s $(U_1, V_1) \sim C_1$ and $(U_2, V_2) \sim C_2$, one has

$$\int_{\mathbb{I}^2} C_1 \, dC_2 = \int_{\mathbb{I}^2} \mathbb{P}(U_1 \leq u, V_1 \leq v) \, dC_2(u, v) = \mathbb{P}\left(U_1 \leq U_2, V_1 \leq V_2\right) \,,$$

but also

$$\begin{aligned}
\int_{\mathbb{I}^2} C_2 \, dC_1 &= \int_{\mathbb{I}^2} \mathbb{P}(U_2 \leq u, V_2 \leq v) \, dC_1(u, v) \\
&= \int_{\mathbb{I}^2} \left(1 - \mathbb{P}(U_2 > u) - \mathbb{P}(V_2 > v) + \mathbb{P}(U_2 > u, V_2 > v)\right) \, dC_1(u, v) \\
&= \mathbb{P}\left(U_2 > U_1, V_2 > V_1\right) .
\end{aligned}$$

Thus, since copulas are continuous, it follows

$$\int_{\mathbb{I}^2} C_1 \, dC_2 = \int_{\mathbb{I}^2} C_2 \, dC_1.$$

The previous expression is quite useful in evaluating integrals involving the Hoeffding–Fréchet bound copulas. In fact, by a direct application of the previous equality and the definition of the copula-based measure, one has

$$\int_{\mathbb{I}^2} C \, dM_2 = \int_{\mathbb{I}^2} M_2 \, dC = \int_0^1 C(t, t) \, dt \,,$$

$$\int_{\mathbb{I}^2} C \, dW_2 = \int_{\mathbb{I}^2} W_2 \, dC = \int_0^1 C(t, 1 - t) \, dt \,,$$

for every copula $C \in \mathscr{C}_2$. ∎

As was seen, the class of all measures induced by $\mathscr{C}_d$, denoted by $\mathscr{P}_C(\mathbb{I}^d)$, is a subset of the space of all probability measures on $(\mathbb{I}^d, \mathscr{B}(\mathbb{I}^d))$, denoted by $\mathscr{P}(\mathbb{I}^d)$. In this latter space, we may consider the so-called *Hutchinson metric* h (sometimes also called *Kantorovich* or *Wasserstein* metric) defined, for all $\mu, \nu \in \mathscr{P}(\mathbb{I}^d)$, by

$$h(\mu, \nu) := \sup \left\{ \left| \int_{\mathbb{I}^d} f \, d\mu - \int_{\mathbb{I}^d} f \, d\nu \right| : f \in Lip_1(\mathbb{I}^d, \mathbb{R}) \right\}, \qquad (3.1.6)$$

where $Lip_1(\mathbb{I}^d, \mathbb{R})$ is the class of all 1-Lipschitz functions $f : \mathbb{I}^d \to \mathbb{R}$, i.e., all the functions fulfilling $|f(\mathbf{x}) - f(\mathbf{y})| \leq \rho(\mathbf{x}, \mathbf{y})$ for all $\mathbf{x}, \mathbf{y} \in \mathbb{I}^d$; ρ is the Euclidean distance.

The following result holds.

Theorem 3.1.6. $\mathscr{P}_C(\mathbb{I}^d)$ *is closed in the metric space* $(\mathscr{P}(\mathbb{I}^d), h)$ *for every* $d \geq 2$.

Proof. Since $\mathbb{I}^d$ is compact, h is a metrisation of the topology of weak convergence on $\mathscr{P}(\mathbb{I}^d)$ (see, e.g., [Dudley, 1989]). If $(\mu_n)_{n \in \mathbb{N}}$ is a sequence in $\mathscr{P}_C(\mathbb{I}^d)$ that converges to $\mu \in \mathscr{P}(\mathbb{I}^d)$ with respect to h, then $\mu_n \to \mu$ weakly as $n \to \infty$. Hence the corresponding sequence of distribution functions $(A_n)_{n \in \mathbb{N}}$ converges to the distribution function A_μ of μ at every continuity point of A_μ. In view of compactness of $\mathscr{C}_d$ with respect to weak convergence (i.e., pointwise convergence), a subsequence $(A_{n_k})_{k \in \mathbb{N}}$ exists that converges uniformly to a (continuous) distribution function $\widetilde{A}$. Weak limits are unique so $\widetilde{A} = A_\mu$ and $\mu \in \mathscr{P}_C(\mathbb{I}^d)$. $\qquad \square$

3.1.1 Doubly stochastic matrices and discrete copulas

The concept of doubly stochastic measure has its roots in matrix theory. In fact, recall that an $n \times n$ matrix $A = (a_{ij})$ is called *doubly stochastic* if all its entries are positive $a_{ij} \geq 0$ and the sum of the entries of each row and each column equals 1, viz., for every $k \in \{1, \ldots, n\}$ one has

$$\sum_{i=1}^{n} a_{ik} = \sum_{j=1}^{n} a_{kj} = 1$$

(see, for instance, [Schur, 1923]). These matrices are used, for instance, in majorisation theory [Marshall and Olkin, 1979] and Markov chains [Seneta, 1981]. One of the most celebrated results concerning these matrices is the Birkhoff-von Neumann Theorem [Birkhoff, 1946; von Neumann, 1953]; this states that each doubly stochastic matrix is a weighted mean of permutation matrices, which are doubly stochastic matrices whose entries are 0 or 1. In a quest to extend this result, G. Birkhoff asked to

extend the result [...] to the infinite-dimensional case, under suitable hypotheses.

See Problem 111 in [Birkhoff, 1940].

The study of doubly stochastic measures originated in the effort to provide an answer to the previous problem (see, e.g., [Peck, 1959]). It is clear that the concept of a copula provides a solution to Birkhoff's quest of infinite-dimensional doubly stochastic matrices. Such a connexion is more evident by using the related notion of a discrete copula, which we now introduce following the work by Kolesárová et al. [2006].

For $n \in \mathbb{N}$ consider the set

$$I_n := \left\{ 0, \frac{1}{n}, \frac{2}{n}, \ldots, \frac{n-1}{n}, 1 \right\}.$$

Definition 3.1.7. Let $n, m \in \mathbb{N}$. A *discrete copula* is a subcopula whose domain is given by $I_n \times I_m$. $\diamond$

Obviously, the restriction of a given copula $C \in \mathscr{C}_2$ to $I_n \times I_m$ is a discrete copula. Moreover, it follows from Lemma 2.3.4 that every discrete copula can be extended to a copula, which in general is not unique, so that every discrete copula $C_{(n,m)}$ on $I_n \times I_m$ is the restriction to $I_n \times I_m$ of some copula $C \in \mathscr{C}_2$,

$$C_{(n,m)} = C|_{I_n \times I_m}.$$

From now on we shall consider only the case $m = n$ and shall write $C_{(n)}$ instead of $C_{(n,n)}$. The set of discrete copulas on I_n^2 will be denoted by $\mathscr{C}_2^{(n)}$.

Given a copula $C \in \mathscr{C}_2$, its restriction $C_{(n)}$ to I_n^2 will be called the *discretisation of order n of C*,

$$C_{(n)} \left(\frac{i}{n}, \frac{j}{n} \right) = C \left(\frac{i}{n}, \frac{j}{n} \right) \quad (i, j = 0, 1, \ldots, n).$$

Theorem 3.1.8. *Let k be a natural number and let $(C_{k^m} : m \in \mathbb{N}, m \geq 2)$ be a sequence of discrete copulas on $I_{k^m}^2$ that satisfy the consistency condition*

$$C_{k^m} \left(\frac{i}{k^{m-1}}, \frac{j}{k^{m-1}} \right) = C_{k^{m-1}} \left(\frac{i}{k^{m-1}}, \frac{j}{k^{m-1}} \right) \quad (i, j = 0, 1, \ldots, k^{m-1}).$$

$$(3.1.7)$$

Then the limit of the sequence

$$\left(C_{k^m} \left(\frac{[k^m u]}{k^m}, \frac{[k^m v]}{k^m} \right) \right)_{m \in \mathbb{N}}$$

$$(3.1.8)$$

exists at every point (u, v) of $\mathbb{I}^2$ (here $[x]$ denotes the integer part of $x \in \mathbb{R}$). Moreover, $C : \mathbb{I}^2 \to \mathbb{I}$ defined by

$$C(u, v) := \lim_{m \to +\infty} C_{k^m} \left(\frac{[k^m u]}{k^m}, \frac{[k^m v]}{k^m} \right)$$

$$(3.1.9)$$

is a copula.

Proof. For every $u \in \mathbb{I}$, one has

$$\frac{[k^m u]}{k^m} \geq \frac{[k^{m-1} u]}{k^{m-1}} ;$$

since $(C_{k^m})_{m \in \mathbb{N}}$ is an increasing sequence, the consistency condition (3.1.7) yields

$$C_{k^m}\left(\frac{[k^m u]}{k^m}, \frac{[k^m v]}{k^m}\right) \geq C_{k^m}\left(\frac{[k^{m-1} u]}{k^{m-1}}, \frac{[k^{m-1} v]}{k^{m-1}}\right)$$

$$= C_{k^{m-1}}\left(\frac{[k^{m-1} u]}{k^{m-1}}, \frac{[k^{m-1} v]}{k^{m-1}}\right).$$

Therefore the sequence (3.1.8) is increasing, but, being a sequence of values taken by a discrete copula, it is also bounded above by 1, so that it converges.

Let $C : \mathbb{I}^2 \to \mathbb{I}$ be defined by (3.1.9); then C clearly satisfies $C(u,v) \in [0,1]$ at every point $(u,v) \in \mathbb{I}^2$. Moreover, since, for every $m \geq 2$, (C_{k^m}) satisfies conditions (a) and (b) of Definition 3.1.7, one immediately has, on taking limits, that, for every $u \in \mathbb{I}$,

$$C(u,0) = C(0,u) = 0, \quad \text{and} \quad C(u,1) = C(1,u) = u.$$

Finally take s, s', t and t' in $\mathbb{I}$ such that $s \leq s'$ and $t \leq t'$. Then (3.1.9) yields

$$
\begin{aligned}
& C(s',t') + C(s,t) - C(s',t) - C(s,t') \\
&= \lim_{m \to +\infty}\left\{ C_{k^m}\left(\frac{[k^m s']}{k^m}, \frac{[k^m t']}{k^m}\right) + C_{k^m}\left(\frac{[k^m s]}{k^m}, \frac{[k^m t]}{k^m}\right) \right. \\
& \left. - C_{k^m}\left(\frac{[k^m s]}{k^m}, \frac{[k^m t']}{k^m}\right) - C_{k^m}\left(\frac{[k^m s']}{k^m}, \frac{[k^m t]}{k^m}\right)\right\},
\end{aligned}
$$

which is positive because of property (b) of Definition 3.1.7. Therefore C is a copula; and this concludes the proof. $\square$

Next we show that there is a one-to-one correspondence between discrete copulas and doubly stochastic matrices, so that these serve to represent discrete copulas.

Theorem 3.1.9. *For a function $C_{(n)} : I_n^2 \to \mathbb{I}$ the following statements are equivalent:*

(a) *$C_{(n)}$ is a discrete copula;*

(b) *there exists an $n \times n$ doubly stochastic matrix $A = (a_{ij})$ such that for $i, j \in \{0, 1, \ldots, n\}$*

$$c_{i,j}^{(n)} := C_{(n)}\left(\frac{i}{n}, \frac{j}{n}\right) = \frac{1}{n}\sum_{k=1}^{i}\sum_{h=1}^{j} a_{kh} . \tag{3.1.10}$$

Proof. (a) $\Longrightarrow$ (b) The 2-increasing property of a discrete copula $C_{(n)}$ is equivalent to the following inequality

$$c_{i,j}^{(n)} - c_{i-1,j}^{(n)} - c_{i,j-1}^{(n)} + c_{i-1,j-1}^{(n)} \geq 0 \quad (i,j = 1, 2, \ldots, n) .$$

Now, let the entries of an $n \times n$ matrix $A = (a_{i,j})$ be defined by

$$a_{ij} := n \left(c_{i,j}^{(n)} - c_{i-1,j}^{(n)} - c_{i,j-1}^{(n)} + c_{i-1,j-1}^{(n)} \right) \quad (i, j = 1, 2, \dots, n).$$

Then $a_{ij} \geq 0$ and, for every $j \in \{1, 2, \dots, n\}$,

$$\sum_{i=1}^{n} a_{ij} = n \sum_{i=1}^{n} \left(c_{i,j}^{(n)} - c_{i-1,j}^{(n)} \right) - n \sum_{i=1}^{n} \left(c_{i,j-1}^{(n)} + c_{i-1,j-1}^{(n)} \right)$$

$$= n \left(c_{n,j}^{(n)} - c_{0,j}^{(n)} - c_{n,j-1}^{(n)} + c_{0,j-1}^{(n)} \right)$$

$$= n \left(C_{(n)} \left(1, \frac{j}{n} \right) - C_{(n)} \left(1, \frac{j-1}{n} \right) \right) = n \frac{1}{n} = 1,$$

since both sums are telescopic.

The proof that $\sum_{j=1}^{n} a_{i,j} = 1$ for every $i \in \{1, \dots, n\}$ is similar.

The implication (b) $\Longrightarrow$ (a) is a matter of a simple straightforward verification. If $C_{(n)}$ is defined by eq. (3.1.10), then

$$C_{(n)} \left(1, \frac{j}{n} \right) = \frac{1}{n} \sum_{h=1}^{j} \sum_{k=1}^{n} a_{k,h} = \frac{1}{n} \sum_{h=1}^{j} 1 = \frac{j}{n},$$

which, together with the analogous property $C_{(n)} \left(\frac{i}{n}, 1 \right) = i/n$, proves the boundary conditions of a subcopula.

As for the 2-increasing property, one has

$$C_{(n,m)} \left(\frac{i}{n}, \frac{j}{m} \right) - C_{(n,m)} \left(\frac{i-1}{n}, \frac{j}{m} \right)$$

$$+ C_{(n,m)} \left(\frac{i-1}{n}, \frac{j-1}{m} \right) - C_{(n,m)} \left(\frac{i}{n}, \frac{j-1}{m} \right)$$

$$= \frac{1}{n} \left\{ \sum_{k=1}^{i} \sum_{h=1}^{j} a_{k,h} - \sum_{k=1}^{i-1} \sum_{h=1}^{j} a_{k,h} - \sum_{k=1}^{i} \sum_{h=1}^{j-1} a_{k,h} + \sum_{k=1}^{i-1} \sum_{h=1}^{j-1} a_{k,h} \right\}$$

$$= \frac{1}{n} \left\{ \sum_{h=1}^{j} a_{i,h} - \sum_{h=1}^{j-1} a_{i,h} \right\} = \frac{1}{n} a_{i,j} \geq 0,$$

which concludes the proof. $\qquad\qquad\square$

Example 3.1.10. The following matrices $A(M_2)$, $A(\Pi_2)$ and $A(W_2)$

$$A(M_2) = \left(\frac{1}{n}\min\{i,j\}\right)_{i,j},$$

$$A(\Pi_2) = \begin{pmatrix} 1/n & 1/n & \dots & 1/n \\ 1/n & 1/n & \dots & 1/n \\ \dots & \dots & \dots & \dots \\ 1/n & 1/n & \dots & 1/n \end{pmatrix},$$

$$A(W_2) = \begin{pmatrix} 0 & 0 & \dots & 0 & 1 \\ 0 & 0 & \dots & 1 & 0 \\ \dots & \dots & \dots & \dots & \dots \\ 0 & 1 & \dots & 0 & 0 \\ 1 & 0 & \dots & 0 & 0 \end{pmatrix}$$

represent the discretisations of the three copulas M_2, Π_2 and W_2. ■

In view of the boundary conditions of a subcopula, the range of every discrete copula $C_{(n)}$ includes I_n. The special case Ran $C_{(n)} = I_n$ deserves a definition.

Definition 3.1.11. A discrete copula $C_{(n)} : I_n^2 \to \mathbb{I}$ will be said to be *irreducible* if Ran $C_{(n)} = I_n$. ◇

Mayor et al. [2005] proved that an irreducible discrete copula is characterised by a doubly stochastic matrix whose entries are all either 0 or 1. In other words, there exists a permutation π of $\{1, \dots, n\}$ such that

$$C_{(n)}\left(\frac{i}{n}, \frac{j}{n}\right) = \frac{1}{n}\sum_{k=1}^{j} \mathbf{1}_{\{1,\dots,i\}}(\pi(k)). \tag{3.1.11}$$

As a consequence, on I_n^2 there are exactly $n!$ irreducible discrete copulas.

Remark 3.1.12. In classical matrix theory [Birkhoff, 1946], it is well known that the set of doubly stochastic matrix is the convex hull of the set of permutation matrix; the following analogous result holds here. In fact, Kolesárová et al. [2006] showed that the class $\mathscr{C}_2^{(n)}$ of discrete copulas on I_n^2 is the smallest convex set containing the set D_n of irreducible discrete copulas on I_n^2. ■

3.2 Absolutely continuous and singular copulas

Since every $C \in \mathscr{C}_d$ is continuous, the probability measure μ_C has no discrete component, namely it does not concentrate any mass on finitely or countably many points of $\mathbb{I}^d$ (see, e.g., [Billingsley, 1979]). Therefore, in view of the Lebesgue decomposition theorem (see, e.g., [Ash, 2000, Theorem 2.2.6]), one has

$$\mu_C = \mu_C^{ac} + \mu_C^s,$$

where:

- μ_C^{ac} is a measure on $\mathscr{B}(\mathbb{I}^d)$ that is absolutely continuous with respect to the d-dimensional Lebesgue measure λ_d, i.e., $\lambda_d(B) = 0$ implies $\mu_C^{ac}(B) = 0$, where $B \in \mathscr{B}(\mathbb{I}^d)$;

- μ_C^s is a measure on $\mathscr{B}(\mathbb{I}^d)$ that is singular with respect to λ^d, i.e., the probability measure is concentrated on a set B such that $\lambda_d(B) = 0$.

Since C has uniform univariate margins, μ_C cannot concentrate probability mass in any hyperplane parallel to a coordinate axis. Thus, $\mu_C(]\mathbf{a}, \mathbf{b}]) = \mu_C([\mathbf{a}, \mathbf{b}])$ for every $\mathbf{a}, \mathbf{b} \in \mathbb{I}^d$. Moreover, for every $\mathbf{u} \in \mathbb{I}^d$, one can write

$$C(\mathbf{u}) = C_{ac}(\mathbf{u}) + C_s(\mathbf{u}),$$

where $C_{ac}(\mathbf{u}) = \mu_C^{ac}(]\mathbf{0}, \mathbf{u}])$ and $C_s(\mathbf{u}) = \mu_C^s(]\mathbf{0}, \mathbf{u}])$ are called, respectively, the *absolutely continuous* and *singular* components of C. Clearly, C_{ac} and C_s need not be copulas. In view of these facts, the following definition may be given.

Definition 3.2.1. A copula $C \in \mathscr{C}_d$ will be said to be *absolutely continuous* (respectively, *singular*) if so is the measure μ_C induced by C with respect to λ_d, viz., if $\mu_C = \mu_C^{ac}$ (respectively, if $\mu_C = \mu_C^s$). ◇

If a copula C is absolutely continuous, then it can be written in the form

$$C(\mathbf{u}) = \int_{[\mathbf{0},\mathbf{u}]^d} c(\mathbf{s}) \, d\mathbf{s},$$

where c is a suitable function called *density* of C. In particular, for almost all $\mathbf{u} \in \mathbb{I}^d$ one has

$$c(\mathbf{u}) = \frac{\partial^d C(\mathbf{u})}{\partial u_1 \cdots \partial u_d}. \tag{3.2.1}$$

As stressed by McNeil and Nešlehová [2009], eq. (3.2.1) is far from obvious. In fact, there are some facts that are implicitly used: first, the mixed partial derivatives of order d of C exist and are equal almost everywhere on $\mathbb{I}^d$; second, each mixed partial derivative is actually almost everywhere equal to the density c. The reader may refer to [Bruckner, 1971, section 4.1] and the references therein, in particular, [Busemann and Feller, 1934; Easton et al., 1967] and [Saks, 1937, page 115], for a proof of (3.2.1).

In general, given a copula C, in view of the Besicovitch derivation theorem, for which see, e.g., [Ash, 2000, Theorem 2.38], one has

$$C(\mathbf{u}) = \int_{[\mathbf{0},\mathbf{u}]^d} c(\mathbf{s}) \, d\mathbf{s} + C_s(\mathbf{u}), \tag{3.2.2}$$

where c, which is the density of the absolutely continuous component of C, coincides almost everywhere with the Radon–Nikodym derivative of μ_C^{ac} with respect to λ_d.

Following [Ash, 2000, Theorem 2.38], the function c of eq. (3.2.2) coincides almost everywhere on $\mathbb{I}^d$ with the derivative of μ_C, given, for almost every $\mathbf{u} \in \mathbb{I}^d$, by:

$$D\mu_C(\mathbf{u}) = \lim_{r \to 0} \frac{\mu_C(B_r)}{\lambda_d(B_r)},$$

where B_r ranges over all the open rectangles of diameter less than r that contain $\mathbf{u}$. Notice that the existence a.e. of $D\mu_C$ is guaranteed in view of Theorem 2.38 in [Ash, 2000]. When the mixed partial derivatives of order d of C exist for almost all $\mathbf{u} \in \mathbb{I}^d$ and are all equal, it can be shown that

$$\frac{\partial^d C(\mathbf{u})}{\partial u_1 \cdots \partial u_d} = \lim_{h_d \to 0} \lim_{h_{d-1} \to 0} \cdots \lim_{h_1 \to 0} \frac{V_C([\mathbf{u}, \mathbf{u} + \mathbf{h}])}{\prod_{i=1}^d h_i},$$

i.e., $\partial_{1 \cdots d}^d C(\mathbf{u}) = D\mu_C(\mathbf{u})$ for almost all $\mathbf{u} \in \mathbb{I}^d$.

The following result easily follows from the representation (3.2.2).

Theorem 3.2.2. *For a copula $C \in \mathscr{C}_d$ the following statements are equivalent:*

(a) *C is singular;*

(b) *$D\mu_C(\mathbf{u}) = 0$ for λ_d-almost all $\mathbf{u}$ in $\mathbb{I}^d$.*

Remark 3.2.3. Suppose that C is a copula such that the following property holds:

(c) The support of μ_C has Lebesgue measure equal to 0, $\lambda_d(\mathrm{supp}(\mu_C)) = 0$.

It can be easily derived from the Lebesgue decomposition of μ_C that such a C is singular. However, if a copula is singular, then it may not satisfy property (c) (a confusion that is sometimes encountered in the literature). The example of a singular bivariate copula whose support is $\mathbb{I}^2$ is given in Example 3.6.15.   ∎

Remark 3.2.4. In some cases, singular copulas can be constructed by assigning a prescribed support (usually, segments or surfaces) on which the probability mass of a given measure is concentrated. For instance, the comonotone copula M_d concentrates the probability mass uniformly on $\{\mathbf{u} \in \mathbb{I}^d : u_1 = \cdots = u_d\}$.   ∎

The example of a singular copula with a prescribed support is given below.

Example 3.2.5. (Tent copula) Choose θ in $]0, 1[$ and consider the probability mass θ uniformly spread on the segment joining the points $(0, 0)$ and $(\theta, 1)$ and the probability mass $1 - \theta$ uniformly spread on the segment joining the points $(\theta, 1)$ and $(1, 1)$. It is now easy to find the expression for the copula C_θ of the resulting probability distribution on the unit square. The two line segments described above divide the unit square into three regions, so that in finding C_θ three cases will have to be considered.

If the point (u, v) is such that $u \leq \theta v$, then one has equality between the two V_{C_θ} volumes

$$V_{C_\theta}([0, u] \times [0, v]) \qquad \text{and} \qquad V_{C_\theta}([0, u] \times [0, 1]) .$$

But $V_{C_\theta}([0, u] \times [0, 1]) = u$ so that $C_\theta(u, v) = u$.

If the point (u, v) is such that u belongs to the interval $]\theta v, 1 - (1 - \theta) v[$, then

$$C_\theta(u, v) = V_{C_\theta}([0, u] \times [0, v]) = V_{C_\theta}([0, \theta v] \times [0, v]) + V_{C_\theta}([\theta v, u] \times [0, v])$$
$$= V_{C_\theta}([0, \theta v] \times [0, v]) = \theta v .$$

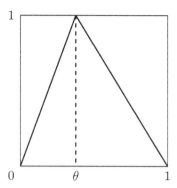

Figure 3.1: *Support of the copula of Example 3.2.5.*

Finally, if the point (u, v) is such that $u \geq 1 - (1 - \theta) v$, notice that the V_{C_θ} volume of any rectangle that is contained in the unit square and which does not intersect any of the two segments is necessarily zero. Therefore

$$1 - u - v + C_\theta(u, v) = V_{C_\theta}([u, 1] \times [v, 1]) = 0 \,,$$

so that $C_\theta(u, v) = u + v - 1$.

Summing up, one has

$$C_\theta(u, v) = \begin{cases} u, & u \in [0, \theta v] \,, \\ \theta v, & u \in]\theta v, 1 - (1 - \theta) v[\,, \\ u + v - 1, & u \in [1 - (1 - \theta) v, 1] \,, \end{cases}$$

which is the expression of the tent copula. ∎

Now, we exhibit the example of a copula that has both non-trivial singular and absolutely continuous components.

Example 3.2.6. Consider the Cuadras-Augé family of copulas (see Example 1.3.7),

$$C_\alpha^{\mathrm{CA}}(u, v) := \begin{cases} uv^{1-\alpha}, & u \leq v \,, \\ u^{1-\alpha}v, & u > v \,. \end{cases}$$

The mixed second partial derivative of C_α^{CA} exists on $]0, 1[^2$ with the exception of the main diagonal $\{(u, u) : u \in \mathbb{I}\}$, and coincides with the density of the absolutely continuous component, which is hence given by

$$c_\alpha^{\mathrm{CA}}(u, v) = (1 - \alpha) \left[\max\{u, v\} \right]^{-\alpha} \,,$$

almost everywhere on $\mathbb{I}^2$. The singular component concentrates the probability mass along the the main diagonal $\{(u, u) : u \in \mathbb{I}\}$. It is given by

$$(C_\alpha^{\mathbf{CA}})_s(u, v) = C_\alpha^{\mathbf{CA}}(u, v) - \int_0^u \int_0^v c_\alpha^{\mathrm{CA}}(s, t)\, \mathrm{d}s\, \mathrm{d}t$$

$$= \frac{\alpha}{2 - \alpha} \left(\min\{u^\alpha, v^\alpha\}\right)^{(2-\alpha)/\alpha}.$$

Thus the singular component of a Cuadras-Augé copula has total mass

$$(C_\alpha^{\mathbf{CA}})_s(1, 1) = \frac{\alpha}{2 - \alpha},$$

as it is immediately checked.   ∎

3.3 Copulas with fractal support

The aim of this section is to present a general construction of copulas whose support is a fractal set in order to help in understanding some real "exotic" dependence structures whose behaviour cannot be captured by standard methods. We should like to avoid the discussion about what a fractal set is, since several definitions are possible and are adopted or suggested in the literature; to this end, it suffices to read the introduction of the book by Edgar [1998]. Here the convention will be adopted that a *fractal* is a set whose topological dimension is strictly smaller than its Hausdorff dimension; see, e.g., [Edgar, 1990]. For the constructions, the concept of *Iterated Functions System* and *Iterated Function System with Probabilities* will be needed; for these we refer to [Edgar, 1990, 1998; Barnsley, 1988]. Here we recall some basic notions that are essentials to make the section self-contained.

Let (Ω, ρ) be a compact metric space. A mapping $w : \Omega \to \Omega$ is called a *contraction* if there exists a constant $L < 1$ such that $\rho(w(x), w(y)) \leq L\rho(x, y)$ holds for all $x, y \in \Omega$. A family $(w_l)_{l=1}^n$ of $n \geq 2$ contractions on Ω is called an *Iterated Function System* (IFS) and will be denoted by $\{\Omega, (w_l)_{l=1}^n\}$. An IFS together with a vector $(p_l)_{l=1}^n \in \,]0, 1[^n$ fulfilling $\sum_{l=1}^n p_l = 1$ is called *Iterated Function System with Probabilities* (IFSP). An IFSP is denoted by $\{\Omega, (w_l)_{l=1}^n, (p_l)_{l=1}^n\}$.

Every IFSP induces the so-called *Hutchinson operator* on the family $\mathscr{K}(\Omega)$ of all non-empty compact subsets of Ω, $\mathscr{H} : \mathscr{K}(\Omega) \to \mathscr{K}(\Omega)$, defined by

$$\mathscr{H}(Z) := \bigcup_{i=1}^n w_i(Z). \tag{3.3.1}$$

It can be shown that $\mathscr{H}$ is a contraction on the compact metric space $(\mathscr{K}(\Omega), \delta_H)$, where δ_H is the Hausdorff metric on $\mathscr{K}(\Omega)$.

Hence Banach's Fixed Point theorem implies the existence of a unique, globally attractive fixed point $Z^\star$ of $\mathscr{H}$, i.e., for every $R \in \mathscr{K}(\Omega)$ one has

$$\lim_{n \to \infty} \delta_H\left(\mathscr{H}^n(R), Z^\star\right) = 0.$$

On the other hand every IFSP also induces an operator on the family $\mathscr{P}(\Omega)$ of all probability measures on $(\Omega, \mathscr{B}(\Omega))$, $\mathscr{V} : \mathscr{P}(\Omega) \to \mathscr{P}(\Omega)$, defined by

$$\mathscr{V}(\mu) := \sum_{i=1}^{n} p_i \, \mu^{w_i} \,, \tag{3.3.2}$$

where μ^{w_i} is the push-forward measure of μ under w_i. It is not difficult to show that $\mathscr{V}$ is a contraction on $(\mathscr{P}(\Omega), h)$, where h is the Hutchinson metric, that h is a metrisation of the topology of weak convergence on $\mathscr{P}(\Omega)$ and that $(\mathscr{P}(\Omega), h)$ is a compact metric space. Consequently, again by Banach's Fixed Point theorem, it follows that there is a unique, globally attractive fixed point $\mu^\star \in \mathscr{P}(\Omega)$ of $\mathscr{V}$, i.e. for every $\nu \in \mathscr{P}(\Omega)$ one has

$$\lim_{n \to \infty} h\left(\mathscr{V}^n(\nu), \mu^\star\right) = 0 \,.$$

Furthermore $Z^\star$ is the support of $\mu^\star$.

IFSPs will now be used in order to construct copulas with fractal support. Fix $d \geq 2$, and d numbers $m_1, \ldots, m_d \in \mathbb{N}$ and set

$$\mathscr{I}_d := \times_{i=1}^{d} I_i \quad \text{where} \quad I_i = \{1, \ldots, m_i\} \text{ for every } i \in \{1, \ldots, d\}. \tag{3.3.3}$$

We shall denote elements of $\mathscr{I}_d$ by $\mathbf{i} = (i_1, \ldots, i_d)$, and, for every probability distribution τ on $(\mathscr{I}_d, 2^{\mathscr{I}_d})$ write $\tau(\mathbf{i}) := \tau(\{\mathbf{i}\})$ for the point mass at $\mathbf{i}$.

Definition 3.3.1. For $d \geq 2$ take $m_1, \ldots, m_d$ in $\mathbb{N}$, $\max_j \{m_j\} \geq 2$, and let $\mathscr{I}_d$ be defined according to (3.3.3). A probability distribution τ on $(\mathscr{I}_d, 2^{\mathscr{I}_d})$ is called a *transformation array* if, for every $j \in \{1, \ldots, d\}$,

$$\sum_{\mathbf{i} \in \mathscr{I}_d: \, i_j = k} \tau(\mathbf{i}) > 0$$

holds for every $k \in I_j$. The class of all transformation arrays for fixed $d \geq 2$ will be denoted by $\mathscr{M}_d$. ◇

Example 3.3.2. Let $n, m \in \mathbb{N}$ such that $\max\{n, m\} \geq 2$. Let $\mathscr{I}_2 = I_n \times I_m$. A transformation array τ on $(\mathscr{I}_2, 2^{\mathscr{I}_2})$ is a probability measure that satisfies $\tau(\{i\} \times I_m) > 0$ for every $i \in I_n$, and $\tau(I_n \times \{j\}) > 0$ for every $j \in I_m$. To each transformation array τ on $(\mathscr{I}_2, 2^{\mathscr{I}_2})$ one can associate a $(n \times m)$-*transformation matrix* $T = (t_{ij})$ with the following properties:

(a) $\max\{n, m\} \geq 2$;

(b) all the entries are positive, $t_{ij} \geq 0$ for all $i \in \{1, \ldots, n\}$ and $j \in \{1, \ldots, m\}$;

(c) $\sum_{i,j} t_{ij} = 1$;

(d) no row and no column has all entries equal to 0.

To this end, it is enough to set $t_{ij} = \tau(i, j)$ for every $i \in I_n$ and $j \in I_m$. Analogously, a transformation array τ on $(\mathscr{I}_2, 2^{\mathscr{I}_2})$ can be constructed from a transformation matrix of the previous type. ■

Every $\tau \in \mathcal{M}_d$ induces a partition of $\mathbb{I}^d$ in the following way: For each $j \in \{1, \dots, d\}$ define $a_0^j := 0$,

$$a_k^j := \sum_{\mathbf{i} \in \mathcal{I}_d : i_j \leq k} \tau(\mathbf{i}),$$

and $E_k^j := [a_{k-1}^j, a_k^j]$ for every $k \in I_j$. Then $\bigcup_{k \in I_j} E_k^j = \mathbb{I}$ and $E_{k_1}^j \cap E_{k_2}^j$ is either empty or consists of exactly one point whenever $k_1 \neq k_2$. Setting $R_{\mathbf{i}} := \times_{j=1}^d E_{i_j}^j$ for every $\mathbf{i} \in \mathcal{I}_d$ therefore yields a family of compact rectangles $(R_{\mathbf{i}})_{\mathbf{i} \in \mathcal{I}_d}$ whose union is $\mathbb{I}^d$ and which is such that $R_{\mathbf{i}_1} \cap R_{\mathbf{i}_1}$ is either empty or a set of zero λ_d-measure whenever $\mathbf{i}_1 \neq \mathbf{i}_2$.

Finally, define an affine contraction $w_{\mathbf{i}} : \mathbb{I}^d \to R_{\mathbf{i}}$ by

$$w_{\mathbf{i}}(x_1, \dots, x_d) = \begin{pmatrix} a_{i_1-1}^1 \\ a_{i_2-1}^2 \\ \vdots \\ a_{i_d-1}^d \end{pmatrix} + \begin{pmatrix} (a_{i_1}^1 - a_{i_1-1}^1) x_1 \\ (a_{i_2}^2 - a_{i_2-1}^2) x_2 \\ \vdots \\ (a_{i_d}^d - a_{i_d-1}^d) x_d \end{pmatrix}.$$

Since the j-th coordinate of $w_{\mathbf{i}}(x_1, \dots, x_d)$ only depends on i_j and x_j we shall also denote it by $w_{i_j}^j$, i.e., $w_{i_j}^j : [0,1] \to E_{i_j}^j$, $w_{i_j}^j(x_j) := a_{i_j-1}^j + (a_{i_j}^j - a_{i_j-1}^j) x_j$. It follows directly from the construction that

$$\left\{ \mathbb{I}^d, (w_{\mathbf{i}})_{\mathbf{i} \in \mathcal{I}_d}, \tau(\mathbf{i})_{\mathbf{i} \in \mathcal{I}_d} \right\} \tag{3.3.4}$$

is an IFSP. Moreover, the following result holds.

Lemma 3.3.3. *If $\tau \in \mathcal{M}_d$, then, for every $\mu_C \in \mathcal{P}_C(\mathbb{I}^d)$, $\mathcal{V}_\tau(\mu_C)$ is a d-fold stochastic measure.*

Proof. Let μ be a d-fold stochastic measure. One only needs to show that $\mathcal{V}_\tau(\mu)$ satisfies eq. (3.1.1); this can be done as follows. For every F in $\mathcal{B}(\mathbb{I})$ consider the rectangle $R := \times_{k=1}^d G_k$ with $G_j = F$ and $G_k = \mathbb{I}$ for every $k \neq j$; then

$$\mu(w_{\mathbf{i}}^{-1}(R)) = \lambda \left((w_{i_j}^j)^{-1}(F) \right) = \lambda^{w_{i_j}^j}(F),$$

and therefore

$$\begin{aligned} (\mathcal{V}_\tau \mu)(R) &= \sum_{\mathbf{i} \in \mathcal{I}_d} \tau(\mathbf{i}) \mu^{w_{\mathbf{i}}}(R) = \sum_{k=1}^{m_j} \sum_{\mathbf{i} \in \mathcal{I}_d : i_j = k} \tau(\mathbf{i}) \lambda^{w_k^j}(F) \\ &= \sum_{k=1}^{m_j} \lambda^{w_k^j}(F) \sum_{\mathbf{i} \in \mathcal{I}_d : i_j = k} \tau(\mathbf{i}) = \sum_{k=1}^{m_j} \lambda^{w_k^j}(F)(a_k^j - a_{k-1}^j) \\ &= \lambda(F), \end{aligned}$$

which completes the proof. $\qquad \square$

As a consequence we shall also write $\mathcal{V}_\tau(A)$ for every $A \in \mathscr{C}_d$. Applying the results on IFSPs mentioned before one has the following result.

Theorem 3.3.4. *Given $\tau \in \mathscr{M}_d$, consider the corresponding IFSP defined by (3.3.4), and let the Hutchinson operator $\mathscr{H}$ and the operator $\mathcal{V}_\tau$ be defined according to (3.3.1) and (3.3.2), respectively. Then there exists a unique compact set $Z^\star \in \mathscr{K}(\mathbb{I}^d)$ (called* attractor*) and a copula $A^\star \in \mathscr{C}_d$ such that the support of $A^\star$ is $Z^\star$, and such that for every $Z \in \mathscr{K}(\mathbb{I}^d)$ and $A \in \mathscr{C}_d$*

$$\lim_{n\to\infty} \delta_H\left(\mathscr{H}^n(Z), Z^\star\right) = 0 \ \text{ and } \ \lim_{n\to\infty} h\left(\mathcal{V}_\tau^n(A), A^\star\right) = 0$$

holds.

Remark 3.3.5. Analogously to the two-dimensional case in [Fredricks et al., 2005] it is straightforward to see that the attractor $Z^\star$ has zero λ_d-measure if, and only if, there is at least one $\mathbf{i} \in \mathscr{I}_d$ such that $\tau(\mathbf{i}) = 0$. Therefore, the limit copula $A^\star \in \mathscr{C}_d$ is singular (w.r.t. the Lebesgue measure λ_d) if $\tau(\mathbf{i}) = 0$ for at least one $\mathbf{i} \in \mathscr{I}_d$. ∎

Having Theorem 3.3.4 at one's disposal, one can easily prove the following result.

Theorem 3.3.6. *For every $d \geq 2$ and every $s \in]1, d[$ there exists a copula $A \in \mathscr{C}_d$ whose support has Hausdorff dimension s.*

Proof. Set $I_j = \{1, 2, 3\}$ for every $j \in \{1, \ldots, d\}$ and, for $r \in]0, 1/2[$, define $\tau_r \in \mathscr{M}_d$ by

$$\tau_r(\mathbf{i}) = \begin{cases} \frac{r}{2^{d-1}}, & \text{if } \mathbf{i} \in \{1, 3\}^d, \\ 1 - 2r, & \text{if } \mathbf{i} = (2, 2, \ldots, 2), \\ 0, & \text{otherwise.} \end{cases} \qquad (3.3.5)$$

Then the IFSP induced by τ_r consists of similarities having contraction factor r or $1 - 2r$ and such that the Morgan's open set condition in [Barnsley, 1988] is obviously fulfilled. Hence, because of Theorem 3.3.4 and of the results of [Edgar, 1990, pp. 107 and 161], the Hausdorff dimension s_r of support of the limit copula $A_r^\star$ is the unique solution of the equality

$$2^d r^s + (1 - 2r)^s = 1 \qquad (3.3.6)$$

in $]1, d[$. Set $f(r, s) := 2^d r^s + (1 - 2r)^s$. By using monotonicity arguments it is straightforward to see that for each $s \in]1, d[$ there exists a unique $r \in]0, 1/2[$ such that $f(r, s) = 1$ holds. On the other hand, for any fixed $r \in]0, 1/2[$ the partial derivative $\partial_2 f(r, s)$ is negative, so that the fact that $f(r, 1) > 1$ and $f(r, d) < 1$ implies that there exists a unique $s_r \in]1, d[$ with $f(r, s_r) = 1$. This completes the proof. □

Obviously there are singular copulas whose support is of Hausdorff dimension 1 (for instance M_d) and no copula has a support of Hausdorff dimension smaller than 1. Below we provide the example of a 2-dimensional singular copula whose fractal support is exactly of Hausdorff dimension 2.

Example 3.3.7. For $i \in \mathbb{N}$ let

$$J_i = \left] \frac{1}{i+1}, \frac{1}{i} \right[,$$

and consider the family of copulas $C^i := C_{T_{r((2i-1)/i)}}$ $(i \in \mathbb{N})$ where T_r is the transformation matrix defined by

$$T_r := \begin{bmatrix} r/2 & 0 & r/2 \\ 0 & 1-2r & 0 \\ r/2 & 0 & r/2 \end{bmatrix} . \tag{3.3.7}$$

Then the copula defined by

$$C(u,v) = \frac{1}{i+1} + \frac{1}{i(i+1)} C^i \left(i(i+1) \left(u - \frac{1}{i+1} \right), i(i+1) \left(v - \frac{1}{i+1} \right) \right),$$

for $(u,v) \in J_i^2$, and by $C(u,v) = \min\{u,v\}$ elsewhere, is singular and has fractal support of Hausdorff dimension equal to 2. It will be seen later (see Section 3.8), that the function defined above is actually a copula. In J_i^2 the similarity

$$F_i(u,v) := \left(\frac{1}{i+1} + \frac{u}{i(i+1)}, \frac{1}{i+1} + \frac{v}{i(i+1)} \right)$$

spreads the mass in the support of C^i. Since F_i is a similarity, and, hence, a bijective Lipschitz function whose inverse is also Lipschitz, it preserves Hausdorff dimension [Falconer, 2003, Corollary 2.4]. Therefore

$$\dim_{\mathscr{H}} F_i \left(S_{r\left(\frac{2i-1}{i}\right)} \right) = \frac{2i-1}{i} .$$

Now, the Hausdorff dimension of the set

$$\bigcup_{i \in \mathbb{N}} F_i \left(S_{r\left(\frac{2i-1}{i}\right)} \right)$$

is the supremum of the above numbers, namely 2; thus the copula C has a support of Hausdorff dimension 2. Interesting extensions of this example to the case when the transformation matrix does not contain zeros have been discussed in [Trutschnig and Fernández Sánchez, 2014]. ∎

Further readings 3.3.8. The content of this section follows mainly the works by Fredricks et al. [2005] and Trutschnig and Fernández Sánchez [2012]. ∎

3.4 Copulas, conditional expectation and Markov kernel

Here, we show how the measure associated with a copula can be conveniently represented in terms of Markov kernels. This result is strictly related to the notion of conditional expectation and, generally, helps to express the relationships among the components of a random vector $\mathbf{X}$ conditioned on the knowledge of the values assumed by some of its components.

We start by recalling some basic notions in Probability. Let $(\Omega, \mathscr{F}, \mathbb{P})$ be a probability space and let $\mathscr{G}$ be a sub-σ-field of $\mathscr{F}$, $\mathscr{G} \subseteq \mathscr{F}$; then, by recourse to the Radon–Nikodym theorem, one sees that, for every random variable Y with finite expectation, there exists a random variable denoted by $\mathbb{E}(Y \mid \mathscr{G})$ (or by $\mathbb{E}_{\mathscr{G}}(Y)$), called *conditional expectation of Y given by $\mathscr{G}$*, that is measurable with respect to $\mathscr{G}$ and which satisfies

$$\int_A Y \, d\mathbb{P} = \int_A \mathbb{E}_{\mathscr{G}}(Y) \, d\mathbb{P} \tag{3.4.1}$$

for every set A in $\mathscr{G}$. Two different r.v.'s $\widetilde{Y}_1$ and $\widetilde{Y}_2$ that satisfy (3.4.1) are called *versions* of the conditional expectation $\mathbb{E}(Y \mid \mathscr{G})$. In such a case, $\widetilde{Y}_1 = \widetilde{Y}_2$ almost surely with respect to $\mathbb{P}$. When $\mathscr{G}$ is the σ-field generated by another r.v. X, i.e., $\mathscr{G} = \mathscr{F}(X)$, one denotes by $E(Y \mid X)$ a version of the conditional expectation of Y given $\mathscr{F}(X)$. It can then be proved that there exists a measurable function $g : \mathbb{R} \to \mathbb{R}$ such that $E(Y \mid X) = g \circ X$ holds $\mathbb{P}$-almost surely (see, e.g., [Ash, 2000]). Thus, if Y is an integrable r.v., then there exists a function denoted by $\mathbb{E}(Y \mid X = \cdot)$ such that for every Borel set B

$$\int_{X^{-1}(B)} Y \, d\mathbb{P} = \int_B \mathbb{E}(Y \mid X = t) \, d\mathbb{P}_X(t), \tag{3.4.2}$$

where $\mathbb{P}_X$ is the probability law of X to $(\mathbb{R}, \mathscr{B})$. Moreover, if h is another function satisfying eq. (3.4.2), then $h = \mathbb{E}(Y \mid X = \cdot)$ $\mathbb{P}_X$-a.e. If $Y = \mathbf{1}_B$ with $B \in \mathscr{F}$, then a.e. with respect to $\mathbb{P}_X$, one defines

$$\mathbb{P}(B \mid X = x) := \mathbb{E}\left(\mathbf{1}_B \mid X = x\right). \tag{3.4.3}$$

It should be recalled that this expression does not define, in general, a conditional probability.

The next result establishes a first link between copulas and conditional expectations.

Theorem 3.4.1. *In the probability space $(\Omega, \mathscr{F}, \mathbb{P})$ let the continuous random vector $(X, \mathbf{Y}) := (X, Y_1, \ldots, Y_{d-1})$ have marginals $F_X, F_{Y_1}, \ldots, F_{Y_{d-1}}$ and copula $C \in \mathscr{C}_d$. Then, for every $\mathbf{y} \in \mathbb{R}^{d-1}$,*

$$\mathbb{P}\left(\mathbf{Y} \leq \mathbf{y} \mid X\right) := \mathbb{E}(\mathbf{1}_{]-\infty, \mathbf{y}]} \circ \mathbf{Y} \mid X)(\omega)$$
$$= \partial_1 C\left(F_X(X(\omega)), F_{Y_1}(y_1), \ldots, F_{Y_{d-1}}(y_{d-1})\right),$$

holds $\mathbb{P}$-almost surely.

Proof. The proof is presented in the case $d = 2$, since the general case may be treated analogously. Specifically, in the probability space $(\Omega, \mathscr{F}, \mathbb{P})$, consider the continuous r.v.'s X and Y whose copula is C. Then, for every $y \in \mathbb{R}$ and for almost all points $\omega \in \Omega$, one should like to prove that

$$\mathbb{E}\left(\mathbf{1}_{\{Y \leq y\}} \mid X\right)(\omega) = \partial_1 C\left(F_X(X(\omega)), F_Y(y)\right) . \tag{3.4.4}$$

Let $\mathscr{F}(X)$ denote the σ-algebra generated by X. Proving the previous assertion is equivalent to establishing the equality $Q_y(A) = \widetilde{Q}_y(A)$, where

$$Q_y(A) = \int_A \mathbf{1}_{\{Y \leq y\}} \, d\mathbb{P}(\omega) ,$$

$$\widetilde{Q}_y(A) = \int_A \partial_1 C\left(F_X(X(\omega)), F_Y(y)\right) \, d\mathbb{P}(\omega) , \tag{3.4.5}$$

for every set $A \in \mathscr{F}(X)$. Both Q_y and $\widetilde{Q}_y$ are well-defined measures on $\mathscr{F}(X)$, since the partial derivative of a copula is integrable. It suffices (see, e.g., [Bauer, 2001, Theorem 5.4] or [Rogers and Williams, 2000, Corollary II.4.7]) to show the equality for sets belonging to a family that generates $\mathscr{F}(X)$ and which is closed under intersections. One such family is $\mathscr{A}' := \{X^{-1}(A) : A \in \mathscr{A}\}$, where the family $\mathscr{A} := \{]-\infty, a] : a \in \mathbb{R}\}$ generates the Borel σ-field $\mathscr{B}$; it is obviously stable under intersection, since if A_1' and A_2' belong to $\mathscr{A}'$, then there are sets A_1 and A_2 in $\mathscr{A}$ such that

$$A_1' \cap A_2' = X^{-1}(A_1) \cap X^{-1}(A_2) = X^{-1}(A_1 \cap A_2) ,$$

which belongs to $\mathscr{A}$. Therefore for a set A of the form $A = X^{-1}(]-\infty, a])$ with $a \in \mathbb{R}$ one has, by the continuity of F_X,

$$\int_A \partial_1 C\left(F_X(X(\omega)), F_Y(y)\right) \, d\mathbb{P}(\omega) = \int_{]-\infty, a]} \partial_1 C\left(F_X(\xi), F_Y(y)\right) \, dF_X(\xi)$$

$$= \int_{]0, F_X(a)]} \partial_1 C\left(\eta, F_Y(y)\right) \, d\eta = \mathbb{P}(X \leq a, Y \leq y)$$

$$= \int_\Omega \mathbf{1}_{\{X \leq a\}} \mathbf{1}_{\{Y \leq y\}} \, d\mathbb{P} = \int_A \mathbf{1}_{\{Y \leq y\}} \, d\mathbb{P} ,$$

which proves that $Q_y(A) = \widetilde{Q}_y(A)$. $\qquad\qquad\qquad\qquad\qquad\qquad\qquad\square$

Now, for a given $\mathbf{y} \in \mathbb{R}^{d-1}$, $\mathbb{P}(\mathbf{Y} \in [-\infty, \mathbf{y}] \mid X = x)$ is well defined for every $x \in \mathbb{R}$ up to a $\mathbb{P}_X$ null set, which depends on $\mathbf{y}$. However, following the classical approach from Probability Theory, one might more conveniently define, for almost every x, a probability measure $\mathbb{P}(\cdot \mid X = x)$. To this end, we adopt the construction of regular conditional distributions (see, e.g., [Kallenberg, 1997; Klenke, 2008]).

Recall the following definition.

Definition 3.4.2. Let $(\Omega_1, \mathscr{F}_1)$ and $(\Omega_2, \mathscr{F}_2)$ be measurable spaces. A mapping $K \colon \Omega_1 \times \mathscr{F}_2 \to \mathbb{R}$ is called a *Markov kernel* (from Ω_1 to $\mathscr{F}_2$) if the following conditions are satisfied:

(a) for every $B \in \mathscr{F}_2$, $\omega_1 \mapsto K(\omega_1, B)$ is $\mathscr{F}_1$-measurable;

(b) for every $\omega_1 \in \Omega_1$, $B \mapsto K(\omega_1, B)$ is a probability measure. $\qquad \diamond$

Now, let $(\Omega, \mathscr{F}, \mathbb{P})$ be a probability space and, on it, let $X \colon \Omega \to \mathbb{R}$ and $\mathbf{Y} \colon \Omega \to \mathbb{R}^{d-1}$ be a random variable and a random vector, respectively. A Markov kernel from $\mathbb{R}$ to $\mathscr{B}(\mathbb{R}^{d-1})$ is said to be a *regular conditional distribution of* $\mathbf{Y}$ *given* X if, for every $B \in \mathscr{B}(\mathbb{R}^{d-1})$,

$$K(X(\omega), B) = \mathbb{E}(\mathbf{1}_B \circ \mathbf{Y} \mid X)(\omega) \tag{3.4.6}$$

holds $\mathbb{P}$-almost surely.

It is well known (see, e.g., [Kallenberg, 1997; Klenke, 2008]) that

- for every d-dimensional random vector $(X, \mathbf{Y})$ a regular conditional distribution $K(\cdot, \cdot)$ of $\mathbf{Y}$ given X exists;
- $K(\cdot, \cdot)$ is unique for $\mathbb{P}_X$-almost all $x \in \mathbb{R}$;
- $K(\cdot, \cdot)$ only depends on the distribution of $(X, \mathbf{Y})$.

Applying the latter classical results to copulas, the following result easily follows.

Theorem 3.4.3. *Let the d-dimensional r.v.* $(X, \mathbf{Y})$ *on* $(\Omega, \mathscr{F}, \mathbb{P})$ *be distributed according to* $C \in \mathscr{C}_d$. *Let* $K_C(\cdot, \cdot)$ *be a version of the regular conditional distribution of* $\mathbf{Y}$ *given* X. *Then, for every Borel set* $B \subseteq \mathbb{I}^d$, *one has*

$$\mu_C(B) = \int_{\mathbb{I}} K_C(x, B_x) \, d\lambda(x), \tag{3.4.7}$$

where B_x *denotes the section of* B *at* x, $B_x := \{\mathbf{y} \in \mathbb{I}^{d-1} : (x, \mathbf{y}) \in B\}$ *and* μ_C *is the measure induced by* C. *In particular, if* $B = \times_{i=1}^d B_i$, *where* $B_i = \mathbb{I}$ *for every* $i \neq j$, *then one has*

$$\lambda(B_j) = \int_{\mathbb{I}} K_C(x, B_x) \, d\lambda(x). \tag{3.4.8}$$

Conversely, every Markov kernel $K \colon \mathbb{I} \times \mathscr{B}(\mathbb{I}^{d-1}) \to \mathbb{I}$ *that satisfies eq.* (3.4.8) *is a version of the regular conditional distribution of* $\mathbf{Y}$ *given* X.

In the remainder of the book, for every $C \in \mathscr{C}_d$, K_C will be simply called the *Markov kernel of* C. As can be seen below, this kernel is related to the derivatives of C.

Theorem 3.4.4 (Markov kernel representation of a copula). *Let* $(\Omega, \mathscr{F}, \mathbb{P})$ *be a probability space and* $(X, \mathbf{Y})$ *be a r.v. distributed according to* $C \in \mathscr{C}_d$. *Let* K_C *be the Markov kernel of* C. *Then, for every* $\mathbf{y} \in \mathbb{I}^{d-1}$, *one has*

$$\partial_1 C(x, \mathbf{y}) = K_C(x, [\mathbf{0}, \mathbf{y}])$$

for λ-*almost every* $x \in \mathbb{I}$.

Proof. It follows from the representation given by (1.6.1).   □

Thus, for every $C \in \mathscr{C}_d$, the Markov kernel of C equals a.e. the first derivative of C. However, the former has the advantage that, for each $x \in \mathbb{I}$, $K_C(x, [\mathbf{0}, \mathbf{y}])$ is a proper distribution function and, hence, it is increasing and right-continuous in $\mathbf{y}$ (see [Ash, 2000, section 5.6]).

Remark 3.4.5. All the results hitherto presented in this section remain valid when one is interested in the conditional expectation of $(X_1, \ldots, X_{j-1}, X_{j+1}, \ldots, X_d)$ given X_j for $j = 1, 2, \ldots, d$, if a r.v. $\mathbf{X} \sim C \in \mathscr{C}_d$ is given.   ∎

Finally, we apply the previous results in order to determine the d.f. of the sum of two continuous random variables X and Y (see also [Cherubini et al., 2011c]).

Theorem 3.4.6. *Let X and Y be two continuous r.v.'s defined on the same probability space $(\Omega, \mathscr{F}, \mathbb{P})$ with marginals F_X and F_Y, respectively. Let C be their copula. Then the d.f. of $X + Y$ is given by*

$$F_{X+Y}(t) = \int_0^1 \partial_1 C\left(w, F_Y\left(t - F_X^{(-1)}(w)\right)\right) \, \mathrm{d}w \, ; \qquad (3.4.9)$$

while the copula of $(X, X + Y)$ is given by

$$C_{X,X+Y}(u, v) = \int_0^u \partial_1 C\left(w, F_Y\left(F_{X+Y}^{(-1)}(v) - F_X^{(-1)}(w)\right)\right) \, \mathrm{d}w \, . \qquad (3.4.10)$$

Proof. Theorem 3.4.1 and eq. (3.4.3) yield

$$F_{X,X+Y}(s, t) = \mathbb{P}(X \le s, X + Y \le t) = \int_{-\infty}^s \mathbb{P}(X + Y \le t \mid X = x) \, \mathrm{d}F_X(x)$$

$$= \int_{-\infty}^s \mathbb{P}(Y \le t - x \mid X = x) \, \mathrm{d}F_X(x)$$

$$= \int_{-\infty}^s \partial_1 C\left(F_X(x), F_Y(t - x)\right) \, \mathrm{d}F_X(x)$$

$$= \int_0^{F_X(s)} \partial_1 C\left(w, F_Y\left(t - F_X^{(-1)}(w)\right)\right) \, \mathrm{d}w \, .$$

Let s tend to $+\infty$ to obtain

$$F_{X+Y}(t) = \int_0^1 \partial_1 C\left(w, F_Y\left(t - F_X^{(-1)}(w)\right)\right) \, \mathrm{d}w \, ,$$

i.e., eq. (3.4.9); moreover, from the expression for $F_{X,X+Y}$ one has eq. (3.4.10).   □

When the random variables X and Y of the previous theorem are independent one recovers the well-known expression for the d.f. of the sum of two independent r.v.'s; it suffices to set $w := F_X(x)$ and $C = \Pi_2$ in eq. (3.4.9).

3.5 Copulas and measure-preserving transformations

It has been traditional in Probability Theory to produce a random vector, which has a given distribution and is defined on the standard space $(\mathbb{I}, \mathscr{B}(\mathbb{I}))$ endowed with Lebesgue measure λ. Here, we revisit this question in the case of copulas. The concept of measure-preserving transformation on the unit interval will be a useful tool. We recall that, given two measure spaces $(\Omega, \mathscr{F}, \mu)$ and $(\Omega', \mathscr{F}', \nu)$, a mapping $f : \Omega \to \Omega'$ is said to be *measure-preserving* if

(a) it is measurable with respect to the σ-fields $\mathscr{F}$ and $\mathscr{F}'$, in the sense that, for every set $B \in \mathscr{F}'$, $f^{-1}(B) \in \mathscr{F}$,

(b) for every set $B \in \mathscr{F}'$, $\mu\left(f^{-1}(B)\right) = \nu(B)$.

We are mainly interested in the case $(\Omega, \mathscr{F}, \mu) = (\Omega', \mathscr{F}', \nu) = (\mathbb{I}, \mathscr{B}(\mathbb{I}), \lambda)$.

The following theorem establishes a connexion between copulas and measure-preserving transformations.

Theorem 3.5.1. *If $f_1, \ldots, f_d$ are measure-preserving transformations on the space $(\mathbb{I}, \mathscr{B}(\mathbb{I}), \lambda)$, then the function $C_{f_1,\ldots,f_d} : \mathbb{I}^d \to \mathbb{I}$ defined by*

$$C_{f_1,\ldots,f_d}(u_1,\ldots,u_d) := \lambda\left(f_1^{-1}[0,u_1] \cap \cdots \cap f_d^{-1}[0,u_d]\right) \qquad (3.5.1)$$

is a d-copula. Conversely, for every d-copula C, there exist d measure-preserving transformations $f_1, \ldots, f_d$ such that C can be expressed in the form (3.5.1).

Proof. The measure-preserving transformations $f_1, \ldots, f_d$ are actually random variables on the standard probability space $(\mathbb{I}, \mathscr{B}(\mathbb{I}), \lambda)$. Therefore the function $C_{f_1,\ldots,f_d} : \mathbb{I}^d \to \mathbb{I}$ defined by eq. (3.5.1) is the restriction to $\mathbb{I}^d$ of a d-dimensional d.f.; thus, in order to prove that it is a copula, it is enough to notice that $\lambda\left(f_j^{-1}([0,u_j])\right) = u_j$, so that the f_j's are uniformly distributed on $\mathbb{I}$ for every $j \in \{1,\ldots,d\}$.

Conversely, let C be a copula in $\mathscr{C}_d$, let μ_C be the probability measure on $\left(\mathbb{I}^d, \mathscr{B}(\mathbb{I}^d)\right)$ induced by C and set $\mathbb{P} = \mu_C$. Then there exist d continuous random variables $X_1, \ldots, X_d$ on the probability space $\left(\mathbb{I}^d, \mathscr{B}(\mathbb{I}^d), \mathbb{P}\right)$, where $\mathbb{P}$ is the Lebesgue measure λ_d on $\mathscr{B}(\mathbb{I}^d)$, such that C is the restriction to $\mathbb{I}^d$ of their joint d.f. A theorem of Kuratowski's (see, e.g., [Royden, 1988]) states that every complete separable uncountable metric space, in particular $\mathbb{I}^d$, is Borel equivalent to $\mathbb{I}$, in the sense that there exists a bijection $\varphi : \mathbb{I}^d \to \mathbb{I}$, such that both φ and φ^{-1} are Borel functions.

Let Z be the random vector on $(\mathbb{I}^d, \mathscr{B}(\mathbb{I}^d), \mathbb{P})$ that takes values in $\mathbb{I}$ and which is defined by $Z := \varphi(X_1, \ldots, X_d)$, and let F be its distribution function; F cannot have any jump, since this would imply that the continuous distribution of the random vector $(X_1, \ldots, X_d)$ has an atom, which would be a contradiction.

Then the random variable $U := F(Z)$ is uniformly distributed on $(0,1)$, and $Z = F^{(-1)}U$, where, as usual, $F^{(-1)}$ denotes the quasi-inverse of F. Thus

$$(X_1, \ldots, X_d) = \varphi^{-1} F^{(-1)} U.$$

Set $(f_1, \ldots, f_d) := \varphi^{-1} F^{(-1)} : \mathbb{I} \to \mathbb{I}^d$. Then, for every $\mathbf{u} = (u_1, \ldots, u_d) \in \mathbb{I}^d$, one has

$$
\begin{aligned}
C(\mathbf{u}) &= \mathbb{P}\left(X_1 \in [0, u_1], \ldots X_d \in [0, u_d]\right) = \mathbb{P}\left(\varphi^{-1} Z \in [0, u_1] \times \cdots \times [0, u_d]\right) \\
&= \mathbb{P}\left(((f_1 \circ U) \in [0, u_1], \ldots, (f_d \circ U) \in [0, u_d]\right) \\
&= \mathbb{P}\left(U \in f_1^{-1}[0, u_1], \ldots, U \in f_d^{-1}[0, u_d]\right) \\
&= \lambda\left(f_1^{-1}[0, u_1] \cap \cdots \cap f_d^{-1}[0, u_d]\right),
\end{aligned}
\tag{3.5.2}
$$

which proves that $C = C_{f_1, \ldots, f_d}$.

It still remains to prove that $f_1, \ldots, f_d$ are measure-preserving. Of course, it is enough to prove that one of them, for instance f_1, is so, since the proof for the remaining ones is similar. To this end, it suffices to check that, for all a and b in $\mathbb{I}$ with $a < b$, one has

$$
\lambda\left(f_1^{-1}\,]a, b]\right) = b - a.
\tag{3.5.3}
$$

Since the random variables $X_1, \ldots, X_d$ have a uniform distribution on $\mathbb{I}$, a computation very similar to that of (3.5.2) yields

$$
\begin{aligned}
\lambda\left(f_1^{-1}\,]a, b]\right) &= \lambda\left(f_1^{-1}\left([0, b] \setminus [0, a]\right)\right) \\
&= \lambda\left(f_1^{-1}\left([0, b] \setminus [0, a]\right) \cap f_2^{-1}(\mathbb{I}) \cap \cdots \cap f_d^{-1}(\mathbb{I})\right) \\
&= \lambda\left(f_1^{-1}[0, b] \cap f_2^{-1}(\mathbb{I}) \cap \cdots \cap f_d^{-1}(\mathbb{I})\right) \\
&\quad - \lambda\left(f_1^{-1}[0, a] \cap f_2^{-1}(\mathbb{I}) \cap \cdots \cap f_d^{-1}(\mathbb{I})\right) \\
&= \mathbb{P}\left(X_1 \in [0, b], X_2 \in \mathbb{I}, \ldots, X_d \in \mathbb{I}\right) \\
&\quad - \mathbb{P}\left(X \in [0, a], X_2 \in \mathbb{I}, \ldots, X_d \in \mathbb{I}\right) \\
&= \mathbb{P}\left(X_1 \in [0, b]\right) - \mathbb{P}\left(X_1 \in [0, a]\right) = b - a,
\end{aligned}
$$

which proves that f_1 is measure-preserving. $\qquad\square$

It may be noticed that if, in the previous proof, a copula C is represented as $C = C_{f_1, \ldots, f_d}$, then $f_1, \ldots, f_d$ are r.v.'s on the standard probability space $(\mathbb{I}, \mathscr{B}(\mathbb{I}), \lambda)$ whose joint d.f. is given by C. In particular the following result easily follows.

Corollary 3.5.2. Let $f_1, \ldots, f_d$ be measure-preserving transformations. Then the following conditions are equivalent for the copula $C_{f_1, \ldots, f_d}$:

(a) $C_{f_1, \ldots, f_d} = \Pi_d$;

(b) $f_1, \ldots, f_d$, when regarded as random variables on the standard probability space $(\mathbb{I}, \mathscr{B}(\mathbb{I}), \lambda)$, are independent.

Useful properties of the representation of a copula through measure-preserving transformations are collected in the next theorem.

Theorem 3.5.3. Let $f_1, \ldots, f_d$ be measure-preserving transformations and let $id_{\mathbb{I}}$ denote the identity mapping on $\mathbb{I}$, viz., $id_{\mathbb{I}}(x) = x$ for every $x \in \mathbb{I}$. Then

(a) *For every measure-preserving transformation φ on $(\mathbb{I}, \mathscr{B}(\mathbb{I}), \lambda)$,*

$$C_{f_1 \circ \varphi, \ldots, f_d \circ \varphi} = C_{f_1, \ldots, f_d} ; \tag{3.5.4}$$

(b) $C_{f_1, \ldots, f_d} = M_d$ *if, and only if, $f_1 = \cdots = f_d$ a.e.;*

(c) *in the case $d = 2$, for every $s \in \mathbb{I}$ and for every measure-preserving transformation f, one has*

$$\partial_2 C_{f, id_{\mathbb{I}}}(s, t) = 1_{f^{-1}[0, s]}(t) = 1_{[0, s]}(f(t)) = 1_{[f(s), 1]}(t) \tag{3.5.5}$$

for almost every $t \in \mathbb{I}$ and, for every $t \in \mathbb{I}$,

$$\partial_1 C_{id_{\mathbb{I}}, f}(s, t) = 1_{f^{-1}[0, t]}(s) = 1_{[0, t]}(f(s)) = 1_{[f(s), 1]}(t) \tag{3.5.6}$$

for almost every $s \in \mathbb{I}$.

Proof. (a) For every $\mathbf{u} \in \mathbb{I}^d$ one has

$$\begin{aligned}
C_{f_1 \circ \varphi, \ldots, f_d \circ \varphi}(\mathbf{u}) &= \lambda \left(\varphi^{-1} f_1^{-1}[0, u_1] \cap \cdots \cap \varphi^{-1} f_d^{-1}[0, u_d] \right) \\
&= \lambda \left(\varphi^{-1} \left(f_1^{-1}[0, u_1] \cap \cdots \cap f_d^{-1}[0, u_d] \right) \right) \\
&= \lambda \left(f_1^{-1}[0, u_1] \cap \cdots \cap f_d^{-1}[0, u_d] \right) = C_{f_1, \ldots, f_d}(\mathbf{u}) .
\end{aligned}$$

(b) Assume $C_{f_1, \ldots, f_d} = M_d$. Then $f_1, \ldots, f_d$ are r.v.'s on the standard probability space $(\mathbb{I}, \mathscr{B}(\mathbb{I}), \lambda)$ whose joint d.f. is given by M_d. From the characterisation of M_d (see Theorem 2.5.7), it follows that $f_1 = \cdots = f_d$ a.e.

Conversely, if $f_1 = \cdots = f_d$ a.e., then for every $t \in \mathbb{I}$ the sets $f_1^{-1}[0, t], \ldots,$ and $f_d^{-1}[0, t]$ differ by a set of zero Lebesgue measure. It follows that there exists a measure-preserving transformation f such that, for every $\mathbf{u} \in \mathbb{I}^d$,

$$\begin{aligned}
C_{f_1, \ldots, f_d}(\mathbf{u}) &= \lambda \left(f^{-1}[0, u_1] \cap f^{-1}[0, u_2], \ldots, \cap f^{-1}[0, u_d] \right) \\
&= \lambda \left(f^{-1}([0, u_1] \cap [0, u_2] \cap \cdots \cap [0, u_d]) \right) \\
&= \lambda \left(f^{-1}[0, \min\{u_1, u_2, \ldots, u_d\}] \right) = \lambda([0, \min\{u_1, u_2, \ldots, u_d\}]) \\
&= M_d(u_1, u_2, \ldots, u_d) .
\end{aligned}$$

(c) One has

$$C_{f, id_{\mathbb{I}}}(s, t) = \lambda \left(f^{-1}[0, s] \cap [0, t] \right) = \int_0^t 1_{f^{-1}[0, s]} \, d\lambda ,$$

from which the first equality in (3.5.5) follows. On the other hand, the three statements

$$t \in f^{-1}[0, s], \qquad f(t) \in [0, s] , \qquad s \in [f(t), 1]$$

are equivalent, as is immediately seen.

The other statement (3.5.6) is proved in a similar manner. $\qquad\square$

Corollary 3.5.4. *If the measure-preserving transformation f is a bijection on $\mathbb{I}$, then there does not exist any measure-preserving transformation g such that f and g are independent random variables on $(\mathbb{I}, \mathscr{B}(\mathbb{I}), \lambda)$.*

Proof. Suppose that there exists a measure-preserving transformation g such that f and g are independent random variables on $(\mathbb{I}, \mathscr{B}(\mathbb{I}), \lambda)$. Then

$$\Pi_2 = C_{f,g} = C_{f \circ f^{-1}, g \circ f^{-1}} = C_{\mathrm{id}_\mathbb{I}, g \circ f^{-1}} .$$

It follows from Theorem 3.5.3(c) that the first derivative of C has a jump discontinuity, which is absurd. □

Further readings 3.5.5. The connexion between copulas and measure-preserving transformations on the unit interval was first established by Olsen et al. [1996] and by Vitale [1996]. In these latter two papers the proof was only sketched; for a detailed proof see [Kolesárová et al., 2008]. A different proof was given by de Amo et al. [2011]. See also [Darsow and Olsen, 2010] for complementary results. ∎

3.6 Shuffles of a copula

In this section, we aim at presenting a possible way for rearranging the probability mass of the measure μ_C induced by $C \in \mathscr{C}_d$. This may be useful, for instance, in order to generate special dependence structures that assign a larger probability to the occurrence of tail events for a random vector $\mathbf{X}$, and/or in approximations. The introduction of these concepts requires tools that will be recalled here.

Denote by $\mathscr{T}$ the set of all measure-preserving transformations of $(\mathbb{I}, \mathscr{B}(\mathbb{I}), \lambda)$ and by $\mathscr{T}_p$ the set of all measure-preserving bijections (automorphisms) of that space. The set $\mathscr{T}$ equipped with the composition of mappings is a semigroup and $\mathscr{T}_p$ is a subgroup of $\mathscr{T}$.

An important subclass of $\mathscr{T}_p$ is formed by the *interval exchange transformations*, see [Cornfeld et al., 1982]. Let $\{J_{1,i}\}$ $(i = 1, \ldots, n)$ be a partition of $\mathbb{I}$ into the non-degenerate intervals $J_{1,i} = [a_{1,i}, b_{1,i}[$ and the singleton $J_{1,n} = \{1\}$. Let $\{J_{2,i}\}$ $(i = 1, \ldots, n)$ be another such partition and suppose that, for all $i \in \{1, \ldots, n\}$, one has $\lambda(J_{1,i}) = \lambda(J_{2,i})$. The interval exchange transformation T associated with these two partitions is the unique permutation of $\mathbb{I}$ which maps every $J_{1,i}$ linearly onto $J_{2,i}$. This map T is given, for every $x \in \mathbb{I}$, by

$$T(x) = \begin{cases} x - a_{1,1} + a_{2,1}, & \text{if } x \in J_{1,i}, \\ \lambda\left((\mathbb{I} \setminus \bigcup_{i=1}^{n} J_{1,i}) \cap [0, x]\right) + \sum_{i=1}^{n}(b_{2,i} - a_{2,i})\, \mathbf{1}_{[a_{2,i}, 1]}(x) & \text{otherwise.} \end{cases} \tag{3.6.1}$$

Next we define the shuffle of a copula.

Definition 3.6.1. Let D be a copula in $\mathscr{C}_d$ and μ_D its induced d-fold stochastic measure. A copula $C \in \mathscr{C}_d$ is a *shuffle of D* if there exist $T_1, \ldots, T_d$ in $\mathscr{T}_p$ such that the d-fold stochastic measure μ_C induced by C is given by

$$\mu_C = \mu_D \circ \left(T_1^{-1}, \ldots, T_d^{-1}\right). \tag{3.6.2}$$

Sometimes one writes $(T_1, \ldots, T_d)_{\#}\, \mu_D$ instead of $\mu_D \circ \left(T_1^{-1}, \ldots, T_d^{-1}\right)$. $\diamond$

It is easily seen that eq. (3.6.2) indeed defines a d-fold stochastic measure; since every T_i is measure-preserving, for every Borel set $A \subseteq \mathbb{I}$ and for every $i \in \{1, \ldots, d\}$, one has

$$\mu_C(\mathbb{I} \times \ldots \times \mathbb{I} \times A \times \mathbb{I} \times \cdots \times \mathbb{I})$$
$$= \left(\mu_D \circ (T_1, \ldots, T_d)^{-1}\right)(\mathbb{I} \times \cdots \times A \times \cdots \times \mathbb{I})$$
$$= \mu_D(\mathbb{I} \times \cdots \times \mathbb{I} \times T_i^{-1}(A) \times \mathbb{I} \times \cdots \times \mathbb{I}) = \lambda(T_i^{-1}(A)) = \lambda(A)\,.$$

Remark 3.6.2. If $T_1, \ldots, T_d$ are in $\mathscr{T}_p$, then $(T_1, \ldots, T_d)$ is a measure-preserving transformation of $(\mathbb{I}^d, \mathscr{B}(\mathbb{I}^d), \lambda_d)$. This can be proved, for instance, by considering that $(T_1, \ldots, T_d)$ preserves the measure of every rectangle $A \in \mathscr{B}(\mathbb{I}^d)$ and, hence, standard arguments of measure theory allow to extend this property to any Borel set in $\mathbb{I}^d$. It follows that, if $C \in \mathscr{S}(D)$, then, in view of (3.6.2), the probability spaces $(\mathbb{I}^d, \mathscr{B}(\mathbb{I}^d), \mu_C)$ and $(\mathbb{I}^d, \mathscr{B}(\mathbb{I}^d), \mu_D)$ are isomorphic since they are connected via the invertible measure-preserving mapping $(T_1, \ldots, T_d)$ (see, e.g., [Walters, 1982]). ∎

Example 3.6.3. Let $T \in \mathscr{T}$ be the transformation given by (3.6.1) that interchanges the intervals $\left[0, \frac{1}{2}\right[$ and $\left[\frac{1}{2}, 1\right[$. Then, for every bivariate copula C, the copula C_T associated with the measure $\mu_C \circ (T^{-1}, \mathrm{id}_\mathbb{I})$ is given by

$$C_T(u, v) = C\left(\left(u + \tfrac{1}{2}\right) \wedge 1, v\right) - C\left(\tfrac{1}{2}, v\right) + C\left(\left(u - \tfrac{1}{2}\right) \vee 0, v\right)\,.$$

For instance, if C is the arithmetic mean of M_2 and W_2, then C_T has support equal to the borders of the square with vertices $(1/2, 0)$, $(0, 1/2)$, $(1/2, 1)$ and $(1, 1/2)$ (see Figure 3.2). This copula corresponds to the circular uniform distribution (see Section 3.1.2 in [Nelsen, 2006]). ∎

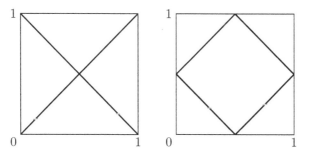

Figure 3.2 *Support of the copula* $C = (M_2 + W_2)/2$ *(left) and support of the related shuffle* C_T *of Example 3.6.3.*

In terms of measure-preserving transformations, a shuffle of a copula may be represented in the following way.

Theorem 3.6.4. *Let $C \in \mathscr{C}_d$ be the copula generated by the measure-preserving transformations $f_1, \ldots, f_d$. Then the shuffle of C given by $\mu_C \circ (T_1^{-1}, \ldots, T_d^{-1})$ is generated by the measure-preserving transformations $T_1 \circ f_1, \ldots, T_d \circ f_d$.*

Proof. If C is the copula generated by $f_1, \ldots, f_d$, then $\mu_C = \lambda \circ (f_1^{-1}, \ldots, f_d^{-1})$. It follows that

$$\mu_C \circ (T_1^{-1}, \ldots, T_d^{-1}) = \left(\lambda \circ (f_1^{-1}, \ldots, f_d^{-1})\right) \circ (T_1^{-1}, \ldots, T_d^{-1})$$
$$= \lambda \circ \left((T_1 \circ f_1)^{-1}, \ldots, (T_d \circ f_d)^{-1}\right),$$

which is the desired assertion. ◻

A shuffle of a copula preserves the absolute continuity of the measure, as shown by the following result.

Theorem 3.6.5. *If $C \in \mathscr{C}_d$ is absolutely continuous then so are all its shuffles.*

Proof. Let C be an absolutely continuous copula, let $T_1, \ldots, T_d$ belong to $\mathscr{T}_p$ and let A be a Borel set of $\mathbb{I}^d$ with $\lambda_d(A) = 0$. Then

$$\lambda_d\left((T_1, \ldots, T_d)^{-1}(A)\right) = \lambda_d(A) = 0,$$

and, by the absolute continuity of C,

$$\mu_C\left((T_1, \ldots, T_d)^{-1}(A)\right) = 0,$$

which is the desired assertion. ◻

In general, the shuffle of a copula produces a copula that is different from the original one. Clearly, one cannot expect that exchangeability is preserved under shuffling. More can be said in this respect.

Theorem 3.6.6. *Every copula $C \in \mathscr{C}_d$ other than Π_d has a non-exchangeable shuffle.*

Proof. We only consider the case $d = 2$, the proof for the general case being analogous. Let ξ be the permutation of $\mathbb{I}^2$ defined by $\xi(x, y) := (y, x)$. Observe, that the exchangeability of a copula C is equivalent to $\mu_C(A) = \mu_C(\xi(A))$ for every Borel set $A \subseteq \mathbb{I}^2$. Let $J_{1x}, J_{1y}, J_{2x}, J_{2y}$ be arbitrary but fixed intervals of type $[a, b[\subseteq \mathbb{I}$ with

$$J_{1x} \cap J_{2x} = J_{1y} \cap J_{2y} = \emptyset \qquad \text{and} \qquad \lambda(J_{1x}) = \lambda(J_{1y}) = \lambda(J_{2x}) = \lambda(J_{2y}).$$

For $i = 1, 2$, define the squares $R_i := J_{ix} \times J_{iy}$ and $R_* := J_{2y} \times J_{1y}$. Further, let T be the interval exchange transformation given by (3.6.1), which, for $i = 1, 2$, sends J_{iy} onto J_{ix}. Let $S_T := (T, \mathrm{id}_{\mathbb{I}})$. Notice that

$$S_T^{-1}(R_*) = R_1 \qquad \text{and} \qquad S_T^{-1}(\xi(R_*)) = R_2.$$

Let C be a copula such that every shuffle of C is exchangeable. Then,

$$\mu_C(R_1) = \left(\mu_C \circ S_T^{-1}\right)(R_*) = \mu_C \circ S_T^{-1}\left(\xi(R_*)\right) = \mu_C(R_2).$$

Thus, if two squares with the same Lebesgue measure have disjoint projections onto the x- and the y-axis, then they have the same C-measure. Now fix a natural number $n \geq 3$ and define, for $i, j \in \{0, 1, \ldots, n\}$,

$$I_n := \left\{ \tfrac{m}{n} \mid m = 0, 1, \ldots, n \right\} \qquad \text{and} \qquad R_{i,j} := \left]\tfrac{i}{n}, \tfrac{i+1}{n}\right[\times \left]\tfrac{j}{n}, \tfrac{j+1}{n}\right[.$$

As a consequence, $\mu_C(R_{i,j}) = \mu_C(R_{k,l})$ whenever $i \neq k$ and $j \neq l$. For $n \geq 3$, this is enough to conclude that the C-measure of any such square is the same. Since the squares $R_{i,j}$ form a partition of the unit square, the C-measure of each of them is n^{-2}. Therefore,

$$C\left(\tfrac{i}{n}, \tfrac{j}{n}\right) = \mu_C\left(]0, \tfrac{i}{n}[\times]0, \tfrac{j}{n}[\right) = \tfrac{ij}{n^2},$$

which can be written also as $C|_{I_n \times I_n} = \Pi_2|_{I_n \times I_n}$. As this can be proved for any arbitrary natural number $n \geq 3$, one has $C = \Pi_2$. $\qquad\square$

The behaviour of the independence copula Π_d with respect to shuffles is quite peculiar, as stated below.

Theorem 3.6.7. *For a copula $C \in \mathscr{C}_d$ the following statements are equivalent:*

(a) $C = \Pi_d$;

(b) *every shuffle of C coincides with C.*

Proof. (a) $\Longrightarrow$ (b) This follows from the fact that $\mu_{\Pi_d} = \lambda_d$ and $(T_1, \ldots, T_d)$ is measure-preserving.

(b) $\Longrightarrow$ (a) Suppose that $C \neq \Pi_d$. Then there exist two rectangles

$$R = [a_1, b_1] \times \cdots \times [a_{d-1}, b_{d-1}] \times [c, d]$$
$$R' = [a_1, b_1] \times \cdots \times [a_{d-1}, b_{d-1}] \times [c', d']$$

such that $[c, d] \neq [c', d']$, $\mu_C(R) \neq \mu_C(R')$ and $d - c = d' - c'$. Consider the interval exchange transformation T that sends $[c, d]$ onto $[c', d']$. Then

$$(\mathrm{id}_{\mathbb{I}}, \ldots, \mathrm{id}_{\mathbb{I}}, T)_{\#}\mu_C(R) = \mu_C(R').$$

Thus, $(\mathrm{id}_{\mathbb{I}}, \ldots, \mathrm{id}_{\mathbb{I}}, T)_{\#}\mu_C(R) \neq \mu_C(R)$ and thus a shuffle of C does not coincide with C. $\qquad\square$

3.6.1   Shuffles of Min

The shuffles of the copula M_d will now be considered in detail; they are called *shuffles of Min* and were introduced by Mikusiński et al. [1992] in the case $d = 2$. Notice that, if a copula $C = (T_1, \ldots, T_d)_{\#}\mu_{M_d}$ is a shuffle of Min, then by Theorem 3.6.4 it follows that C is generated by $T_1, \ldots, T_d$. Moreover, since T_1 is a bijection, we may express C in an equivalent way in terms of the measure-preserving transformations $T_1 \circ T_1^{-1}, T_2 \circ T_1^{-1}, \ldots, T_d \circ T_1^{-1}$. In other words, C can be written in the form

$$C(\mathbf{u}) = \lambda \left([0, u_1] \cap T_2^{-1} [0, u_2] \cap \cdots \cap T_d^{-1} [0, u_d] \right), \qquad (3.6.3)$$

for suitable λ-measure-preserving $T_i \in \mathscr{T}_{\mathrm{p}}$.

The following characterisation holds.

Theorem 3.6.8. *Let C be a copula. The following statements are equivalent:*

(a) $\mu_C = (T_1, \ldots, T_d)_{\#} \mu_{M_d}$ *for suitable $T_1, \ldots, T_d$ in $\mathscr{T}_p$;*

(b) C *concentrates the probability mass on the set*

$$\{\mathbf{x} \in \mathbb{I}^d \mid \mathbf{x} = (x_1, f_2(x_1), \ldots, f_d(x_1))\}$$

   for suitable $f_i \in \mathscr{T}_p$;

(c) *there exists a random vector $\mathbf{U} = (U_1, \ldots, U_d)$ on the probability space $(\Omega, \mathscr{F}, \mathbb{P})$ having distribution function equal to C for which there exist $g_{ij} \in \mathscr{T}_{\mathrm{p}}$ for all $i, j \in \{1, \ldots, d\}$, $i \neq j$, such that $U_i = g_{ij}(U_j)$ almost surely.*

Proof. (a) $\Longrightarrow$ (b) Let $\mathbf{T} = (T_1, \ldots, T_d)$. Since μ_{M_d} concentrates the probability mass on the set $\Gamma = \{\mathbf{u} \in \mathbb{I}^d : u_1 = u_2 = \cdots = u_d\}$, one has

$$\mu_C(\mathbf{T}(\Gamma)) = \mathbf{T}_{\#} \mu_{M_d}(\mathbf{T}(\Gamma)) = \mu_{M_d}(\Gamma) = 1,$$

from which it follows that C concentrates the probability mass on

$$\{\mathbf{u} = (T_1(t), \ldots, T_d(t)), t \in \mathbb{I}\} = \{\mathbf{u} = (s, T_2(T_1^{-1}(s)), \ldots, T_d(T_1^{-1}(s))), s \in \mathbb{I}\},$$

which is the desired assertion.

(b) $\Longrightarrow$ (c) Let $\mathbf{U}$ be a random vector distributed according to the copula C. If C satisfies (b), then

$$\mathbb{P} \left(\mathbf{U} \in \{\mathbf{x} \in \mathbb{I}^d \mid \mathbf{x} = (x_1, f_2(x_1), \ldots, f_d(x_1))\} \right) = 1,$$

from which it follows that each component of $\mathbf{U}$ is almost surely a function of the other via a transformation in $\mathscr{T}_p$.

(c) $\Longrightarrow$ (a) Because of (c), if the random vector $\mathbf{U}$ is distributed according to the copula C, then it is almost surely equal to the vector $(U_1, f_2(U_1), \ldots, f_d(U_1))$ for suitable $f_2, \ldots, f_d$ in $\mathscr{T}_p$, $f_j = g_{j1}$ $(j = 2, \ldots, d)$. It follows that

$$\begin{aligned} \mu_C &= (U_1, f_2(U_1), \ldots, f_d(U_1))_{\#} \mathbb{P} \\ &= ((\mathrm{id}_{\mathbb{I}}, f_2, \ldots, f_d) \circ (U_1, U_1, \ldots, U_1))_{\#} \mathbb{P} \\ &= (\mathrm{id}_{\mathbb{I}}, f_2, \ldots, f_d)_{\#} \mu_{M_d}, \end{aligned}$$

which implies (a). $\qquad \square$

Property (c) of Theorem 3.6.8 is equivalent to the fact that all the bivariate random pairs extracted from $(U_1, \ldots, U_d)$ are mutually complete dependent in the sense of Lancaster [1963]. This latter characterisation is, in some sense, an extension of the comonotonicity property of a random vector. In fact, while the comonotonicity of a vector is equivalent to saying that each of its components is a monotone increasing function of the others, mutually complete dependence implies that each component of a vector is a bijective transformation of the others.

The notion of shuffles of Min has been first introduced by Mikusiński et al. [1992] for the bivariate case in a slightly different setting. Here we recall their definition and we show how it is connected with the framework previously considered.

Definition 3.6.9. A bivariate copula is said to be a *shuffle of Min à la* Mikusiński et al. [1992] if it is obtained through the following procedure:

(s1) the probability mass is spread uniformly on the support of the copula M_2, namely on the main diagonal of the unit square;

(s2) then the unit square is cut into a finite number of vertical strips;

(s3) these vertical strips are permuted ("shuffled") and, possibly, some of them are flipped about their vertical axes of symmetry;

(s4) finally, the vertical strips are reassembled to form the unit square again;

(s5) to the probability mass thus obtained there corresponds a unique copula C, which is a shuffle of Min.   ◇

The procedure is summarised in Figure 3.3.

A shuffle of Min *à la* Mikusiński et al. [1992] is hence characterised by

- a number $n \in \mathbb{N}$, $n \geq 2$;
- $n - 1$ numbers $x_1 < x_2 < \cdots < x_{n-1}$ in $]0, 1[$ that give rise to the partition

$$\mathbb{I} = [0, x_1] \cup \,]x_1, x_2] \cup \cdots \cup \,]x_{n-1}, 1] \, ;$$

- a permutation σ_n of $\{1, \ldots, n\}$;
- a function $\rho : \{1, \ldots, n\} \to \{-1, 1\}$.

The j-strip is, or is not, flipped about its vertical axis of symmetry, according to whether $\rho(j)$ equals -1 or 1, respectively. If $\rho(j) = 1$ for every j, then the shuffle of Min is called *straight*; if, instead, $\rho(j) = -1$ for every j, then the shuffle is called *flipped*.

As an immediate consequence of Theorem 2.5.7, one can state the following result. Here, we assume that a function is *piecewise continuous* if it is defined on a non-degenerate interval and has at most finitely many discontinuities, all of them being jumps.

Theorem 3.6.10. *Let the copula C of two continuous r.v.'s X and Y on the probability space $(\Omega, \mathscr{F}, \mathbb{P})$ be a shuffle of Min à la Mikusiński et al. [1992]. Then there exists a bijective and piecewise continuous function $\varphi : \mathbb{R} \to \mathbb{R}$ for which $\mathbb{P}(X = \varphi \circ Y) = 1$.*

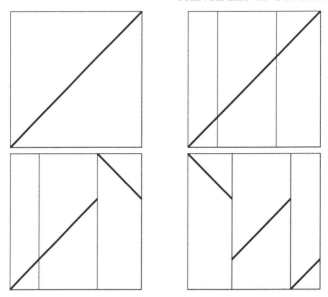

Figure 3.3 *Graphical procedure to determine a shuffle of Min. (s1): upper left figure. (s2): upper right figure. (s3): lower left figure. (s4): lower right figure.*

By using the previous arguments, a more concise way of introducing shuffles of Min is derived.

Theorem 3.6.11. *The following statements are equivalent:*

(a) *a copula $C \in \mathscr{C}_2$ is a shuffle of Min à la Mikusiński et al. [1992];*

(b) *there exists a piecewise continuous $T \in \mathscr{T}$ such that $\mu_C = \mu_M \circ S_T^{-1}$, where $S_T = (T, id_{\mathbb{I}})$.*

Proof. (b) $\Longrightarrow$ (a) Let Y be a random variable uniformly distributed on $\mathbb{I}$. Then $(Y, Y) \sim M_2$ or, equivalently, $\mathbb{P}_{(Y,Y)} = \mu_{M_2}$. Thus, by the definitions of S_T and of image measure, one can derive

$$\mathbb{P}_{(T \circ Y, Y)} = \mathbb{P}_{(S_T \circ (Y,Y))} = \mathbb{P}_{(Y,Y)} \circ S_T = \mu_{M_2} \circ S_T^{-1} \qquad (3.6.4)$$

for every measurable (but not necessarily measure-preserving) transformation T of the unit interval.

Now, let T be a piece-wise continuous function in $\mathscr{T}$. Then the measure μ_{S_T} is doubly stochastic, and, hence, corresponds to a copula C. By (3.6.4) the random vector $(T \circ Y, Y)$ is distributed according to C, which then is a shuffle of Min.

(a) $\Longrightarrow$ (b) Let C be a shuffle of Min. Then there exists a probability space $(\Omega, \mathscr{F}, \mathbb{P})$ and a random vector (X, Y) defined on it such that $(X, Y) \sim C$. Moreover, by Theorem 3.6.10 there exists a bijective and piece-wise continuous function T for which $\mathbb{P}(X = T(Y)) = 1$. Moreover, T is easily seen to be Borel-measurable, and, as a consequence, also $(T \circ Y, Y)$ is a random vector. This random vector differs from (X, Y) on a set of zero $\mathbb{P}$-measure, which proves that

$$\mathbb{P}_{(T \circ Y, Y)} = \mathbb{P}_{(X,Y)} = \mu_C \, .$$

Thus, invoking (3.6.4) allows to derive the representation $\mu_C = \mu_{M_2} \circ S_T^{-1}$. In order to conclude the proof, it is enough to notice that T is measure-preserving. $\square$

Shuffles of Min *à la* Mikusiński et al. [1992] are not allowed to have a support with *countably* many discontinuity points, as instead is possible by Definition 3.6.1.

Example 3.6.12. For every $i \in \mathbb{N}$ define

$$J_{1,i} := \left[\frac{1}{i+1}, \frac{1}{i} \right[\qquad \text{and} \qquad J_{2,i} := \left[1 - \frac{1}{i+1}, 1 - \frac{1}{i} \right[\, .$$

Clearly, the indexed systems $\{J_{1,i}\}_{i \in \mathbb{N}}$ and $\{J_{2,i}\}_{i \in \mathbb{N}}$ consist of nonoverlapping intervals such that $\lambda(J_{1,i}) = \lambda(J_{2,i})$ for every natural i. Let $\widetilde{T}$ be the interval exchange transformation given by (3.6.1). Also $T(x) := \widetilde{T}(1 - x)$ belongs to $\mathscr{T}$. Indeed, T is a composition of $\widetilde{T}$ and $x \mapsto 1 - x$, which are both measure-preserving. Clearly, T has countably many discontinuities. Therefore, the push-forward measure $\mu_{M_2} \circ (T, id_{\mathbb{I}})^{-1}$ is not a shuffle of Min *à la* Mikusiński et al. [1992]. $\blacksquare$

From the above representation of bivariate shuffles of Min, it not not hard to see that shuffles of Min are extremal copulas, since they are obtained by shuffling in a convenient way the probability mass distribution of the extremal copula M_2. In view of Remark 1.7.10, it is not surprising that these copulas appear in various optimization problems. We illustrate this in two examples.

Example 3.6.13. Consider the functional $\varphi : \mathscr{C}_2 \to \mathbb{R}$ given by

$$\varphi(C) = \max_{(u,v) \in \mathbb{I}^2} |C(u,v) - C(v,u)| \, .$$

This functional expresses the lack of exchangeability of C and has been employed to measure the non-exchangeability of a copula [Durante et al., 2010c; Genest and Nešlehová, 2013]. Now, as shown by Klement and Mesiar [2006] and Nelsen [2007], the maximum of this function is reached by the shuffles of Min whose mass distribution is depicted in Figure 3.4. They spread the probability mass, respectively, on the segments joining $(2/3, 0)$ and $(1, 1/3)$, $(0, 1/3)$ and $(2/3, 1)$ (Figure 3.4, left); the segments joining $(0, 2/3)$ and $(1/3, 1/3)$, $(1/3, 1)$ and $(2/3, 2/3)$, $(2/3, 1/3)$ and $(1, 0)$ (Figure 3.4, right). $\blacksquare$

Example 3.6.14. For fixed continuous r.v.'s $X_1 \sim F_1$ and $X_2 \sim F_2$, let F_C be the d.f. of $X_1 + X_2$ (dependent on the copula C of (X_1, X_2)). For a fixed $p \in (0, 1)$, consider the functional $\varphi : \mathscr{C}_2 \to \mathbb{R}$ given by

$$\varphi(C) = F_C^{(-1)}(p) \, ,$$

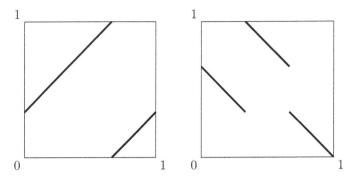

Figure 3.4: *Support of the copulas related to Example 3.6.13.*

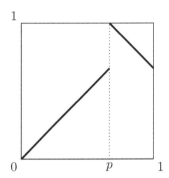

Figure 3.5: *Support of the copula related to Example 3.6.14.*

which provides the quantile (at level p) of the sum of $X_1 + X_2$. Then, it has been proved by Makarov [1981] and Rüschendorf [1982] (see also [Frank et al., 1987]) that the supremum of this functional is reached by the shuffle of Min whose support is depicted in Figure 3.5. This problem has relevant applications in risk management, for instance in the determination of worst-case Value-at-Risk scenario for a given portfolio: see, e.g., Embrechts et al. [2013] and Rüschendorf [2013] and references quoted in these papers.   ∎

We conclude with the example of a singular copula with full support, which is constructed via shuffles of Min.

Example 3.6.15. Let C_n be the bivariate shuffle of Min obtained in the following way: consider a partition of the unit interval $\mathbb{I}$ into n^2 subintervals of length $1/n^2$ and the permutation π of $\{1, 2 \dots, n^2\}$ defined by $\pi(n^2(j-1)+k) = n^2(k-1)+j$ where $j, k = 1, 2, \ldots, n$. The copula C_n approximates the independence copula Π_2 in the sense that

$$\sup_{(u,v) \in \mathbb{I}^2} |C_n(u, v) - \Pi_2(u, v)| < \varepsilon,$$

for every $n \geq 4/\varepsilon$. Consider the copula

$$C = \sum_{n \in \mathbb{N}} \frac{1}{2^n} C_n .$$

Let T_n be the support of C_n and set $T := \cup_{n \in \mathbb{N}} T_n$; then $\mu_C(T) = 1$, viz., the probability mass of μ_C is concentrated on T. On the other hand, one has $\lambda_2(T_n) = 0$, so that $\lambda_2(T) = 0$. This implies that C is singular.

Since the closure of T is $\mathbb{I}^2$, the support of μ_C is $\mathbb{I}^2$. ∎

Further readings 3.6.16. The concept of shuffle was originally introduced for the copula M_2 by Mikusiński et al. [1992], who gave it the name of Shuffles of Min. Shuffles in the generalised sense of this book were introduced by Durante et al. [2009b] in the case $d = 2$ and extended to the multivariate case by Durante and Fernández-Sánchez [2010]. The extension of the definition of shuffles of Min to the d-dimensional case can be found in Section 6 of [Mikusiński and Taylor, 2009]. Further generalisations have been presented by Trutschnig and Fernández Sánchez [2013]. Shuffles of Min have been used in the approximation of solutions of several optimization problems (under discrete marginal distributions); see, for instance, the rearrangement algorithm [Puccetti and Rüschendorf, 2012] and [Hofer and Iacò, 2014]. ∎

3.7 Sparse copulas

The way of constructing multidimensional copulas we present here is related in spirit to that of shuffles of the previous section, and, in the literature, is also given the name of "multidimensional shuffle"; the copulas thus constructed will be called *sparse copulas*. Their construction exploits the one-to-one correspondence between the set $\mathscr{C}_d$ of d-dimensional copulas and the set of d-fold stochastic measures established in Section 3.1.

The construction of a d-dimensional sparse copula is based on the following structure.

Definition 3.7.1. Let $d \in \mathbb{N}$, $d \geq 2$. For every $i \in \{1, \ldots, d\}$, let $\mathscr{J}^{i,N}$ be a system of closed and non-empty intervals of $\mathbb{I}$,

$$\mathscr{J}^{i,N} = \left(J_n^i = \left[a_n^i, b_n^i \right] \right)_{n \in N}$$

such that:

(S1) N represents a finite or countable index set, i.e., either $N = \{0, 1, \ldots, \tilde{n}\}$ or $N = \mathbb{Z}_+$;

(S2) for every $i \in \{1, \ldots, d\}$ and $n, m \in N$, $n \neq m$, J_n^i and J_m^i have at most one endpoint in common;

(S3) for every $i \in \{1, \ldots, d\}$, $\sum_{n \in N} \lambda \left(J_n^i \right) = 1$;

(S4) for every $n \in N$, $\lambda \left(J_n^1 \right) = \lambda \left(J_n^2 \right) = \cdots = \lambda \left(J_n^d \right)$.

The structure $\widetilde{\mathscr{J}} = \left(\mathscr{J}^{1,N}, \mathscr{J}^{2,N} \ldots, \mathscr{J}^{d,N} \right)$ satisfying the above properties is called a *sparse structure*.   ◇

Definition 3.7.2. Let N be an index set as in (S1). Let $(\mu_n)_{n \in N}$ be a sequence of probability measures on $\left(\mathbb{I}^d, \mathscr{B}(\mathbb{I}^d) \right)$ that are supported on $]0, 1[^d$. Let $\widetilde{\mathscr{J}}$ be a sparse structure. Let $\mu : \mathbb{I}^d \to \mathbb{R}_+$ be the set function defined, for every $A \in \mathscr{B}(\mathbb{I}^d)$, by

$$\mu(A) := \sum_{(n_1, \ldots, n_d) \in N^d} \mu_{n_1, \ldots, n_d} \left(A \bigcap \left(J_{n_1}^1 \times J_{n_2}^2 \times \cdots \times J_{n_d}^d \right) \right),$$

where, for every $(n_1, \ldots, n_d) \in N^d$, the set function $\mu_{n_1, \ldots, n_d}$ is defined on the Borel sets of

$$J_{n_1}^1 \times J_{n_2}^2 \times \cdots \times J_{n_d}^d$$

in the following way:

(M1) $\mu_{n_1, \ldots, n_d} = 0$ if $n_k \neq n_{k'}$ for some k and k';

(M2) for every Borel set $A \subseteq J_n^1 \times J_n^2 \times \cdots \times J_n^d$,

$$\mu_{n, \ldots, n}(A) = \lambda(J_n^1) \, \mu_n \left(\varphi_n(A) \right),$$

where $\varphi_n : J_n^1 \times J_n^2 \times \cdots \times J_n^d \to \mathbb{I}^d$ is defined by

$$\varphi_n(x_1, \ldots, x_d) = \left(\frac{x_1 - a_n^1}{\lambda(J_n^1)}, \ldots, \frac{x_d - a_n^d}{\lambda(J_n^d)} \right) = \left(\frac{x_1 - a_n^1}{\lambda(J_n^1)}, \ldots, \frac{x_d - a_n^d}{\lambda(J_n^1)} \right).$$

The function μ is called a *sparse set function* related to $\widetilde{\mathscr{J}}$ and to $(\mu_n)_{n \in N}$. It is denoted by the symbol $\mu = \langle \widetilde{\mathscr{J}}, (\mu_n)_{n \in N} \rangle$.   ◇

A sparse set function is actually a probability measure, as the following result shows. Notice that, here, we use the convention that a d-fold stochastic measure (respectively, copula) may be considered as a measure (respectively, d.f.) on $\mathbb{R}^d$ that concentrates all the probability mass on $\mathbb{I}^d$.

Theorem 3.7.3. *Every sparse set function* $\mu = \langle \widetilde{\mathscr{J}}, (\mu_n)_{n \in N} \rangle$ *is a probability measure. Moreover, if each μ_n is d-fold stochastic, then so is μ.*

Proof. Let $\mu = \langle \widetilde{\mathscr{J}}, (\mu_n)_{n \in N} \rangle$ be a sparse set function. That μ is a measure is clear since every $\mu_{n_1, n_2, \ldots, n_d}$ is a measure. Since

$$\bigcup_{(n_1, n_2, \ldots, n_d) \in N^d} (J_{n_1}^1 \times J_{n_2}^2 \cdots \times J_{n_d}^d) = \mathbb{I}^d,$$

then

$$\mu\left(\mathbb{I}^d\right) = \sum_{(n_1,\ldots,n_d)\in N^d} \mu_{n_1,\ldots,n_d} \left(\bigcup_{(n_1,\ldots,n_d)\in N^d} \left(J_{n_1}^1 \times \cdots \times J_{n_d}^d\right) \right)$$

$$= \sum_{n\in N} \mu_{n,\ldots,n} \left(J_{n_1}^1 \times J_{n_2}^2 \times \cdots \times J_{n_d}^d\right)$$

$$= \sum_{n\in N} \lambda(J_n^1)\,\mu_n \left(\varphi_n \left(J_{n_1}^1 \times J_{n_2}^2 \times \cdots \times J_{n_d}^d\right)\right) = 1\,,$$

so that μ is a probability measure.

Now, suppose that each μ_n is a d-fold stochastic measure. In order to prove that so is μ, one has to show that, for every $i \in \{1,\ldots,d\}$ and for every $x_i \in \mathbb{I}$,

$$\mu\left(\mathbb{I}^{i-1} \times [0,x_i] \times \mathbb{I}^{d-i}\right) = \lambda([0,x_i])\,.$$

Suppose $i = 1$ (the other cases are treated analogously) and let x_1 belong to $\mathbb{I}$. Since $\mu\left(\{x_1\} \times \mathbb{I}^{d-1}\right) = 0$, one can assume, without loss of generality, that x_1 does not belong to $\{a_n^1, b_n^1\}$ for every $n \in N$. Let N_1 be the following index set

$$N_1 := \{n \in N : J_n^1 \cap [0,x_1] \neq \emptyset \text{ and } x_1 \notin J_n^1\}.$$

Let $n(0)$ be a natural number such that x_1 belongs to $\left[a_{n(0)}^1, b_{n(0)}^1\right]$ and notice that $\sum_{n\in N_1}\left(b_n^1 - a_n^1\right) = a_{n(0)}$. It follows that

$$\mu([0,x_1] \times \mathbb{I}^{d-1}) = \left(\sum_{n\in N_1} \lambda(J_n^1)\,\mu_n \left(\varphi_n \left(J_n^1 \times \cdots \times J_n^d\right)\right)\right)$$

$$+ \lambda(J_{n(0)}^1)\,\mu_{n(0)} \left(\varphi_{n(0)} \left(\left(J_{n(0)}^1 \cap [0,x_1]\right) \times J_{n(0)}^2 \times \cdots \times J_{n(0)}^d\right)\right)$$

$$= \sum_{n\in N_1} \lambda(J_n^1) + \lambda(J_{n(0)}^1)\mu_{n(0)} \left(\left[0, \frac{x_1 - a_{n(0)}^1}{b_{n(0)}^1 - a_{n(0)}^1}\right] \times \mathbb{I}^{d-1}\right)$$

$$= \sum_{n\in N_1}\left(b_n^1 - a_n^1\right) + \lambda(J_{n(0)}^1)\frac{x_1 - a_{n(0)}^1}{b_{n(0)}^1 - a_{n(0)}^1} = x_1\,,$$

which is the desired assertion. $\square$

Remark 3.7.4. Because of its very definition the support of a sparse probability measure is on the sets of type $J_n^1 \times \cdots \times J_n^d$ for every $n \in N$, while the probability mass is equal to 0 in all the sets of type $J_{n_1}^1 \times \cdots \times J_{n_d}^d$, when $n_i \neq n_j$ for at least one index $i \neq j$. This feature also motivates the adoption of the adjective "sparse" to denote such measures, since "sparsity" indicates a mathematical structure that is null in most of its components. ∎

We are now in the position of defining a multidimensional sparse copula.

Definition 3.7.5. A copula $C \in \mathscr{C}_d$ is a *sparse copula* if the measure μ_C it induces can be represented as a sparse measure. It will be denoted by $\langle \widetilde{\mathscr{J}}, (C_n)_{n \in N} \rangle$, where C_n is the copula induced by μ_{C_n}. $\diamond$

This leads to the following procedure (see Figure 3.6) in order to construct a sparse copula $C \in \mathscr{C}_d$:

(a) define a suitable partition $\{ J_{n_1}^1 \times J_{n_2}^2 \times \cdots \times J_{n_d}^d \}_{(n_1, \ldots, n_d) \in N^d}$ of $\mathbb{I}^d$ formed by d-dimensional boxes, namely the sparse structure $\widetilde{\mathscr{J}}$;

(b) given a system of copulas $(C_n)_{n \in N}$, plug a transformation of the probability mass of C_n into $J_n^1 \times J_n^2 \times \cdots \times J_n^d$;

(c) define C as the copula whose associated measure coincides with the previous transformation of μ_{C_n} in each box of type $J_n^1 \times J_n^2 \times \cdots \times J_n^d$.

An explicit expression of a sparse copula is given by the following result.

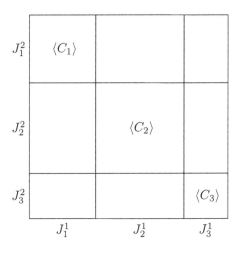

Figure 3.6: *Graphical representation of a sparse copula.*

Theorem 3.7.6. *Let* $C = \langle \widetilde{\mathscr{J}}, (C_n)_{n \in N} \rangle$ *be a sparse copula. Then, for every* $\mathbf{u} \in \mathbb{I}^d$,

$$C(\mathbf{u}) = \sum_{n \in N} \lambda(J_n^1) \, C_n \left(\frac{u_1 - a_n^1}{\lambda(J_n^1)}, \ldots, \frac{u_d - a_n^d}{\lambda(J_n^1)} \right). \qquad (3.7.1)$$

Proof. It follows from Definition 3.7.2 that, for every $\mathbf{u} \in \mathbb{I}^d$,

$$C(\mathbf{u}) = \mu_C([0, \mathbf{u}]) = \sum_{n \in N} \lambda(J_n^1) \, \mu_{C_n} \left(\varphi_n([0, \mathbf{u}] \cap (J_n^1 \times \cdots \times J_n^d)) \right)$$

$$= \sum_{n \in N} \lambda(J_n^1) \, C_n \left(\frac{u_1 - a_n^1}{\lambda(J_n^1)}, \ldots, \frac{u_d - a_n^d}{\lambda(J_n^1)} \right),$$

which is the desired assertion.   □

In eq. (3.7.1) we use the convention that the values assumed by C_n outside $\mathbb{I}^d$ can be uniquely determined by the usual properties of a distribution function.

Thanks to eq. (3.7.1), any sparse copula may be interpreted as a convex combination (with coefficients $\lambda(J_n^1)$) of the d-variate distribution functions

$$F_n(\mathbf{x}) = C_n \left(\frac{x_1 - a_n^1}{\lambda(J_n^1)}, \ldots, \frac{x_d - a_n^d}{\lambda(J_n^1)} \right).$$

Shuffles of Min *à la* Mikusiński et al. [1992] are sparse copulas.

Example 3.7.7. Consider a bivariate shuffle of Min (Section 3.6). Let the points

$$0 = s_0 < s_1 < \cdots < s_{\tilde{n}} = 1 \quad \text{and} \quad 0 = t_0 < t_1 < \cdots < t_{\tilde{n}} = 1$$

determine two partitions of $\mathbb{I}$ and let ς denote a permutation of $\{1, \ldots, n\}$ such that each

$$[s_{i-1}, s_i] \times [t_{\varsigma(i)-1}, t_{\varsigma(i)}]$$

is a square on which the copula C spreads uniformly a mass $s_i - s_{i-1}$ either on the main diagonal or on the opposite diagonal.

This shuffle of Min may be represented as a sparse copula of the form $\langle (\mathscr{J}^{1,N}, \mathscr{J}^{2,N}), (C_n)_{n \in N} \rangle$, where:

- $N = \{0, 1, \ldots, \tilde{n} - 1\}$;
- $\mathscr{J}^{1,N} = (J_i^1)_{i \in N}$ is the partition given by $J_i^1 = [s_{i-1}, s_i]$;
- $\mathscr{J}^{2,N} = (J_i^2)_{i \in N}$ is the partition given by $J_i^2 = [t_{\varsigma(i-1)}, t_{\varsigma(i)}]$;
- $(C_n)_{n \in N}$ is a system of copulas where $C_n = M_2$, if the mass is distributed along the main diagonal of $[s_{i-1}, s_i] \times [t_{\varsigma(i)-1}, t_{\varsigma(i)}]$, $C_i = W_2$, otherwise.

The extension of the definition of shuffles of Min to the d-dimensional case is readily seen to be a special case of the construction we have presented above.   ■

Sparse copulas can be characterised in terms of measure-preserving transformations, in the spirit of equation (3.5.1).

Theorem 3.7.8. *Let* $C = \langle \widetilde{\mathscr{J}}, (C_n)_{n \in N} \rangle$ *be a sparse copula. If every* C_n *can be represented in the form* (3.5.1) *by means of suitable measure-preserving transformations* $f_1^{(n)}, \ldots, f_d^{(n)}$, *then* C *can be represented in the form* (3.5.1) *where* f_i, *for* $i = 1, \ldots, d$, *is given by*

$$f_i(t) = \begin{cases} f_i^{(n)} \left(\dfrac{t - a_n^i}{b_n^i - a_n^i} \right) (b_n^i - a_n^i) + a_n^i, & t \in \,]a_n^i, b_n^i[\,, \\ 0, & \text{otherwise.} \end{cases} \quad (3.7.2)$$

Proof. Let $C = \langle \widetilde{\mathscr{J}}, (C_n)_{n \in N} \rangle$ be a sparse copula. Let C^* be the copula that can be represented in the form (3.5.1) where, for $i = 1, \ldots, d$, f_i is given by (3.7.2). In order to prove the assertion, it is enough to show that μ_C coincides with μ_{C^*} on every box of type $J_n^1 \times J_n^2 \times \cdots \times J_n^d$. Now, let $\mathbf{u}$ be in $J_n^1 \times J_n^2 \times \cdots \times J_n^d$. Then, it follows that:

$$\mu_{C^*}\left((a_n^1, u_1) \times \cdots \times (a_n^d, u_d)\right)$$
$$= \lambda\left(\{t \in \mathbb{I} : f_1(t) \in (a_n^1, u_1), \ldots, f_d(t) \in (a_n^d, u_d)\}\right)$$
$$= (b_n^1 - a_n^1)\, \lambda\left(\left\{t \in \mathbb{I} : f_n^j(t) \in \left(0, \frac{u_j - a_n^j}{\lambda(J_n^1)}\right), (j = 1, \ldots, d)\right\}\right)$$
$$= (b_n - a_n)\, \mu_{C_n}\left(\left(0, \frac{u_1 - a_n^1}{\lambda(J_n^1)}\right) \times \cdots \times \left(0, \frac{u_d - a_n^d}{\lambda(J_n^1)}\right)\right)$$
$$= \mu_C\left((a_n^1, u_1) \times \cdots \times (a_n^d, u_d)\right).$$

Therefore $\mu_{C^*}(A) = \mu_C(A)$ for every Borel subset A of $J_n^1 \times \cdots \times J_n^d$, which is the desired assertion. $\qquad \square$

3.8   Ordinal sums

In this section, we present another construction of copulas that is related to the idea of a sparse copula.

Definition 3.8.1. Let N be a finite or countable subset of the natural numbers $\mathbb{N}$. Let $(]a_k, b_k[)_{k \in N}$ be a family of non-overlapping intervals of $\mathbb{I}$ with $0 \leq a_k < b_k \leq 1$ for each $k \in N$. Let $(C_k)_{k \in N}$ be a family of copulas. Then the *ordinal sum* C of $(C_k)_{k \in N}$ with respect to $(]a_k, b_k[)_{k \in N}$ is defined, for every $\mathbf{u} \in \mathbb{I}^d$, by

$$C(\mathbf{u}) = \sum_{k \in N} (b_k - a_k) C_k\left(\frac{u_1 - a_k}{b_i - a_k}, \ldots, \frac{u_d - a_k}{b_k - a_k}\right)$$
$$+ \lambda\left([0, \min\{u_1, \ldots, u_d\}] \setminus \bigcup_{k \in N}]a_k, b_k[\right), \qquad (3.8.1)$$

where λ is the Lebesgue measure. One writes $C = (\langle]a_k, b_k[, C_k \rangle)_{k \in N}$. $\qquad \diamond$

Notice that, in eq. (3.8.1), each C_k is supposed to be extended on $\mathbb{R}^d$ via standard arguments.

Theorem 3.8.2. *The ordinal sum C of type* (3.8.1) *is a d-copula.*

Proof. First, it is not difficult to prove that C has uniform univariate margins (the proof can be done via similar arguments as in Theorem 3.7.3). Moreover, in order to prove that C is d-increasing, it is enough to notice that C is the sum of two terms that are d-increasing functions. The first term is the weighted sum of d.f.'s F_k that have copula C_k and univariate marginals that are uniform on $]a_k, b_k[$ for every $k \in N$. The second term is the measure-generating function (see, e.g., [Bauer, 1996]) associated with a measure on $\mathbb{I}^d$. Thus, C is d-increasing. $\qquad \square$

Remark 3.8.3. The ordinal sum in the case $d = 2$ has a simpler expression. Let $(]a_k, b_k[)_{k \in N}$ be as in Definition 3.8.1 and C_k be copulas in $\mathscr{C}_2$; then

$$C(u_1, u_2) = \begin{cases} a_k + (b_k - a_k) C_k \left(\frac{u_1 - a_k}{b_k - a_k}, \frac{u_2 - a_k}{b_k - a_k} \right), & (u_1, u_2) \in]a_k, b_k[^2, \\ \min\{u_1, u_2\}, & \text{elsewhere,} \end{cases}$$

is a copula. It is important to stress that for $d \geq 3$ the similar expression

$$C(\mathbf{u}) = \begin{cases} a_k + (b_k - a_k) C_k \left(\frac{u_1 - a_k}{b_k - a_k}, \dots, \frac{u_d - a_k}{b_k - a_k} \right), & \mathbf{u} \in]a_k, b_k[^d, \\ \min\{u_1, \dots, u_d\}, & \text{elsewhere,} \end{cases}$$

need not be a copula. As an example, consider the function $C : \mathbb{I}^3 \to \mathbb{I}$ defined by

$$C(u_1, u_2, u_3) := \begin{cases} \frac{1}{3} \Pi_3(3u_1, 3u_2, 3u_3), & (u_1, u_2, u_3) \in [0, 1/3]^3, \\ \min\{u_1, u_2, u_3\}, & \text{elsewhere.} \end{cases}$$

At the point $(1/3, 1/4, 1/4)$, which lies on the boundary of the box $[0, 1/3]^3$, one has

$$\frac{1}{3} \Pi_3 \left(1, \frac{3}{4}, \frac{3}{4} \right) = \frac{3}{16} \neq \frac{1}{4} = \min \left\{ \frac{1}{3}, \frac{1}{4}, \frac{1}{4} \right\},$$

so that C is not even continuous, and, hence, is not a copula. ■

Example 3.8.4. Let $(]a_k, b_k[)_{k \in N}$ be a family of non-overlapping intervals of $\mathbb{I}$ with $0 \leq a_k < b_k \leq 1$ for each $k \in N$ such that $\bigcup_{k \in N} [a_k, b_k] = \mathbb{I}$. Consider the ordinal sum $C = (\langle]a_k, b_k[, C_k \rangle)_{k \in N}$. For every $i = 1, \dots, d$ and for every $k \in N$, set $J_k^i = [a_k, b_k]$. Moreover, for $i = 1, \dots, d$ set $\mathscr{J}^{i,N} = (J_k^i)_{k \in N}$. Let C' be the sparse copula of type

$$\langle (\mathscr{J}^{i,N})_{i=1,\dots,d}, (C_k)_{k \in N} \rangle.$$

Then by using eq. (3.7.1), it is easy to check that C' coincides with C.

In such a case, thanks to Theorem 3.7.6, C can be easily simulated in view of the following mechanism.

1. Generate a random number $v \in [0, 1]$.
2. If $v \in]a_i, b_i[$ for some $i \in N$, then we generate $\mathbf{u}$ from a distribution with uniform margins on $[a_i, b_i]$ and copula C_i.
3. Return $\mathbf{u}$.

Notice that, with probability zero, v can be equal to a_i or b_i for some $i \in N$. Examples generated from previous algorithm are presented in Figure 3.7. ■

Example 3.8.5. Let C' be the ordinal sum of type $\langle]a, b[, C \rangle$ for a suitable $C \in \mathscr{C}_d$ and $0 < a < b < 1$. Then, if one sets

$$\widetilde{F}(t) = \frac{1}{b - a} \max\{\min\{b, t\} - a, 0\},$$

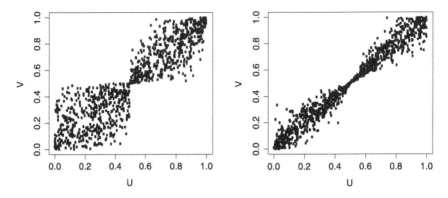

Figure 3.7 *Bivariate sample clouds of 1000 points from the ordinal sum* $(\langle 0, 0.5, C_1\rangle, \langle 0.5, 1, C_2\rangle)$, *where (left) C_1 and C_2 are Gaussian 2-copulas with parameters equal to 0.5 and 0.75, respectively; (right) C_1 is the Gumbel copula and C_2 is the Clayton copula, both with Kendall's τ equal to 0.75.*

for every $t \in \mathbb{I}$,

$$C(\mathbf{u}) = (b - a) C\left(\widetilde{F}(u_1), \dots, \widetilde{F}(u_d)\right) + \mu_{M_d}\left([\mathbf{0}, \mathbf{u}] \cap B^c\right),$$

where $B =]a, b[^d$. This copula concentrates the probability mass on the segments joining $\mathbf{0}$ and $\mathbf{a}$, and $\mathbf{b}$ and $\mathbf{1}$. ∎

Example 3.8.6. Consider an ordinal sum C of type (3.8.1), where $0 < \lambda(\cup_{k \in N}]a_k, b_k[) < 1$. Suppose that $A = \mathbb{I} \setminus (]a_k, b_k[)_{k \in N}$ is a Smith-Volterra-Cantor set (or generalised Cantor set), i.e., a closed set in $\mathbb{I}$ that is nowhere dense and has positive measure [Royden, 1988]. Then, C is an ordinal sum, but it is not a sparse copula since A cannot be expressed as a countable union of intervals. ∎

Example 3.8.7. Let C be an ordinal sum of type

$$C = \left(\left\langle \frac{1}{2^k}, \frac{1}{2^k - 1}, C_k \right\rangle\right)_{k \in \mathbb{N}},$$

where $C_k = W_2$, if k is odd, while $C_k = \Pi_2$ if k is even. Then C does not admit a lower tail dependence coefficient, since $C(t, t)/t$ does not have a limit as t goes to 0. Analogously, the survival copula $\widehat{C}$ associated to C does not admit an upper tail dependence coefficient. ∎

It can be shown that an ordinal sum C of type (3.8.1) can be rewritten as

$$C(\mathbf{u}) := \begin{cases} a_k + (b_k - a_k) C_k\left(\frac{\min\{u_1, b_k\} - a_k}{b_k - a_k}, \dots, \frac{\min\{u_d, b_k\} - a_k}{b_k - a_k}\right), \\ \qquad \text{if } \min\{u_1, u_2, \dots, u_d\} \in \cup_{k \in N}]a_k, b_k[, \\ \min\{u_1, u_2, \dots, u_d\}, \qquad \text{elsewhere.} \end{cases} \tag{3.8.2}$$

Historical remark 3.8.8. The concept of ordinal sum was introduced in an algebraic framework, first by Birkhoff [1940] in dealing with posets, and then by Climescu [1946] and Clifford [1954] in the theory of semigroups. It has been later applied to copulas by Schweizer and Sklar [1983]. The presentation of ordinal sums of this section follows essentially the works by Jaworski [2009] and Mesiar and Sempi [2010].

Other methods have been presented in the literature in order to extend ordinal sum-type constructions, mainly based on patchwork techniques. See, for instance, [De Baets and De Meyer, 2007; Siburg and Stoimenov, 2008a; Durante et al., 2009a, 2013a; González-Barrios and Hernández-Cedillo, 2013]. ∎

3.9 The Kendall distribution function

It is well known that if X is a random variable on the probability space $(\Omega, \mathscr{F}, \mathbb{P})$ and if its d.f. F is continuous, then the random variable $F(X)$ is uniformly distributed on $\mathbb{I}$ (see Theorem 1.2.6). This is called the *probability integral transform*. In this section, following the work of Genest and Rivest [2001] and of Nelsen et al. [2001, 2003], we study the extension of the probability integral transform to the multivariate case. To this end the Kendall d.f. will be needed. As will be seen in the following, it allows each measure μ_C induced by a copula C to be associated with a probability measure on $\mathbb{I}$.

We start with a definition.

Definition 3.9.1. Let H_1 and H_2 be d-dimensional d.f.'s belonging to the same Fréchet class $\Gamma(F_1, \ldots, F_d)$, where each F_i is a continuous (one-dimensional) distribution function. Let $(\Omega, \mathscr{F}, \mathbb{P})$ be a probability space and, on this, let $X_1, \ldots, X_d$ be random variables with joint d.f. given by H_2. Denote by $\langle H_1 \mid H_2 \rangle (X_1, \ldots, X_d)$ the random variable $H_1(X_1, \ldots, X_d)$,

$$\langle H_1 \mid H_2 \rangle (X_1, \ldots, X_d) := H_1(X_1, \ldots, X_d).$$

The H_2 *distribution function of* H_1 is defined, for $t \in \mathbb{I}$, by

$$
\begin{aligned}
(H_1 \mid H_2)(t) &:= \mathbb{P}\left(\langle H_1 \mid H_2 \rangle (X_1, \ldots, X_d) \le t\right) \\
&= \mu_{H_2}\left(\{\mathbf{x} \in \mathbb{R}^d : H_1(\mathbf{x}) \le t\}\right),
\end{aligned}
\tag{3.9.1}
$$

where μ_{H_2} is the measure induced on $\mathscr{B}(\mathbb{R}^d)$ by H_2. ◇

In other words, $(H_1 \mid H_2)$ is a distribution function on $\mathbb{I}$ such that, to every $t \in \mathbb{I}$, $(H_1 \mid H_2)(t)$ is equal to the H_2-measure of the set of all points $\mathbf{x}$ in $\mathbb{R}^d$ lying on the level curves $H_1(\mathbf{x}) = s$ for some $s \in [0, t]$.

Obviously, Definition 3.9.1 applies to copulas, since these are special d.f.'s. More interestingly, $(H_1 \mid H_2)$ depends on the copulas of H_1 and H_2, and not on the respective marginals, as is shown below.

Theorem 3.9.2. *Let H_1, H_2, F_1, ..., F_d, X_1, ..., X_d be as in Definition 3.9.1; if C_1 and C_2 are the copulas of H_1 and H_2, then $(H_1 \mid H_2) = (C_1 \mid C_2)$.*

Proof. For $t \in \mathbb{I}$ one has

$$
\begin{aligned}
(H_1 \mid H_2)(t) &= \mu_{H_2}\left(\left\{\mathbf{x} \in \overline{\mathbb{R}}^d : H_1(\mathbf{x}) \le t\right\}\right) \\
&= \mu_{H_2}\left(\left\{\mathbf{x} \in \overline{\mathbb{R}}^d : C_1(F_1(x_1), \dots, F_d(x_d)) \le t\right\}\right) \\
&= \mu_{C_2}\left(\left\{\mathbf{u} \in \mathbb{I}^d : C_1(\mathbf{u}) \le t\right\}\right) = (C_1 \mid C_2)(t) \,,
\end{aligned}
$$

by recourse to the substitution $u_i = F_i(x_i)$ for every $i \in \{1, \dots, d\}$. $\qquad\square$

The equality provided by Theorem 3.9.2 allows restricting the consideration to the distribution functions of copulas.

Example 3.9.3. From (3.9.1) one readily computes, for $t \in \mathbb{I}$,

$$
(M_2 \mid M_2)(t) = t \,, \quad (\Pi_2 \mid M_2)(t) = \sqrt{t} \,, \quad (W_2 \mid M_2)(t) = \frac{1+t}{2} \,,
$$

$$
(M_2 \mid \Pi_2)(t) = 1 - (1-t)^2 = 2t - t^2 \,, \quad (\Pi_2 \mid \Pi_2)(t) = t - t \ln t \,,
$$

$$
(W_2 \mid \Pi_2)(t) = t + \frac{1}{2}\left(1 - t^2\right) \,, \quad (W_2 \mid W_2)(t) = 1 \,,
$$

$$
(M_2 \mid W_2)(t) = \min\{2t, 1\} \,, \quad (\Pi_2 \mid W_2)(t) = 1 - \sqrt{\max\{0, 1 - 4t\}} \,.
$$

Notice that $(M_2 \mid M_2)$ is the uniform law on $\mathbb{I}$; $(M_2 \mid W_2)$ is the uniform law on $(0, 1/2)$. Recall that the beta law of parameters $\alpha > 0$ and $\beta > 0$ has density

$$
f_{\alpha,\beta}(x) = \frac{x^{\alpha-1}(1-x)^{\beta-1}}{B(\alpha, \beta)} \, \mathbf{1}_{]0,1[}(x) \,,
$$

where $B(\alpha, \beta)$ is the beta function, defined for $\alpha > 0$ and $\beta > 0$ by

$$
B(\alpha, \beta) := \int_0^1 x^{\alpha-1}(1-x)^{\beta-1} \, \mathrm{d}x \,.
$$

Its d.f. is given, for $t \in \mathbb{I}$, by

$$
F_{\alpha,\beta}(t) = \frac{1}{B(\alpha, \beta)} \int_0^t x^{\alpha-1}(1-x)^{\beta-1} \, \mathrm{d}x \,.
$$

Thus one has $(M_2 \mid \Pi_2) = F_{1,2}$ and $(\Pi_2 \mid M_2) = F_{1/2,1}$, where the equality is limited to $\mathbb{I}$. $\qquad\blacksquare$

Example 3.9.4. It is easily seen that one has, for every $t \in \mathbb{I}$ and for every copula $C \in \mathscr{C}_2$,

$$
(M_2 \mid C)(t) = 2t - C(t, t) \,.
$$

In fact, the set $\{(u,v) \in \mathbb{I}^2 : M_2(u,v) \le t\}$ is the whole unit square $\mathbb{I}^2$ deprived of the square $[t,1] \times [t,1]$. The measure μ_C of this region is the sum of the C-volumes of the two disjoint rectangles $[0,t] \times \mathbb{I}$ and $[t,1] \times [0,t]$. Therefore

$$(M_2 \mid C)(t) = C(t,1) - C(t,0) - C(0,1)$$
$$+ C(0,0) + C(1,t) - C(1,0) - C(t,t) + C(t,0) = 2t - C(t,t).$$

∎

A relevant instance of the distribution function of d.f.'s discussed above is given by the Kendall d.f. that we now introduce.

Definition 3.9.5. Let $(\Omega, \mathscr{F}, \mathbb{P})$ be a probability space and, on this, let $\mathbf{X}$ be a r.v. with joint d.f. given by H and with univariate marginals equal to $F_1, \ldots, F_d$, respectively. Then the *Kendall distribution function of* $\mathbf{X}$ is the d.f. of the random variable $H(\mathbf{X})$, namely, for $t \in \mathbb{R}$,

$$t \mapsto \kappa_H(t) := \mathbb{P}\left(H(X_1, \ldots, X_d) \le t\right) = \mu_H\left(\{\mathbf{x} \in \mathbb{R}^d : H(\mathbf{x}) \le t\}\right). \quad (3.9.2)$$

◇

As a consequence of Theorem 3.9.2, it follows that

$$\kappa_C(t) := \mu_C\left(\{\mathbf{u} \in \mathbb{I}^d : C(\mathbf{u}) \le t\}\right). \quad (3.9.3)$$

Example 3.9.6. It follows from (3.9.3) and from Example 3.9.3 that, for $t \in \mathbb{I}$, $\kappa_{M_2}(t) = t$, $\kappa_{\Pi_2} = t - t \ln t$ and $\kappa_{W_2}(t) = 1$. Thus, the Kendall d.f. need not be uniform on $(0,1)$. ∎

Let C be a bivariate copula, $C \in \mathscr{C}_2$; because of the Hoeffding–Fréchet bounds, $W_2 \le C \le M_2$, one has, for every $t \in \mathbb{I}$,

$$\{(u,v) \in \mathbb{I}^2 : M_2(u,v) \le t\} \subseteq \{(u,v) \in \mathbb{I}^2 : C(u,v) \le t\}$$
$$\subseteq \{(u,v) \in \mathbb{I}^2 : W_2(u,v) \le t\},$$

which, in its turn, implies

$$t = \kappa_{M_2}(t) \le \kappa_C(t) \le \kappa_{W_2}(t) = 1.$$

This yields part (a) of the following theorem.

Theorem 3.9.7. *For every copula* $C \in \mathscr{C}_2$, κ_C *is a d.f. concentrated on* $\mathbb{I}$ *such that, for every* $t \in \mathbb{I}$,

(a) $t \le \kappa_C(t) \le 1$;

(b) $\ell^- \kappa_C(0) = 0$.

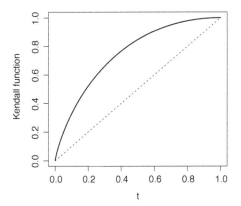

Figure 3.8: *Visualisation of the Kendall distribution function associated with Π_2.*

Moreover the bounds of (a) *are attained, since* $\kappa_{M_2}(t) = t$ *and* $\kappa_{W_2}(t) = 1$ *for every* $t \in \mathbb{I}$.

Proof. Since $C(0,0) = 0$, it follows that $\ell^- \kappa_C(0) = 0$. The remaining assertions were established in Example 3.9.3 and recalled in Example 3.9.6. $\qquad\square$

An example of a Kendall d.f. is depicted in Figure 3.8. Notice that the Kendall d.f. does not uniquely determine the related copula, as the following example shows.

Example 3.9.8. Consider the copulas

$$C_1(u,v) = \min\{u, \max\{v/2, u + v - 1\}\}$$

and

$$C_2(u,v) = \max\{0, u + v - 1, \min\{u, v - 1/2\}, \min\{u - 1/2, v\}\},$$

respectively. After some calculation one obtains, for $t \in \mathbb{I}$,

$$\kappa_{C_1}(t) = \kappa_{C_2}(t) = \min\{2t, 1\},$$

so that C_1 and C_2 have the same Kendall distribution function. ∎

Properties (a) and (b) of Theorem 3.9.7 characterise Kendall d.f.'s. One needs first a preliminary result.

Lemma 3.9.9. *Let δ be a diagonal and let $C_\delta^{\mathbf{Ber}}$ be the Bertino copula of (2.6.5) associated with it. Then, for every $t \in \mathbb{I}$, $\kappa_{C_\delta^{\mathbf{Ber}}}(t) = 2\,\delta^{(-1)}(t) - t$.*

Proof. For $t \in \mathbb{I}$, define the sets

$$S_B := \left\{ (u,v) \in \mathbb{I}^2 : C_\delta^{\mathbf{Ber}}(u,v) \le t \right\},$$

$$S_{M_2} := \left\{ (u,v) \in \mathbb{I}^2 : \min\{u,v\} \le \delta^{(-1)}(t) \right\}.$$

It follows from Example 3.9.4 that

$$\mu_{C_\delta^{\mathbf{Ber}}}(S_{M_2}) = (M_2 \mid C_\delta^{\mathbf{Ber}})\left(\delta^{(-1)}(t)\right) = 2\,\delta^{(-1)}(t) - t.$$

Thus it only remains to prove that $\mu_{C_\delta^{\mathbf{Ber}}}(S_B) = \mu_{C_\delta^{\mathbf{Ber}}}(S_{M_2})$, which will be achieved by showing

(a) $S_B \subseteq S_{M_2}$;

(b) $\mu_{C_\delta^{\mathbf{Ber}}}(S_B \setminus S_{M_2}) = 0$.

If $u \le v$, then $C_\delta^{\mathbf{Ber}}(u,v) \le t$ implies

$$t \ge C_\delta^{\mathbf{Ber}}(u,v) = u - \min_{s \in [u,v]}(s - \delta(s)) \ge u - u + \delta(u) = \delta(u).$$

Since an analogous inequality holds when $u > v$, one has that $C_\delta^{\mathbf{Ber}}(u,v) \le t$ implies $\delta(\min\{u,v\}) \le t$, and, hence, $\min\{u,v\} \le \delta^{(-1)}(t)$, which establishes (a).

Now, assume, first, $u \le v$. The set

$$S := \,]0,1[^2 \bigcap \left\{ (u,v) \in \mathbb{I}^2 : C_\delta^{\mathbf{Ber}}(u,v) < t \right\}$$

is decreasing, in the sense that, for every pair of points (x,y) and (u,v) in S, the inequality $x < u$ implies $y \ge v$. As a consequence, in order to show that $\mu_{C_\delta^{\mathbf{Ber}}}(S_B \setminus S_{M_2}) = 0$, it suffices to show

$$\mu_{C_\delta^{\mathbf{Ber}}}\left(\left[u, \delta^{(-1)}(t)\right] \times [v,1]\right) = 0$$

at all the points (u,v) with

$$u \in \left[t, \delta^{(-1)}(t)\right] \qquad \text{and} \qquad v = \min\left\{ s : C_\delta^{\mathbf{Ber}}(u,s) = t \right\}.$$

If u and v satisfy the above restrictions, then one has

$$\left[u, \delta^{(-1)}(t)\right] \times [v,1] \subseteq [u,v] \times [v,1].$$

We claim that $\mu_{C_\delta^{\mathbf{Ber}}}([u,v] \times [v,1]) = 0$.

$$\mu_{C_\delta^{\mathbf{Ber}}}([u,v] \times [v,1]) = C_\delta^{\mathbf{Ber}}(v,1) - C_\delta^{\mathbf{Ber}}(v,v) + C_\delta^{\mathbf{Ber}}(u,v) - C_\delta^{\mathbf{Ber}}(u,1)$$

$$= v - \delta(v) + t - u.$$

It follows from the definition of v that $C_\delta^{\mathrm{Ber}}(u, h) < t$ for every $h \in [u, v[$, so that

$$\min\{s - \delta(s) : s \in [u, h]\} > u - t.$$

But then $h - \delta(h) > u - t$ for every $h \in [u, v[$; and, since $C_\delta^{\mathrm{Ber}}(u, v) = t$, one has $\min\{s - \delta(s) : s \in [u, v]\} = u - t$. This implies $v - \delta(v) = u - t$, and, as a consequence, $\mu_{C_\delta^{\mathrm{Ber}}}([u, v] \times [v, 1]) = 0$, which establishes (b) when $u \leq v$. A similar argument holds when $u > v$. □

We are in the position of proving

Theorem 3.9.10. *For every d.f. F that satisfies properties* (a) *and* (b) *of Theorem 3.9.7 there exists a copula $C \in \mathscr{C}_2$ such that F coincides with the Kendall d.f. associated with C, namely $F = \kappa_C$.*

Proof. Define two functions on $\mathbb{I}$ via

$$\alpha(t) := \frac{t + F(t)}{2} \qquad \text{and} \qquad \delta(t) := \sup\{s \in \mathbb{I} : \alpha(s) \leq t\}.$$

It follows immediately from the definition that $\alpha \leq 1$, that α is increasing and that $\alpha(\delta(t)) \geq t$. Moreover, it is obvious that $\delta(0) = 0$, $\delta(1) = 1$ and that $\delta^{(-1)}(t) = \alpha(t)$ and $\delta(t) \leq t$ for every $t \in \mathbb{I}$; it is also obvious that δ is increasing. Only the Lipschitz property, $\delta(t_2) - \delta(t_1) \leq 2(t_2 - t_1)$ when $t_1 < t_2$, has to be proved. Since $\delta(t_1) \leq \delta(t_2)$, there is nothing to prove when $\delta(t_1) = \delta(t_2)$. Assume, next, $\delta(t_1) < \delta(t_2)$. For every $s \in [\delta(t_1), \delta(t_2)]$ one has $\alpha(s) \leq t_2$. Then

$$s - \delta(t_1) \leq s - \delta(t_1) + F(s) - F(\delta(t_1)) = 2\{\alpha(s) - \alpha(\delta(t_1))\} \leq 2(t_2 - t_1),$$

which establishes the Lipschitz property for δ. Then consider the Bertino copula C_δ^{Ber} associated with δ. Lemma 3.9.9 yields, for $t \in \mathbb{I}$,

$$\kappa_{C_\delta^{\mathrm{Ber}}}(t) = 2\,\delta^{(-1)}(t) - t = 2\,\alpha(t) - t = F(t),$$

which concludes the proof. □

Remark 3.9.11. The Kendall's tau related to $C \in \mathscr{C}_2$ may be expressed in a different way by using the Kendall distribution of C. In fact, if U and V are uniformly distributed on $(0, 1)$, the integral in equation (2.4.4) is the expectation of the probability transform $C(U, V)$, so that

$$\tau(C) = 4\,\mathbb{E}[C(U, V)] - 1.$$

But it is known that the expectation of a positive random variable may be expressed as an integral of its d.f., in this case of κ_C; thus,

$$\mathbb{E}(C(U, V)) = \int_0^1 t\,\mathrm{d}\kappa_C(t) = \int_0^1 (1 - \kappa_C(t))\,\mathrm{d}t,$$

whence

$$\tau(C) = 3 - \int_0^1 \kappa_C(t) \, \mathrm{d}t \,. \tag{3.9.4}$$

■

Further readings 3.9.12. The content of this section is drawn from the works by Nelsen et al. [2001], Nelsen et al. [2003], Nelsen et al. [2009] and Genest and Rivest [2001].

The Kendall d.f. was introduced by Genest and Rivest [1993] and by Capéraà et al. [1997] who called it a "decomposition of Kendall's tau"; it was given the name that it now bears, and which we have adopted, by Nelsen et al. [2003]. In Nelsen et al. [2009], it is presented as a one-to-one correspondence between Kendall d.f.'s and associative copulas.

Kendall d.f.'s have been used in various problems in risk quantification; see, e.g., [Nappo and Spizzichino, 2009; Salvadori et al., 2011; Di Bernardino and Prieur, 2014]. ■

Chapter 4

Copulas and approximation

In this chapter, we study the approximation of copulas by means of different sub-classes and by using different kinds of convergence. Pros and cons of each method will be discussed.

4.1 Uniform approximations of copulas

Here we present three different methods to approximate (in the L^∞-norm) a copula. Specifically, given a copula $C \in \mathscr{C}_d$, possible ways are determined to find a sequence $(C_n)_{n \in \mathbb{N}} \subset \mathscr{C}_d$ such that, as n goes to ∞, C_n tends to C uniformly. We recall that, under this kind of convergence, the space of copulas $\mathscr{C}_d$ is complete and compact (Theorem 1.7.7).

4.1.1 Checkerboard copulas

In this subsection we shall introduce the checkerboard approximation of copulas. To this end we need some preliminary notations. Given a fixed natural number $d \geq 2$, let $n_1, \ldots, n_d$ be d natural numbers, and, for every $k \in \{1, \ldots, d\}$, let $a_0^k, a_1^k, \ldots, a_{n_k}^k$ be numbers in $\mathbb{I}$ such that

$$0 = a_0^k < a_1^k < \cdots < a_{n_k}^k = 1.$$

Consider the mesh

$$D :- \times_{k=1}^d \{a_0^k, a_1^k, \ldots, a_{n_k}^k\} = \times_{k=1}^d I_k,$$

which divides $\mathbb{I}^d$ into $\prod_{i=1}^d n_i$ d-boxes

$$B_{(i_1,i_2,\ldots,i_d)} := \left[a_{i_1}^1, a_{i_1+1}^1\right] \times \left[a_{i_2}^2, a_{i_2+1}^2\right] \cdots \times \left[a_{i_d}^d, a_{i_d+1}^d\right], \qquad (4.1.1)$$

where, for every $k \in \{1, 2, \ldots, d\}$, $i_k \in \{0, \ldots, n_k - 1\}$. The norm of the mesh D is defined by $\|D\| = \max_{k,i_k} \left(a_{i_k+1}^k - a_{i_k}^k\right)$. A mesh D' is a *refinement* of D when $D \subseteq D'$.

Moreover, for every $k \in \{1, 2, \ldots, d\}$ and every $i_k \in \{0, 1, \ldots, n_k - 1\}$, consider the function $f_{i_k}^k : \mathbb{I} \to \mathbb{I}$ defined by

$$
f_{i_k}^k(x) := \begin{cases} 0, & x < a_{i_k}^k, \\[2mm] \dfrac{x - a_{i_k}^k}{a_{i_k+1}^k - a_{i_k}^k}, & a_{i_k}^k \le x \le a_{i_k+1}^k, \\[2mm] 1, & x > a_{i_k+1}^k. \end{cases} \tag{4.1.2}
$$

For the multi-index

$$
\mathbf{i} = (i_1, \dots, i_d) \in \mathscr{I} = \{0, 1, \dots, n_1 - 1\} \times \cdots \times \{0, 1, \dots, n_d - 1\},
$$

we define $f_{\mathbf{i}} : \mathbb{I}^d \to \mathbb{I}^d$ via

$$
f_{\mathbf{i}}(\mathbf{u}) = (f_{i_1}^1(u_1), \dots, f_{i_d}^d(u_d)).
$$

Definition 4.1.1. Let S be a subcopula with domain given by the closed set D. For each $\mathbf{i} \in \mathscr{I}$, consider a measure $\mu_{\mathbf{i}}$ on $(\mathbb{I}^d, \mathscr{B}(\mathbb{I}^d))$. Define the measure μ on $\mathscr{B}(\mathbb{I}^d)$ such that, for every Borel set $B \in \mathscr{B}(\mathbb{I}^d)$, one has

$$
\mu(B) = \sum_{\mathbf{i} \in \mathscr{I}} \beta_{\mathbf{i}} \, \mu_{\mathbf{i}}(f_{\mathbf{i}}(B \cap B_{\mathbf{i}})), \tag{4.1.3}
$$

where $\beta_{\mathbf{i}}$ is equal to the S-volume of $B_{\mathbf{i}}$, i.e. $\beta_{\mathbf{i}} = V_S(B_{\mathbf{i}})$.

Notice that the set-function μ is a *bona fide* measure because each $\mu_{\mathbf{i}}$ is a measure. Moreover, the following result holds.

Theorem 4.1.2. *Under the notations of Definition 4.1.1, suppose that, for each $\mathbf{i} \in \mathscr{I}$, $\mu_{\mathbf{i}}$ is a probability measure such that $\mu_{\mathbf{i}}(]0, 1[^d) = 1$. Then μ defined by (4.1.3) is a probability measure. Moreover, if each $\mu_{\mathbf{i}}$ is d-fold stochastic, then so is μ.*

Proof. Obviously, μ is a measure and, since each $\mu_{\mathbf{i}}$ has no mass on the border of $\mathbb{I}^d$, it follows easily that the total mass of μ is 1.

Moreover, consider the case when each $\mu_{\mathbf{i}}$ is d-fold stochastic. Let $u_1 \in \mathbb{I}$ such that $a_{i_1}^1 \le u_1 \le a_{i_1+1}^1$. Let

$$
B = [0, u_1] \times \mathbb{I}^{d-1} = \left([0, a_{i_1}^1] \times \mathbb{I}^{d-1} \right) \bigcup \left(\cup_{\mathbf{j} \in \mathscr{I}'} (B_{\mathbf{j}} \cap B) \right),
$$

where the set $\mathscr{I}'$ is formed by all indices $\mathbf{j} = (j_1, \dots, j_d)$ with $j_1 = i_1$. Thus

$$
\mu\left([0, u_1] \times \mathbb{I}^{d-1}\right) = S(a_{i_1}^1, 1, \dots, 1) + \sum_{\mathbf{j} \in \mathscr{I}'} \beta_{\mathbf{j}} \frac{u_1 - a_{i_1}^1}{a_{i_1+1}^1 - a_{i_1}^1}
$$

$$
= a_{i_1}^1 + \frac{u_1 - a_{i_1}^1}{a_{i_1+1}^1 - a_{i_1}^1} \left(a_{i_1+1}^1 - a_{i_1}^1\right) = u_1,
$$

because, for every $\mathbf{j} \in \mathscr{I}'$,

$$\mu_{\mathbf{j}}\left(f_{\mathbf{j}}(B \cap B_{\mathbf{j}})\right) = \mu_{\mathbf{j}}\left(\left[0, \frac{u_1 - a_{i_1}^1}{a_{i_1+1}^1 - a_{i_1}^1}\right] \times \mathbb{I}^{d-1}\right) = \frac{u_1 - a_{i_1}^1}{a_{i_1+1}^1 - a_{i_1}^1}.$$

The assertion follows by repeating the above procedure for each coordinate. □

From now on, we devote our attention to measures of the type (4.1.3) that are generated by d-fold stochastic measures $(\mu_{\mathbf{i}})_{\mathbf{i} \in \mathscr{I}}$. Moreover, we denote by C and $C_{\mathbf{i}}$ the copulas defined from the measures μ and $\mu_{\mathbf{i}}$, respectively. When the explicit expression of C is required, the following result can be helpful.

Theorem 4.1.3. *For every* $\mathbf{u} \in B_{\mathbf{i}}$ *with* $\mathbf{i} \in \mathscr{I}$, *one has*

$$C(\mathbf{u}) = S(a_{i_1}^1, \ldots, a_{i_d}^d) + \sum_{\mathbf{j} \in \mathscr{J}} \beta_{\mathbf{j}} C_{\mathbf{j}}(f_{\mathbf{j}}(\mathbf{u})), \tag{4.1.4}$$

where the set $\mathscr{J}$ *is formed by all indices* $\mathbf{j} = (j_1, \ldots, j_d)$ *with* $j_k \leq i_k$ *for every* $k = 1, 2, \ldots, d$, *and* $j_k = i_k$ *for at least one index* k.

Proof. Let $\mathbf{u} \in B_{\mathbf{i}}$. Then $[\mathbf{0}, \mathbf{u}]$ can be decomposed into the following union of boxes whose interiors are disjoint:

$$[\mathbf{0}, \mathbf{u}] = \left([0, a_{i_1}^1] \times \cdots \times [0, a_{i_d}^d]\right) \bigcup \left(\cup_{\mathbf{j} \in \mathscr{J}} (B_{\mathbf{j}} \cap [\mathbf{0}, \mathbf{u}])\right). \tag{4.1.5}$$

Thus, we have

$$C(\mathbf{u}) = \mu\left([0, a_{i_1}^1] \times \cdots \times [0, a_{i_d}^d]\right) + \sum_{\mathbf{j} \in \mathscr{J}} \beta_{\mathbf{j}} \mu_{\mathbf{j}}(f_{\mathbf{j}}([a_{j_1}^1, u_1] \times \cdots \times [a_{j_d}^d, u_d]))$$

$$= S(a_{i_1}^1, \ldots, a_{i_d}^d) + \sum_{\mathbf{j} \in \mathscr{J}} \beta_{\mathbf{j}} C_{\mathbf{j}}(f_{\mathbf{j}}(\mathbf{u})),$$

which is the desired assertion. □

It easily follows that $C = S$ on the mesh D. Moreover, let $\mathbf{u} = (u_1, a_{i_2}^2, \ldots, a_{i_d}^d)$ be in $B_{\mathbf{i}}$. Let $[\mathbf{0}, \mathbf{u}]$ be expressed as in (4.1.5). If $\mathbf{j} \in \mathscr{I}_{\mathbf{i}}^1$, then

$$\beta_{\mathbf{j}} C_{\mathbf{j}}(f_{j_1}^1(u_1), f_{j_2}^2(a_{i_2}^2), \ldots, f_{j_d}^d(a_{i_d}^d)) = \beta_{\mathbf{j}} \frac{u_1 - a_{i_1}^1}{a_{i_1+1}^1 - a_{i_1}^1}.$$

Hence

$$C(\mathbf{u}) - C(a_{i_1}^1, a_{i_2}^2, \ldots, a_{i_d}^d) = \mu([\mathbf{0}, \mathbf{u}]) - S(a_{i_1}^1, a_{i_2}^2, \ldots, a_{i_d}^d)$$

$$= \frac{u_1 - a_{i_1}^1}{a_{i_1+1}^1 - a_{i_1}^1} \sum_{\mathbf{j} \in \mathscr{I}_{\mathbf{i}}^1} \beta_{\mathbf{j}}$$

$$= \frac{u_1 - a_{i_1}^1}{a_{i_1+1}^1 - a_{i_1}^1} \mu\left([a_{i_1}^1, u_{i_1+1}^1] \times [0, a_{i_2}^2] \times \cdots \times [0, a_{i_d}^d]\right)$$

$$= \frac{u_1 - a_{i_1}^1}{a_{i_1+1}^1 - a_{i_1}^1} \left(S(a_{i_1+1}^1, a_{i_2}^2, \ldots, a_{i_d}^d) - S(a_{i_1}^1, a_{i_2}^2, \ldots, a_{i_d}^d)\right).$$

By repeating the above procedure for each coordinate, it follows that the extension of the subcopula S to the copula C provided by (4.1.4) is linear along segments joining two points in the mesh D such that $(d-1)$ of their components are equal.

For this reason, the copula C defined by (4.1.4) is called the *L-extension of S by means of* $(C_i)_{i \in \mathscr{I}}$, where the prefix "L" indicates that the extension is linear on specific segments of the copula domain. We write $C = \langle D, C_i \rangle_{i \in \mathscr{I}}^S$ or, with the measure notation, $\mu = \langle D, \mu_i \rangle_{i \in \mathscr{I}}^S$.

Now, let $C = \langle D, C_i \rangle_{i \in \mathscr{I}}^S$. Obviously, if every C_i is absolutely continuous (respectively, singular), then C is absolutely continuous (respectively, singular).

The construction just introduced may also be used as a tool to approximate a given copula C with respect to a mesh D and a sequence of copulas $(C_i)_{i \in \mathscr{I}}$. In fact, if C is a copula, one may consider the subcopula S_C that coincides with the restriction of C to D. Starting from it, one may obtain the copula C_D that is the L-extension of S_C by means of $(C_i)_{i \in \mathscr{I}}$. Then, C_D is called the $(D, (C_i)_{i \in \mathscr{I}})$-*approximation* of C. In particular, the following approximation has proved to be useful in the literature.

Definition 4.1.4. Let $n \in \mathbb{N}$. Let $C \in \mathscr{C}_d$ be a copula. We call the *n-th checkerboard approximation* of C the L-extension of the subcopula $S : D \to \mathbb{I}$ by means of $(C_i)_{i \in \mathscr{I}} = (\Pi_d)_{i \in \mathscr{I}}$, where $D = \{0, 1/n, \ldots, (n-1)/n, 1\}^d$. $\diamond$

The checkerboard copulas are dense in $(\mathscr{C}_d, d_\infty)$ as shown below.

Theorem 4.1.5. *Given a copula $C \in \mathscr{C}_d$, the sequence (C_n) of its checkerboard approximations converges uniformly to C.*

Proof. For $\mathbf{u} \in \mathbb{I}^d$ there are $j_1, \ldots, j_d$ in $\{1, \ldots, n\}$ such that $\mathbf{u}$ belongs to the box B_j defined as in (4.1.1). Then

$$C(\mathbf{u}) - C_n(\mathbf{u}) \leq C\left(\frac{j_1}{n}, \ldots, \frac{j_d}{n}\right) - C_n\left(\frac{j_1-1}{n}, \ldots, \frac{j_d-1}{n}\right)$$
$$= C\left(\frac{j_1}{n}, \ldots, \frac{j_d}{n}\right) - C\left(\frac{j_1-1}{n}, \ldots, \frac{j_d-1}{n}\right) \leq \frac{d}{n}.$$

Similarly

$$C_n(\mathbf{u}) - C(\mathbf{u}) \leq C_n\left(\frac{j_1}{n}, \ldots, \frac{j_d}{n}\right) - C\left(\frac{j_1-1}{n}, \ldots, \frac{j_d-1}{n}\right)$$
$$= C\left(\frac{j_1}{n}, \ldots, \frac{j_d}{n}\right) - C\left(\frac{j_1-1}{n}, \ldots, \frac{j_d-1}{n}\right) \leq \frac{d}{n}.$$

Hence

$$|C(\mathbf{u}) - C_n(\mathbf{u})| \leq \frac{d}{n},$$

from which the assertion immediately follows. $\square$

In particular, since checkerboard copulas are absolutely continuous, the following result easily follows.

Theorem 4.1.6. *The set of absolutely continuous d-copulas is dense in* $(\mathscr{C}_d, d_\infty)$.

Thanks to previous approximation one is now able to prove the sharpness of the lower Hoeffding–Fréchet bound, a result which was announced in Subsection 1.7.1.

Theorem 4.1.7 (Sharpness of Hoeffding–Fréchet bounds). *For every* $d \geq 3$ *and for every* $\mathbf{u} \in \mathbb{I}^d$ *there exists a d-copula C such that*

$$C(\mathbf{u}) = W_d(\mathbf{u}),$$

so that the lower bound in (1.7.2) *is the best possible even though* W_d *is not a d-copula.*

Proof. Let a point $\mathbf{u} \in \mathbb{I}^d$, different from $\mathbf{0} = (0, \ldots, 0)$ and $\mathbf{1} = (1, \ldots, 1)$, be fixed.

Assume first $W_d(\mathbf{u}) = 0$, so that $0 < u_1 + \cdots + u_d \leq d - 1$ and consider the points $\mathbf{v}$ such that each component v_j is in the set $\{0, 1, t_j\}$, where

$$t_j := \min\left\{\frac{(d-1)\,u_j}{u_1 + \cdots + u_d}, 1\right\} \qquad (j = 1, \ldots, d).$$

On these points define $C'(\mathbf{v}) = W_d(\mathbf{v})$; C' thus defined is obviously a subcopula. Let C be the checkerboard copula that extends it to the whole of $\mathbb{I}^d$. For every $\mathbf{x}$ in the d-box $[\mathbf{0}, \mathbf{t}]$, where $\mathbf{t} := (t_1, \ldots, t_d)$, one has $C(\mathbf{x}) = W_d(\mathbf{x}) = 0$. Since $u_j \leq \min\{1, (d-1)/(u_1 + \cdots + u_d)\}$, the point $\mathbf{u}$ belongs to $[\mathbf{0}, \mathbf{t}]$, so that $C(\mathbf{u}) = W_d(\mathbf{u}) = 0$.

On the other hand, if $W_d(\mathbf{u}) > 0$, so that $d - 1 < u_1 + \cdots + u_d < d$, consider the points $\mathbf{v}$ for which each component v_j is in the set $\{0, 1, s_j\}$, where

$$s_j := 1 - \frac{1 - u_j}{d - (u_1 + \cdots + u_d)} \qquad (j = 1, \ldots, d).$$

On these points define $C'(\mathbf{v}) = W_d(\mathbf{v})$ As above C' is a subcopula that can be extended to a checkerboard copula C. Let $\mathbf{s} := (s_1, \ldots, s_d)$. For every $\mathbf{x} \in [\mathbf{s}, \mathbf{1}]$ one has

$$C(\mathbf{x}) = W_d(\mathbf{x}) = x_1 + \cdots + x_d - d + 1.$$

Since $u_j \leq 1$, and, hence, $u_j\,((d-1) - (u_1 + \cdots + u_d)) \geq (d-1) - (u_1 + \cdots + u_d)$, one has $u_j \geq s_j$ for every index $j = 1, \ldots, d$, so that $\mathbf{u}$ belongs to the box $[\mathbf{s}, \mathbf{1}]$; therefore, $C(\mathbf{u}) = W_d(\mathbf{u})$.

In either case there is a copula C that agrees with W_d at $\mathbf{u}$. $\qquad\square$

Historical remark 4.1.8. To the best of our knowledge, the term "checkerboard copula" appeared for the first time in [Li et al., 1997], and was hence considered by Li et al. [1998]; Kulpa [1999]. It has been employed to prove Sklar's theorem by Carley and Taylor [2003]. Notice that these copulas are also called multilinear copula by Genest et al. [2014]. It should be also stressed that, in different contexts, similar constructions have been obtained by Johnson and Kotz [2007]; Ghosh and Henderson [2009] (see also [Durante et al., 2015b]).   ∎

4.1.2 Bernstein copulas

Since a copula is a continuous function on the compact domain $\mathbb{I}^d$, it is possible to approximate it by Bernstein polynomials. As will be clarified below, such an approximation turns out to be quite convenient, since the approximating functions are themselves copulas.

Definition 4.1.9. Let $n \in \mathbb{N}$. Given a d-copula C, the function defined, for $\mathbf{u} \in \mathbb{I}^d$, by

$$B_n^C(\mathbf{u}) := \sum_{j_1,\ldots,j_d=0}^n C\left(\frac{j_1}{n},\ldots,\frac{j_d}{n}\right) p_{j_1,n}(u_1)\ldots p_{j_d,n}(u_d), \qquad (4.1.6)$$

where

$$p_{j,n}(t) := \binom{n}{j} t^j (1-t)^{n-j}, \qquad (4.1.7)$$

is called its *n-th order Bernstein approximation* of C.   ◇

The probabilistic interpretation of (4.1.6) will make the proof that the function it defines is indeed a copula easy.

Let $F_{\mathrm{Bin}(n,t)}$ be the binomial distribution with parameters (n,t), $n \in \mathbb{N}$ and $t \in \mathbb{I}$, and denote by $F_{\mathrm{Bin}(n,t)}^{(-1)}$ its (left-continuous) quasi-inverse function. Note that, in view of Theorem 1.2.6, if U on $(\Omega, \mathscr{F}, \mathbb{P})$ is uniformly distributed in $\mathbb{I}$, $F_{\mathrm{Bin}(n,t)}^{(-1)}(U)$ follows a binomial distribution whose probability mass function coincides with $p_{j,n}(t)$ of eq. (4.1.7). Moreover, eq. (4.1.6) can be rewritten as

$$B_n^C(\mathbf{u}) = \mathbb{E}\left(C\left(\frac{F_{\mathrm{Bin}(n,u_1)}^{(-1)}(U_1)}{n},\ldots,\frac{F_{\mathrm{Bin}(n,u_d)}^{(-1)}(U_d)}{n}\right)\right), \qquad (4.1.8)$$

for $U_1,\ldots,U_d$ independent r.v.'s that are uniformly distributed on $\mathbb{I}$. Thanks to this interpretation, one can prove the following.

Theorem 4.1.10. *Given $C \in \mathscr{C}_d$, and a natural number $n \geq 2$, the function $B_n^C = B_n : \mathbb{I}^d \to \mathbb{I}$ defined by (4.1.6) is a d-copula.*

Proof. First, B_n satisfies the boundary conditions. In fact, let $u_i = 0$. Then $p_{j_i,n}(0) = 0$ if $j_i \neq 0$, while, if $j_i = 0$, then the j_i-th argument in the coefficient $C(\dots)$ in eq. (4.1.6) equals zero; in either case one has

$$B_n(u_1, \dots, u_{i-1}, 0, u_{i+1}, \dots, u_d) = 0.$$

Moreover, for every $u \in \mathbb{I}$ one has

$$B_n(1, \dots, 1, u, 1, \dots, 1)$$
$$= \sum_{j=0}^{n} p_{j,n}(u) \sum_{\substack{j_1,\dots,j_d=0 \\ j_i \neq j}}^{n} C\left(\frac{j_1}{n}, \dots, \frac{j_{i-1}}{n}, \frac{j}{n}, \frac{j_{i+1}}{n}, \frac{j_d}{n}\right) p_{j_1,n}(1) \dots p_{j_d,n}(1).$$

But $p_{j_i,n}(1) = 0$ for $j_i \neq n$ and $p_{j_i,n}(1) = 1$, for $j_i = n$, so that

$$B_n(1, \dots, 1, u, 1, \dots, 1) = \sum_{j=1}^{n} p_{j,n}(u) C\left(1, \dots, 1, \frac{j}{n}, 1, \dots, 1\right)$$
$$= \frac{1}{n} \sum_{j=0}^{n} j \, p_{j,n}(u) = \frac{nu}{n} = u,$$

since the last sum represents the mean value nu of a binomial distribution of parameter u.

Then, consider the d-box $[\mathbf{a}, \mathbf{b}]$ in $\mathbb{I}^d$. In view of eq. (4.1.8), it follows that

$$V_{B_n}([\mathbf{a}, \mathbf{b}]) = \mathbb{E}\left(V_C([\mathbf{a}', \mathbf{b}'])\right),$$

where, since $F_{Bin(n,u)}(t)$ is decreasing in u for a fixed t,

$$a'_j = \frac{F_{Bin(n,a_j)}^{(-1)}(U_j)}{n}, \quad b'_j = \frac{F_{Bin(n,b_j)}^{(-1)}(U_j)}{n}$$

for $j = 1, \dots, d$. Hence, $V_{B_n}([\mathbf{a}, \mathbf{b}]) \geq 0$ because C is a copula. $\qquad\square$

The announced approximation by Bernstein polynomials, which we state below as a theorem, is now an immediate consequence of the multidimensional Weierstrass Theorem (see, e.g., Theorem 3.1 in Chapter 1 of [DeVore and Lorentz, 1993]).

Theorem 4.1.11. *Given $C \in \mathscr{C}_d$ the sequence $(B_n^C)_{n \geq 2}$ converges uniformly to C.*

Example 4.1.12. Theorem 4.1.11 may be used to prove that the result of certain transformations of copulas is again a copula. For instance, let $\Xi^2(\mathbb{I}^2)$ be the space of all functions in $\mathbb{I}^2$ that admit continuous partial derivatives up to second order. Consider, for $C \in \mathscr{C}_2$, the mapping $\varphi : \mathscr{C}_2 \to \Xi^2(\mathbb{I}^2)$, defined by $\varphi(C) = C'$, where

$$C'(u, v) := C(u, v)(u + v - C(u, v)).$$

Now, it can be easily checked that if C is in $\Xi^2(\mathbb{I}^2)$, then C' is also in $\Xi^2(\mathbb{I}^2)$ and, moreover, is a copula, since

$$
\begin{aligned}
\partial_{12}C'(u,v) = {}& \partial_{12}C(u,v)(u+v-2C(u,v)) \\
& + \partial_1 C(u,v)(1-\partial_2 C(u,v)) + \partial_2 C(u,v)(1-\partial_1 C(u,v))
\end{aligned}
$$

is positive. Now, given a (general) copula C, Theorem 4.1.11 ensures that there exists a sequence of copulas $(C_n)_{n\in\mathbb{N}}$ in $\Xi^2(\mathbb{I}^2)$ that converges uniformly to C as $n\to\infty$. Moreover, φ is continuous (with respect to the L^∞-norm). Thus, $\varphi(C_n)$ converges uniformly to $\varphi(C)$, that is a copula since it is a limit of a sequence of copulas. A probabilistic interpretation of the previous construction has been presented in Example 1.3.8. ∎

Bernstein copulas may also be used in order to prove an integration-by-parts formula that is very useful when an integration with respect to the measure induced by a copula is performed.

Theorem 4.1.13. *Let A and B be 2-copulas, and let the function $\varphi : \mathbb{I} \to \mathbb{R}$ be continuously differentiable. Then*

$$
\int_{\mathbb{I}^2} \varphi \circ A \, dB = \int_0^1 \varphi(t)\, dt - \int_{\mathbb{I}^2} \varphi'(A(u,v))\, \partial_1 A(u,v)\, \partial_2 B(u,v)\, du\, dv \quad (4.1.9)
$$

$$
= \int_0^1 \varphi(t)\, dt - \int_{\mathbb{I}^2} \varphi'(A(u,v))\, \partial_2 A(u,v)\, \partial_1 B(u,v)\, du\, dv. \tag{4.1.10}
$$

Proof. The proof will be given first under the more restrictive assumption that both A and B admit continuous partial derivatives up to second order. Then one has

$$
\int_{\mathbb{I}^2} \varphi \circ A \, dB = \int_{\mathbb{I}^2} \varphi(A(u,v))\, \partial_{12} B(u,v)\, du\, dv.
$$

Integrating by parts with respect to u yields

$$
\begin{aligned}
\int_{\mathbb{I}^2} \varphi \circ A \, dB = {}& \int_0^1 \left[\varphi(A(u,v))\, \partial_2 B(u,v) \right]_{u=0}^{u=1} dv \\
& - \int_0^1 dv \int_0^1 \varphi'(A(u,v))\, \partial_1 A(u,v)\, \partial_2 B(u,v)\, du \\
= {}& \int_0^1 \left(\varphi(A(1,v))\, \partial_2 B(1,v) - \varphi(A(0,v))\, \partial_2 B(0,v) \right) dv \\
& - \int_0^1 dv \int_0^1 \varphi'(A(u,v))\, \partial_1 A(u,v)\, \partial_2 B(u,v)\, du.
\end{aligned}
$$

Now $A(1,v) = v$, $\partial_2 B(1,v) = 1$ and $\partial_2 B(0,v) = 0$, so that

$$
\int_{\mathbb{I}^2} \varphi \circ A \, dB = \int_0^1 \varphi(v)\, dv - \int_{\mathbb{I}^2} \varphi'(A(u,v))\, \partial_1 A(u,v)\, \partial_2 B(u,v)\, du\, dv;
$$

this proves (4.1.9) in this case.

In the general case, let (A_n) and (B_n) be sequences of Bernstein copulas that converge uniformly to A and B, respectively. Since Bernstein copulas admit continuous partial derivatives of second order, one has

$$\int_{\mathbb{I}^2} \varphi \circ A_n \, dB_n$$

$$= \int_0^1 \varphi(t) \, dt - \int_{\mathbb{I}^2} \varphi'(A_n(u,v)) \, \partial_1 A_n(u,v) \, \partial_2 B_n(u,v) \, du \, dv. \quad (4.1.11)$$

Now, as n goes to $+\infty$, the dominated convergence theorem yields that the right-hand side of (4.1.11) converges to

$$\int_0^1 \varphi(t) \, dt - \int_{\mathbb{I}^2} \varphi'(A(u,v)) \, \partial_2 A(u,v) \, \partial_1 B(u,v) \, du \, dv.$$

The left-hand side of (4.1.11) can be instead rewritten as

$$\int_{\mathbb{I}^2} \varphi \circ A_n \, dB_n = \int_0^1 (\varphi \circ A_n - \varphi \circ A) \, dB_n - \int_0^1 \varphi \circ A \, dB_n,$$

which, as n goes to $+\infty$, converges to $\int_{\mathbb{I}^2} \varphi \circ A \, dB$. Thus, the desired assertion follows.

Integrating by parts with respect to v rather than u leads to eq. (4.1.10). $\qquad \square$

Further readings 4.1.14. Nowadays, Bernstein copulas are used in various problems in order to approximate dependence structure due to their flexibility. See, for instance, recent contributions by Sancetta and Satchell [2004]; Mikusiński and Taylor [2010]; Janssen et al. [2012]; Dou et al. [2015]; Yang et al. [2015]. The integral formula in Theorem 4.1.13 has been presented in Li et al. [2003]. $\qquad \blacksquare$

4.1.3 Sparse copulas

Sparse copulas of Section 3.7 provide another kind of approximation (see [Durante and Fernández-Sánchez, 2010]).

Theorem 4.1.15. *For every* $C \in \mathscr{C}_d$ *and for every* $\varepsilon > 0$, *there exists a sparse copula* C' *such that* $\|C - C'\|_\infty < \varepsilon$. *In other words, the set of all sparse copulas is dense in* $\mathscr{C}_d$ *with respect to uniform convergence.*

Proof. Let C be in $\mathscr{C}_d$. For $p \in N$ with $p \geq 2$, divide $\mathbb{I}^d$ into p^d boxes of the type

$$B_m^p = \left[\frac{a_m^{(1)}}{p}, \frac{a_m^{(1)} + 1}{p} \right] \times \cdots \times \left[\frac{a_m^{(d)}}{p}, \frac{a_m^{(d)} + 1}{p} \right],$$

where $a_m^{(1)}$ is in $\{0, 1, \ldots, p-1\}$ for every $i \in \{1, \ldots, d\}$ and m can be written in the form

$$m = a_m^{(1)} + a_m^{(2)} p + \cdots + a_m^{(d)} p^{d-1}, \quad (4.1.12)$$

so that it belongs to $M(p) := \{0, 1, \ldots, p^d - 1\}$. Denote by $M_{k,i}(p)$ the subset of $M(p)$ composed by all the m's that can be written in the form (4.1.12) with $a_m^{(i)} = k$.

Since μ_C is a d-fold stochastic measure, one has, for every $m \in M_{k,i}(p)$,

$$\sum_{m \in M_{k,i}(p)} \mu_C\left(B_m^p\right) = \lambda_d\left(\mathbb{I}^{i-1} \times \left[\frac{k}{p}, \frac{k+1}{p}\right] \times \mathbb{I}^{d-i}\right) = \frac{1}{p}.$$

For $i \in \{1, \ldots, d\}$, choose the i-th axis and divide each interval of the type $[k/p, (k+1)/p]$ into a system of closed intervals $\left(J_m^i\right)_{m \in M_{k,i}(p)}$ such that

- $\lambda\left(J_m^i\right) = \mu_C\left(B_m^p\right)$ for every $m \in M_{k,i}(p)$;
- $\sup\left(J_{m(1)}\right) \leq \inf\left(J_{m(2)}\right)$, whenever $a_{m(1)}^{(i)} < a_{m(2)}^{(i)}$;
- $\sup\left(J_{m(1)}\right) \leq \inf\left(J_{m(2)}\right)$, whenever $a_{m(1)}^{(i)} = a_{m(2)}^{(i)}$ and $m(1) < m(2)$.

According to the above procedure the unit interval $\mathbb{I}$ is partitioned as follows

$$\mathbb{I} = \bigcup_{k=0}^{p-1}\left[\frac{k}{p}, \frac{k+1}{p}\right] = \bigcup_{k=0}^{p-1}\left(\bigcup_{m \in M_{k,i}(p)} J_m^i\right).$$

Set $\mathscr{J}^{i,M(p)} := \left(J_m^i\right)_{m \in M(p)}$; then it is easily proved that $\left(\mathscr{J}^{(i,M(p))}\right)_{i=1}^d$ forms a sparse structure that depends on p.

Let C_p be the sparse copula given by

$$\left\langle \left(\mathscr{J}^{i,M(p)}\right)_{i=1}^d, (C_m = C)_{m \in M(p)}\right\rangle.$$

Obviously, $\mu_{C_p}\left(B_m^p\right) = \mu_C\left(B_m^p\right)$.

Now, let f be a continuous function on $\mathbb{I}^d$. Then, for every $\varepsilon > 0$ and for every $m \in M(p)$, there exists $q \in \mathbb{N}$ such that

$$|f(\mathbf{x}) - f(\mathbf{x}')| < \varepsilon$$

for every $p > q$, and for all $\mathbf{x}$ and $\mathbf{x}'$ in B_m^p. Therefore

$$\sum_{m \in M(p)} \min_{\mathbf{x} \in B_m^p} f(\mathbf{x}) \mu_C\left(B_m^p\right) = \sum_{m \in M(p)} \min_{\mathbf{x} \in B_m^p} f(\mathbf{x}) \mu_{C_p}\left(B_m^p\right)$$

$$\leq \int_{\mathbb{I}^d} f(\mathbf{x}) d\mu_{C_p} \leq \sum_{m \in M(p)} \max_{\mathbf{x} \in B_m^p} f(\mathbf{x}) \mu_{C_p}\left(B_m^p\right)$$

$$= \sum_{m \in M(p)} \max_{\mathbf{x} \in B_m^p} f(\mathbf{x}) \mu_C\left(B_m^p\right) < \sum_{m \in M(p)} \min_{\mathbf{x} \in B_m^p} f(\mathbf{x}) \mu_C\left(B_m^p\right) + \varepsilon.$$

It follows that

$$\left|\int_{\mathbb{I}^d} f(\mathbf{x}) d\mu_{C_p} - \int_{\mathbb{I}^d} f(\mathbf{x}) d\mu_C\right| < \varepsilon,$$

when $p > q$. The arbitrariness of f allows to conclude that (C_p) converges pointwise to C as p tends to $+\infty$. $\qquad\square$

In view of Example 3.7.7 the following corollary is now obvious.

Corollary 4.1.16. *The shuffles of Min are dense in $\mathscr{C}_2$, with respect to the topology of uniform convergence.*

The next result, in its simplicity, will have a perhaps surprising consequence (see, also, Kimeldorf and Sampson [1978] and Vitale [1990]).

Theorem 4.1.17. *Let X and Y be continuous random variables on the same probability space $(\Omega, \mathscr{F}, \mathbb{P})$, let F and G be their marginal d.f.'s and H their joint d.f. Then, for every $\varepsilon > 0$, there exist two random variables X_ε and Y_ε on the same probability space and a piecewise linear function $\varphi : \mathbb{R} \to \mathbb{R}$ such that*

(a) $Y_\varepsilon = \varphi \circ X_\varepsilon$;

(b) $F_\varepsilon := F_{X_\varepsilon} = F$ and $G_\varepsilon := F_{Y_\varepsilon} = G$;

(c) $\|H - H_\varepsilon\|_\infty < \varepsilon$,

where H_ε is the joint d.f. of X_ε and Y_ε, and $\|\cdot\|_\infty$ denotes the L^∞-norm on $\mathbb{R}^2$.

Proof. Let C be the copula of (X, Y) and let $B \in \mathscr{C}_2$ be a shuffle of Min such that, for all u and v in $\mathbb{I}$,
$$|C(u, v) - B(u, v)| < \varepsilon.$$
By virtue of Theorem 2.1.1, the function $K : \mathbb{R}^2 \to \mathbb{R}$ defined via
$$K(x, y) := B(F(x), G(y))$$
is a d.f. in $\mathbb{R}^2$. Endow $\mathbb{R}^2$ with the Stieltjes measure μ_K induced by K, and let X_ε and Y_ε be the orthogonal projections of $\mathbb{R}^2$ onto the coordinate axes. Then K is the joint d.f. of the pair $(X_\varepsilon, Y_\varepsilon)$. By construction, the marginal d.f.'s of X_ε and Y_ε are F and G, respectively. As a consequence of Corollary 2.5.10 and of the construction of shuffles of Min, there exists a piecewise linear function $\varphi : \mathbb{R} \to \mathbb{R}$ such that $Y_\varepsilon = \varphi \circ X_\varepsilon$. Finally, for all $(x, y) \in \mathbb{R}^2$,
$$|H(x, y) - K(x, y)| = |C(F(x), G(y)) - B(F(x), G(y))| < \varepsilon,$$
so that setting $K = H_\varepsilon$ completes the proof. $\square$

As was announced, the last result has a surprising consequence. Let X and Y be independent (and continuous) random variables on the same probability space, let F and G be their marginal d.f.'s and H their joint d.f. Then, according to the previous theorem, it is possible to construct two sequences (X_n) and (Y_n) of random variables such that, for every $n \in \mathbb{N}$, their joint d.f. H_n approximates H to within $1/n$ in the L^∞-norm, but Y_n is almost surely a (piecewise linear) function of X_n.

This fact shows that with uniform convergence it becomes difficult, in a certain sense, to distinguish between types of statistical dependence that seem, intuitively, to be very different. This motivates the search for other types of convergences, a programme that will pursued in the following.

4.2   Application to weak convergence of multivariate d.f.'s

In this section we consider the relationship between the weak convergence of sequences $(F_n^{(j)})_{n \in \mathbb{N}}$, $(j = 1, \ldots, d)$, of univariate d.f.'s, the sequence of the d-dimensional d.f.'s (H_n) of which they are the marginals and the sequence (C_n) of the connecting d-copulas. The results are presented in the framework of the space $\mathcal{D}^d$. For the extension of these results beyond the setting of proper d.f.'s we refer to [Sempi, 2004].

We recall that, given F_n ($n \in \mathbb{N}$) and F_0 d-dimensional d.f.'s in $\mathcal{D}^d$, F_n converges weakly to F_0 (as $n \to +\infty$) if $\lim_{n \to +\infty} F_n(\mathbf{x}) = F_0(\mathbf{x})$ at all continuity points $\mathbf{x}$ of F_0. Equivalent formulations are presented, e.g., in Billingsley [1979].

Theorem 4.2.1. *Let $(F_n^{(j)})_{n \in \mathbb{N}}$ and $(C_n)_{n \in \mathbb{N}}$ be d sequences of univariate d.f.'s and a sequence of d-copulas, respectively, and let a sequence (H_n) of d-dimensional d.f.'s be defined through*

$$H_n := C_n \left(F_n^{(1)}, \ldots, F_n^{(d)} \right).$$

If every sequence $(F_n^{(j)})$ $(j = 1, \ldots, d)$ converges weakly to a d.f. $F_0^{(j)}$ and if the sequence (C_n) of d-copulas converges pointwise to the copula C_0 in $\mathbb{I}^d$, then the sequence (H_n) converges to

$$H_0 := C_0 \left(F_0^{(1)}, \ldots, F_0^{(d)} \right)$$

in the weak topology of $\mathcal{D}^d$.

Proof. Let $F_n^{(j)}$ be continuous at $x_j \in \mathbb{R}$ ($j = 1, \ldots, d$) and put $\mathbf{x} = (x_1, \ldots, x_d)$. Then

$$\lim_{n \to +\infty} F_n^{(j)}(x_j) = F_0^{(j)}(x_j) \qquad (j = 1, \ldots, d).$$

Since C_0 is continuous at every point $\mathbf{t} \in \mathbb{I}^d$, H_0 is continuous at $\mathbf{x}$. The points $\mathbf{x} \in \mathbb{R}^d$ at which H_0 is continuous are dense in $\overline{\mathbb{R}}^d$, since the closure of the Cartesian product of all the sets of continuity points of $F_0^{(j)}$ is dense in $\overline{\mathbb{R}}^d$. Therefore, the

Lipschitz condition (1.2.7) yields

$$|H_0(\mathbf{x}) - H_n(\mathbf{x})|$$

$$= \left| C_0\left(F_0^{(1)}(x_1),\ldots,F_0^{(d)}(x_d)\right) - C_n\left(F_n^{(1)}(x_1),\ldots,F_n^{(d)}(x_d)\right)\right|$$

$$\leq \left| C_0\left(F_0^{(1)}(x_1),\ldots,F_0^{(d)}(x_d)\right) - C_n\left(F_0^{(1)}(x_1),\ldots,F_0^{(d)}(x_d)\right)\right|$$

$$+ \left| C_n\left(F_0^{(1)}(x_1)\ldots,F_0^{(d)}(x_d)\right) - C_n\left(F_n^{(1)}(x_1),F_0^{(2)}(x_2),\ldots,F_0^{(d)}(x_d)\right)\right|$$

$$+ \sum_{j=2}^{d}\left| C_n\left(F_n^{(1)}(x_1),\ldots,F_n^{(j-1)}(x_{j-1}),F_0^{(j)}(x_j),F_0^{(j+1)}(x_{j+1}),\ldots,F_0^{(d)}(x_d)\right)\right.$$

$$\left. - C_n\left(F_n^{(1)}(x_1),\ldots,F_n^{(j-1)}(x_{j-1}),F_0^{(j)}(x_j),\ldots,F_0^{(d)}(x_d)\right)\right|$$

$$\leq \left| C_0\left(F_0^{(1)}(x_1),\ldots,F_0^{(d)}(x_d)\right) - C_n\left(F_0^{(1)}(x_1),\ldots,F_0^{(d)}(x_d)\right)\right|$$

$$+ \sum_{j=1}^{d}\left| F_0^{(j)}(x_j) - F_n^{(j)}(x_j)\right| \xrightarrow[n\to+\infty]{} 0.$$

This proves the assertion. $\qquad\square$

In the opposite direction one has the following result.

Theorem 4.2.2. *Let $(H_n)_{n\in\mathbb{N}}$ be a sequence of d-dimensional d.f.'s, and, for every $n \in \mathbb{N}$, let $F_n^{(j)}$ $(j = 1,\ldots,d)$ be the marginals of H_n. For every $n \in \mathbb{N}$, let C_n be a connecting d-copula, so that, for every $\mathbf{x} \in \mathbb{R}^d$,*

$$H_n(\mathbf{x}) = C_n\left(F_n^{(1)}(x_1),\ldots,F_n^{(d)}(x_d)\right).$$

If (H_n) converges to the d.f. H_0 in the weak topology of $\mathscr{D}^d$, and, if C_0 is a d-copula such that

$$H_0(\mathbf{x}) = C_0\left(F_0^{(1)}(x_1),\ldots,F_0^{(d)}(x_d)\right),$$

then for every point $\mathbf{t} \in \times_{j=1}^{d}\mathrm{Ran}\,F_0^{(j)}$, one has

$$\lim_{n\to+\infty} C_n(\mathbf{t}) = C_0(\mathbf{t}).$$

Proof. First of all, notice that the Continuous Mapping Theorem [Billingsley, 1968] guarantees that if (H_n) converges to the d.f. H_0 in the weak topology of $\mathscr{D}^d$, then each of the sequences of marginals $(F_n^{(j)})$ $(j = 1,\ldots,d)$ converges weakly to the d.f. $F_0^{(j)}$ $(j = 1,\ldots,d)$.

Now, the set A_j of continuity points of $F_0^{(j)}$ $(j = 1,\ldots,d)$ is, as is well known, the whole extended real line $\bar{\mathbb{R}}$, with the exception of at most a countable infinity of points, say

$$A_j = \bar{\mathbb{R}} \setminus \left(\bigcup_n \{x_j^{(n)}\}\right).$$

If $\overline{B}$ denotes the closure of a subset B of $\overline{\mathbb{R}}$, then one has, for $j = 1, \ldots, d$,

$$\operatorname{Ran} F_0^{(j)} = F_0^{(j)}(\overline{A_j}) = F_0^{(j)}(\overline{\mathbb{R}})$$

$$= \mathbb{I} \setminus \left(\bigcup_n \left[\ell^- F_0^{(j)}(x_j^{(n)}), \ell^+ F_0^{(j)}(x_j^{(n)}) \right[\right) = F_0^{(j)}(A_j).$$

Thus, for every $\mathbf{t} \in F_0^{(1)}(A_1) \times \cdots \times F_0^{(d)}(A_d)$, there is $x_j \in A_j$ $(j = 1, \ldots, d)$ such that $t_j = F_0^{(j)}(x_j)$. Therefore

$$|C_n(\mathbf{t}) - C_0(\mathbf{t})|$$

$$= \left| C_n \left(F_0^{(1)}(x_1), \ldots, F_0^{(d)}(x_d) \right) - C_0 \left(F_0^{(1)}(x_1), \ldots, F_0^{(d)}(x_d) \right) \right|$$

$$\leq \left| C_n \left(F_0^{(1)}(x_1), \ldots, F_0^{(d)}(x_d) \right) - C_n \left(F_n^{(1)}(x_1), \ldots, F_n^{(d)}(x_d) \right) \right|$$

$$+ \left| C_n \left(F_n^{(1)}(x_1), \ldots, F_n^{(d)}(x_d) \right) - C_0 \left(F_0^{(1)}(x_1), \ldots, F_0^{(d)}(x_d) \right) \right|.$$

Since $A_1 \times \cdots \times A_d$ is included in the subset of $\overline{\mathbb{R}}^d$ consisting of the continuity points of H_0, one has

$$\left| C_n \left(F_n^{(1)}(x_1), \ldots, F_n^{(d)}(x_d) \right) - C_0 \left(F_0^{(1)}(x_1), \ldots, F_0^{(d)}(x_d) \right) \right|$$

$$= |H_n(\mathbf{x}) - H_0(\mathbf{x})| \xrightarrow[n \to +\infty]{} 0.$$

On the other hand, the Lipschitz condition and the weak convergence of the marginals imply

$$\left| C_n \left(F_0^{(1)}(x_1), \ldots, F_0^{(d)}(x_d) \right) - C_n \left(F_n^{(1)}(x_1), \ldots, F_n^{(d)}(x_d) \right) \right|$$

$$\leq \sum_{j=1}^{d} \left| F_n^{(j)}(x_j) - F_0^{(j)}(x_j) \right| \xrightarrow[n \to +\infty]{} 0.$$

The assertion is now proved. $\square$

Corollary 4.2.3. *If the d-dimensional d.f.'s H_0 and H_n, for every $n \in \mathbb{N}$, have continuous marginals, $F_0^{(j)}$ and $F_n^{(j)}$ $j \in \{1, \ldots, d\}$, and (unique) copulas C_0 and C_n, respectively, then the following are equivalent:*

(a) *the sequence (H_n) converges to H_0 in the weak topology of $\mathscr{D}^d$;*

(b) *the sequence (C_n) converges pointwise to C_0 and, for every $j \in \{1, \ldots, d\}$, $(F_n^{(j)})$ converges to $F_0^{(j)}$ in the weak topology of $\mathscr{D}$ (or, equivalently, pointwise).*

Although the sequence $(C_n(\mathbf{t}))$ converges to $C_0(\mathbf{t})$ at all the points $\mathbf{t}$ of a suitable set D, according to Theorem 4.2.2, it need not converge to $C_0(\mathbf{t})$ in $\mathbb{I}^d$.

Example 4.2.4. Let (α_n) be a sequence of real numbers in $]0, 1[$ that converges to $1/2$, let Φ be the d.f. of the standard normal law $N(0, 1)$ and consider the d.f.'s

$$F_n := \alpha_n \Phi + (1 - \alpha_n) \varepsilon_{1/n}, \qquad\qquad G_n := \Phi,$$

$$F_0 := \frac{1}{2} (\Phi + \varepsilon_0), \qquad\qquad\qquad G_0 := \Phi,$$

where $\varepsilon_{1/n}(t) = 0$, if $t < 1/n$, $\varepsilon_{1/n}(t) = 1$, if $t \geq 1/n$. Then, one has $F_n \to F_0$ and $G_n \to G_0$ in the sense of weak convergence as $n \to +\infty$. Let $\widetilde{C}$ be the copula given by

$$\widetilde{C}(u, v) = \begin{cases} 4uv, & (u, v) \in [0, 1/4]^2, \\ \frac{3}{4} + \left(u - \frac{3}{4}\right) \left(v - \frac{3}{4}\right), & (u, v) \in [3/4, 1]^2, \\ M_2(u, v), & \text{elsewhere,} \end{cases}$$

i.e., it is obtained by ordinal sum of type $(\langle]0, 1/4[, \Pi_2, \rangle, \langle]3/4, 1[, \Pi_2\rangle)$. Notice that

$$\overline{D} = \overline{\operatorname{Ran} F_0} \times \overline{\operatorname{Ran} G_0} = \left(\left[0, \frac{1}{4}\right] \cup \left[\frac{3}{4}, 1\right]\right) \times \mathbb{I}.$$

Consider now the copula C_0 that coincides with $\widetilde{C}$ on $\overline{D}$, while on the complement of D, namely on the set $[1/4, 3/4] \times \mathbb{I}$, C_0 is defined as the bilinear interpolation of the restriction of $\widetilde{C}$ to $\overline{D}$. Thus, if $u \in]1/4, 3/4[$, then

$$C_0(u, v) = \left(\frac{2}{3} - 2u\right) \widetilde{C}\left(\frac{1}{4}, v\right) + \left(2u - \frac{1}{2}\right) \widetilde{C}\left(\frac{3}{4}, v\right).$$

Define the following d.f.'s in $\mathscr{D}^2$ via

$$H_0 := C_0(F_0, G_0), \qquad H_n := C_n(F_n, G_n) \quad (n \in \mathbb{N}),$$

where $C_n = \widetilde{C}$ for every $n \in \mathbb{N}$. By construction, the sequence (H_n) converges weakly to H_0, and, by Theorem 4.2.2, (C_n) converges to C_0 at the points of $\overline{D}$. However, if u belongs to the open interval $]1/4, 3/4[$, then $\widetilde{C}(u, v) \neq C_0(u, v)$, so that $(C_n(u, v))$ does not converge to $C_0(u, v)$. ∎

4.3 Markov kernel representation and related distances

In this section, we exploit the possibility of introducing metrics in $\mathscr{C}_d$, which are based on the Markov kernel representation of a copula from Section 3.4.

For every $A, B \in \mathscr{C}_d$, let K_A and K_B be (versions of) the Markov kernels associated with A and B, respectively. For every $(x, \mathbf{y}) \in \mathbb{I}^d$, set

$$\psi(x, [\mathbf{0}, \mathbf{y}]) := K_A(x, [\mathbf{0}, \mathbf{y}]) - K_B(x, [\mathbf{0}, \mathbf{y}]).$$

Moreover, consider the real functions defined on $\mathscr{C}_d \times \mathscr{C}_d$ by:

$$D_1(A, B) := \int_{\mathbb{I}^{d-1}} d\lambda_{d-1}(\mathbf{y}) \int_{\mathbb{I}} |\psi(x, [\mathbf{0}, \mathbf{y}])| \, d\lambda(x) \,, \tag{4.3.1}$$

$$D_2(A, B) := \left(\int_{\mathbb{I}^{d-1}} d\lambda_{d-1}(\mathbf{y}) \int_{\mathbb{I}} |\psi(x, [\mathbf{0}, \mathbf{y}])|^2 \, d\lambda(x) \right)^{1/2} \,, \tag{4.3.2}$$

$$D_\infty(A, B) := \sup_{\mathbf{y} \in \mathbb{I}^{d-1}} \int_{\mathbb{I}} |\psi(x, [\mathbf{0}, \mathbf{y}])| \, d\lambda(x) \,. \tag{4.3.3}$$

It can be proved that these functions are metrics on $\mathscr{C}_d$.

Lemma 4.3.1. *The functions D_1, D_2 and D_∞ defined by equations (4.3.1), (4.3.2) and (4.3.3) are metrics on $\mathscr{C}_d$.*

Proof. We first prove that the integrands in equations (4.3.1), (4.3.2) and (4.3.3) are measurable. The function H defined on $\mathbb{I}^d$ by

$$H(x, \mathbf{y}) := K_A(x, [\mathbf{0}, \mathbf{y}])$$

is measurable in x, increasing and right-continuous in $\mathbf{y}$. Fix $z \in \mathbb{I}$; then, for every vector with rational components, $\mathbf{q} \in \mathbb{Q}^{d-1} \cap \mathbb{I}^{d-1}$, define

$$E_\mathbf{q} := \{x \in \mathbb{I} : H(x, \mathbf{q}) < z\}$$

in $\mathscr{B}(\mathbb{I})$, and set

$$E := \bigcup_{\mathbf{q} \in \mathbb{Q}^{d-1} \cap \mathbb{I}^{d-1}} (E_\mathbf{q} \times [\mathbf{0}, \mathbf{q}])$$

belonging to $\mathscr{B}(\mathbb{I}^d)$. Since H is right-continuous in $\mathbf{y}$, it follows immediately that $H^{(-1)}([0, z[) = E$, so that H is measurable and, hence, so are the integrands in question.

Now, we shall only prove that D_1 is a metric, as the proof for both D_2 and D_∞ is very similar.

Both the symmetry of D_1 and the fact that it satisfies the triangle inequality are obvious. Assume now that $D_1(A, B) = 0$; then there exists a Borel subset F of $\mathbb{I}^d$ with $\lambda_d(F) = 1$ such that, for every $(x, \mathbf{y})$ in F, one has $K_A(x, [\mathbf{0}, \mathbf{y}]) = K_B(x, [\mathbf{0}, \mathbf{y}])$. As a consequence, one has $\lambda_{d-1}(F_x) = 1$ for almost every $x \in \mathbb{I}$; for every such x the kernels K_A and K_B coincide on a dense set, and, hence, they are identical. It now follows from (3.4.7) that $A = B$. $\qquad\square$

Moreover, for every $A, B \in \mathscr{C}_d$, consider the function $\Phi_{A,B} : \mathbb{I}^{d-1} \to \mathbb{R}$ given by

$$\Phi_{A,B}(\mathbf{y}) := \int_{\mathbb{I}} \psi(x, [\mathbf{0}, \mathbf{y}]) \, d\lambda(x) = \int_{\mathbb{I}} |K_A(x, [\mathbf{0}, \mathbf{y}]) - K_B(x, [\mathbf{0}, \mathbf{y}])| \, d\lambda(x) \,.$$

The following result holds.

Lemma 4.3.2. *The function $\Phi_{A,B}$ is Lipschitz continuous (with Lipschitz constant equal to 2) with respect to the norm $\| \cdot \|_1$ on $\mathbb{I}^{d-1}$. Moreover,*

$$\Phi_{A,B}(\mathbf{y}) \leq 2 \min \left\{ \min_{j=1,\dots,d-1} y_j, \sum_{j=1}^{d-1}(1 - y_j) \right\}, \qquad (4.3.4)$$

for every $\mathbf{y} \in \mathbb{I}^{d-1}$.

Proof. Since $|x - y| \leq |x| + |y|$, if $F = \times_{j=1}^{d-1} F_j$ is a measurable box in $\mathbb{I}^{d-1}$, then

$$\int_{\mathbb{I}} |K_A(x, F) - K_B(x, F)| \, d\lambda(x) \leq 2 \min_{j=1,\dots,d-1} \lambda(F_j),$$

and, since $B \mapsto K(x, B)$ is a probability measure,

$$\int_{\mathbb{I}} |K_A(x, F) - K_B(x, F)| \, d\lambda(x) = \int_{\mathbb{I}} |K_A(x, F^c) - K_B(x, F^c)| \, d\lambda(x)$$
$$\leq \int_{\mathbb{I}} (K_A(x, F^c) + K_B(x, F^c)) \, d\lambda(x).$$

Now, set $R := \int_{\mathbb{I}} K_A(x, F^c) \, d\lambda(x)$, and use the fact that

$$F^c = \bigcup_{j=1}^{d-1} \left(F_1 \times F_2 \times \cdots \times F_{j-1} \times F_j^c \times \mathbb{I} \times \cdots \times \mathbb{I} \right).$$

Then

$$R - \sum_{j=1}^{d-1} \int_{\mathbb{I}} K_A \left(x, F_1 \times F_2 \times \cdots \times F_{j-1} \times F_j^c \times \mathbb{I} \times \cdots \times \mathbb{I} \right) d\lambda(x)$$
$$\leq \lambda(F_1^c) + \min\{\lambda(F_1), \lambda(F_2^c)\} + \min\{\lambda(F_1), \lambda(F_2), \lambda(F_3^c)\} + \cdots +$$
$$+ \min\{\lambda(F_1), \lambda(F_2), \dots, \lambda(F_{d-2}), \lambda(F_{d-1}^c)\} \leq \sum_{j=1}^{d-1}(1 - \lambda(F_j)).$$

Eq. (4.3.4) is now proved.

Let $\mathbf{y}$ and $\mathbf{z}$ belong to $\mathbb{I}^{d-1}$ and set $U := \times_{j=1}^{d-1} [0, \min\{y_j, z_j\}]$. As above, U^c is the union of $d - 1$ sets. Then

$$|\Phi_{A,B}(\mathbf{y}) - \Phi_{A,B}(\mathbf{z})|$$

$$= \left| \int_{\mathbb{I}} |K_A(x, [\mathbf{0}, \mathbf{y}]) - K_B(x, [\mathbf{0}, \mathbf{y}])| \ d\lambda(x) \right.$$

$$\left. - \int_{\mathbb{I}} |K_A(x, [\mathbf{0}, \mathbf{z}]) - K_B(x, [\mathbf{0}, \mathbf{z}])| \ d\lambda(x) \right|$$

$$\leq \int_{\mathbb{I}} |K_A(x, [\mathbf{0}, \mathbf{y}]) - K_B(x, [\mathbf{0}, \mathbf{y}]) - K_A(x, [\mathbf{0}, \mathbf{z}]) + K_B(x, [\mathbf{0}, \mathbf{z}])| \ d\lambda(x)$$

$$\leq \int_{\mathbb{I}} |K_A(x, [\mathbf{0}, \mathbf{y}] \cap U^c) - K_B(x, [\mathbf{0}, \mathbf{y}] \cap U^c)| \ d\lambda(x)$$

$$+ \int_{\mathbb{I}} |K_A(x, [\mathbf{0}, \mathbf{z}] \cap U^c) - K_B(x, [\mathbf{0}, \mathbf{z}] \cap U^c)| \ d\lambda(x)$$

$$\leq 2 \ (y_1 - \min\{y_1, z_1\} + y_2 - \min\{y_2, z_2\} + \cdots + y_{d-1} - \min\{y_{d-1}, z_{d-1}\})$$

$$+ 2 \ (z_1 - \min\{y_1, z_1\} + z_2 - \min\{y_2, z_2\} + \cdots + z_{d-1} - \min\{y_{d-1}, z_{d-1}\})$$

$$\leq 2 \sum_{j=1}^{d-1} |y_j - z_j| = 2 \ \|\mathbf{y} - \mathbf{z}\|_1 \,,$$

which concludes the proof.   □

Remark 4.3.3. In the case $d = 2$, eq. (4.3.4) reads $\Phi_{A,B}(y) \leq 2 \ \min\{y, 1 - y\}$. Moreover, since

$$\Phi_{M_2, W_2}(y) = 2 \ \min\{y, 1 - y\} \,,$$

the upper bound is attained.   ■

The relationships between the new metrics introduced on $\mathscr{C}_d$ and the metric d_∞ of uniform convergence are established in the next result.

Theorem 4.3.4. *For all copulas A and B in $\mathscr{C}_d$ one has*

$$D_2(A, B) \geq D_1(A, B) \geq D_2^2(A, B) \,, \tag{4.3.5}$$

$$D_\infty(A, B) \geq D_1(A, B) \geq \left(\frac{1}{2 \ (d - 1)} \right)^{d-1} \frac{D_\infty^d(A, B)}{d} \,. \tag{4.3.6}$$

Moreover, $D_\infty(A, B) \geq d_\infty(A, B)$.

Proof. The inequality $D_1(A, B) \leq D_2(A, B)$ is a direct consequence of the Schwarz inequality, while $D_1(A, B) \geq D_2^2(A, B)$ is obvious. This proves (4.3.5).

Moreover, for given $A, B \in \mathscr{C}_d$, let $\mathbf{z} \in \mathbb{I}^{d-1}$ such that $\Phi_{A,B}(\mathbf{z}) = D_\infty(A, B)$. Then there exists a hypercube Q contained in $\mathbb{I}^{d-1}$ having $\mathbf{z}$ as one of its 2^{d-1} vertices

$$\left(\mathbf{v}^{(1)}, \ldots, \mathbf{v}^{(2^{d-1})} \right)$$

and having side length equal to $\Phi_{A,B}(\mathbf{z})/(2(d-1))$. For $j \in \{1,\ldots,2^{d-1}\}$ define a point in $\mathbb{I}^d$ via $\mathbf{w}^{(j)} := (w_1^{(j)},\ldots,w_{d-1}^{(j)},0)$ and put $\mathbf{w}^{2^{d-1}+1} := (z_1,\ldots,z_{d-1},\Phi_{A,B}(\mathbf{z}))$. The convex hull P in $\mathbb{I}^d$ of the points $\mathbf{w}^{(1)},\ldots,\mathbf{w}^{2^{d-1}+1}$ is contained in the region between the hyperplane $x_d = 0$ and the graph of $\Phi_{A,B}$. Therefore

$$D_\infty(A,B) \geq D_1(A,B) = \int_{\mathbb{I}^{d-1}} \Phi_{A,B}(\mathbf{y}) \, d\lambda_{d-1}(\mathbf{y}) \geq \lambda_d(P)$$

$$= \int_{[0,\Phi_{A,B}(\mathbf{z})]} \left(\frac{x}{2(d-1)}\right)^{d-1} d\lambda(x) = \left(\frac{1}{2(d-1)}\right)^{d-1} \frac{\Phi_{A,B}^d(\mathbf{z})}{d}.$$

As for the last assertion, one has, for every point $(x,\mathbf{y}) \in \mathbb{I}^d$,

$$|A(x,\mathbf{y}) - B(x,\mathbf{y})| = \left|\int_{[0,x]} (K_A(t,[0,\mathbf{y}]) - K_B(t,[0,\mathbf{y}])) \, d\lambda(t)\right|$$

$$\leq \int_{[0,x]} |K_A(t,[0,\mathbf{y}]) - K_B(t,[0,\mathbf{y}])| \, d\lambda(t)$$

$$\leq \Phi_{A,B}(\mathbf{y}) \leq D_\infty(A,B),$$

which concludes the proof. $\qquad\qquad\qquad\qquad\qquad\qquad\qquad\qquad\square$

The following consequence is now immediate.

Corollary 4.3.5. *For a sequence $(C_n)_{n\in\mathbb{N}}$ of copulas in $\mathscr{C}_d$ and for a copula $C \in \mathscr{C}_d$, the following statements are equivalent:*

(a) $\lim_{n\to+\infty} D_\infty(C_n,C) = 0$;

(b) $\lim_{n\to+\infty} D_1(C_n,C) = 0$;

(c) $\lim_{n\to+\infty} D_2(C_n,C) = 0$.

Moreover, each of the previous statements implies that C_n tends to C uniformly, as $n \to +\infty$.

Starting with the metric D_1, one may also consider another kind of convergence for copulas.

Definition 4.3.6. A sequence $(C_n)_{n\in\mathbb{N}}$ in $\mathscr{C}_d$ is said to converge in the ∂ sense to the copula $C \in \mathscr{C}_d$, and one writes

$$C_n \xrightarrow[n\to\infty]{\partial} C \qquad \text{or} \qquad \partial - \lim_{n\to\infty} C_n = C,$$

if, for all $k \in \{1,\ldots,d\}$ and $(u_1,\ldots,u_{k-1},u_{k+1},\ldots,u_d) \in \mathbb{I}^{d-1}$, one has

$$\lim_{n\to\infty} \int_0^1 |\partial_k C_n(u_k(t)) - \partial_k C(u_k(t))| \, dt = 0,$$

where $u_k(t) := (u_1,\ldots,u_{k-1},t,u_{k+1},\ldots,u_d)$. $\qquad\qquad\qquad\qquad\diamond$

As can be easily seen, uniform convergence is implied by ∂-convergence.

Theorem 4.3.7. *Given a sequence (C_n) of copulas in $\mathscr{C}_d$, if $C_n \xrightarrow[n\to\infty]{\partial} C$, then $C_n \xrightarrow[n\to\infty]{} C$ uniformly on $\mathbb{I}^d$.*

Proof. For every $\mathbf{u} = (u_1, \ldots, u_d) \in \mathbb{I}^d$ one has

$$|C_n(\mathbf{u}) - C(\mathbf{u})| = \left| \int_0^{u_j} \left(\partial_k C_n(u_k(t)) - \partial_k C(u_k(t)) \right) \mathrm{d}t \right|$$

$$\leq \int_0^1 |\partial_k C_n(u_k(t)) - \partial_k C(u_k(t))| \, \mathrm{d}t,$$

which concludes the proof. $\square$

Now, for all $j \in \{1, \ldots, d\}$ and $A \in \mathscr{C}_d$, let A^{τ_j} denote the copula obtained by the permutation of $\mathbf{x} \in \mathbb{I}^d$ that swaps the components at places 1 and j, keeping fixed the other components, i.e.

$$A^{\tau_j}(u_1, \ldots, u_d) := A(u_j, u_2, \ldots, u_{j-1}, u_1, u_{j+1}, \ldots, x_d). \tag{4.3.7}$$

The following result holds.

Theorem 4.3.8. *The metric on $\mathscr{C}_d$ defined by*

$$D_\partial(A, B) := \sum_{j=1}^d D_1(A^{\tau_j}, B^{\tau_j})$$

is a distance that induces the ∂-convergence.

Proof. It easily follows from the definition of D_1 by using Theorem 3.4.4. $\square$

While $(\mathscr{C}_d, d_\infty)$ is a compact space (Theorem 1.7.7), $(\mathscr{C}_d, D_1)$ fails to be so, as the following example shows (see [Trutschnig, 2012]).

Example 4.3.9. Let $(A_n)_{n\in\mathbb{N}}$ be a sequence of shuffles of Min that converges uniformly to Π_2. Suppose, if possible, that $(\mathscr{C}_2, D_1)$ is compact. Then there would exist a subsequence (A_{n_k}) that converges to Π_2 in the metric D_1. Since every A_{n_k} is a shuffle of Min, it can be written as $A_{n_k}(x, y) = \lambda\left([0, x] \cap f_{n_k}^{-1}([0, y])\right)$ for a measure-preserving transformation f_{n_k}. It follows that, for every $y \in \mathbb{I}$,

$$\int_0^1 |K_{A_{n_k}}(x, [0, y]) - K_{\Pi_2}(x, [0, y])| \, \mathrm{d}x = \int_0^1 |K_{A_{n_k}}(x, [0, y]) - y| \, \mathrm{d}x$$

$$= \int_0^1 |\mathbf{1}_{[0,y]}(f_{n_k}(x)) - y| \, \mathrm{d}x = \int_0^1 |\mathbf{1}_{[0,y]}(x) - y| \, \mathrm{d}x = 2y(1 - y).$$

Thus, $D_1(A_{n_k}, \Pi_2) = 1/3$ for every n_k, which is a contradiction.   ∎

However, $(\mathscr{C}_d, D_1)$ is still topologically rich as the following result asserts.

Theorem 4.3.10. *The metric spaces $(\mathscr{C}_d, D_1)$, $(\mathscr{C}_d, D_2)$ and $(\mathscr{C}_d, D_\infty)$ are complete and separable.*

Proof. Since the three metrics D_1, D_2 and D_∞ are equivalent by Theorem 4.3.4, it suffices to prove the assertion only for D_1.

Let (C_n) be a Cauchy sequence of copulas in $\mathscr{C}_d$ with respect to D_1 and, for every $n \in \mathbb{N}$, let K_n denote the Markov kernel related to C_n; on $\mathbb{I}^d$ define the function $H_n(x, \mathbf{y}) := K_n(x, [\mathbf{0}, \mathbf{y}])$. One has

$$D_1(C_n, C_m) = \int_{\mathbb{I}^{d-1}} d\lambda_{d-1}(\mathbf{y}) \int_{\mathbb{I}} |H_n(x, \mathbf{y}) - H_m(x, \mathbf{y})| \, d\lambda(x) = \|H_n - H_m\|_1 \, ;$$

here $\| \cdot \|_1$ denotes the norm of the space $L^1(\mathbb{I}^d, \mathscr{B}(\mathbb{I}^d), \lambda_d)$. Thus (H_n) is a Cauchy sequence in this latter space, which is complete, and, as a consequence, it has a limit $H \in L^1(\mathbb{I}^d, \mathscr{B}(\mathbb{I}^d), \lambda_d)$. Now, as is well known, convergence in L^1 implies convergence in probability, in this case in λ_d-probability; in its turn, this means that there exists a subsequence $(H_{n(k)})$ that converges to H λ_d-a.e., namely that there exists a measurable subset F of $\mathbb{I}^d$ with $\lambda_d(F) = 1$ such that

$$\lim_{k \to +\infty} H_{n(k)}(x, \mathbf{y}) = H(x, \mathbf{y})$$

at every point $(x, \mathbf{y}) \in F$. There is no loss of generality in assuming $H(x, \mathbf{1}) = 1$ for every $x \in \mathbb{I}$.

We wish to show that there exists a measurable function $G : \mathbb{I}^d \to \mathbb{I}$ such that

- $G = H$ λ_d-a.e.;
- $K(x, \mathbf{y}) := G(x, \mathbf{y})$ is a regular conditional distribution of a copula $C \in \mathscr{C}_d$.

For the section $F_{\mathbf{y}} = \{x \in \mathbb{I} : (x, \mathbf{y}) \in F\}$ one has from eq. (3.4.8)

$$\lambda(F_{\mathbf{y}}) = \int_{\mathbb{I}} K(x, \mathbf{1}) \, d\lambda(x) = 1 \, ,$$

for almost every $\mathbf{y} \in \mathbb{I}^{d-1}$ with respect to λ_{d-1}. Therefore, it is possible to find a countable dense subset $Q = \{\mathbf{q}_n : n \in \mathbb{N}\} \subseteq \mathbb{I}^{d-1}$ that contains $\mathbf{1}$ and a measurable subset F_0 of $\mathbb{I}$ such that, for every $\mathbf{q} \in Q$ and for every $x \in F_0$,

$$\lim_{k \to +\infty} H_{n(k)}(x, \mathbf{q}) = H(x, \mathbf{q}) \, .$$

As above, it is possible to find a measurable subset $F_1 \subseteq F_0$ such that $\lambda(F_x) = 1$ for every $x \in F_1$. Fix $x \in F_1$ and define two new functions h_x and g_x from $\mathbb{I}^{d-1}$ into $\mathbb{I}$ via

$$h_x(\mathbf{y}) := \begin{cases} 0, & \text{if } \min\{y_1, \ldots, y_{d-1}\} = 0, \\ \sup_{\mathbf{q} \in Q : \mathbf{y} \prec \mathbf{q}} H(x, \mathbf{q}), & \text{otherwise,} \end{cases}$$

and

$$g_x(\mathbf{y}) := \begin{cases} 1, & \text{if } \mathbf{y} = 1, \\ \inf_{\mathbf{q} \in \mathbb{Q} \cap \mathbb{I}^{d-1} : \mathbf{q} \gg \mathbf{y}} h_x(\mathbf{q}), & \text{otherwise.} \end{cases}$$

Here $\mathbf{y} \prec \mathbf{q}$ means $y_j \le q_j$ for every $j \in \{1, \ldots, d-1\}$, while $\mathbf{q} \gg \mathbf{y}$ means $\mathbf{y} \prec \mathbf{q}$ and, furthermore, $q_j > y_j$ when $y_j < 1$.

It follows from the definition that, for every $j \in \{1, \ldots, d-1\}$, the function

$$t \mapsto h_x(y_1, \ldots, y_{j-1}, t, y_{j+1}, \ldots, y_{d-1})$$

is increasing, bounded above by 1 and left-continuous. At each $\mathbf{y}$ that is a continuity point of h_x one has

$$\lim_{k \to +\infty} H_{n(k)}(x\,\mathbf{y}) = h_x(\mathbf{y}).$$

But, with respect to λ_{d-1}, almost all points $\mathbf{y} \in \mathbb{I}^{d-1}$ are points of continuity for h_x, so that $H(x, \mathbf{y}) = h_x(\mathbf{y})$ for almost all points $\mathbf{y} \in [0, 1[^{d-1}$. The same argument allows to conclude that g_x is increasing in each place, bounded above by 1, right-continuous and that $H(x, \mathbf{y}) = g_x(\mathbf{y})$ for almost all points $\mathbf{y} \in [0, 1[^{d-1}$. By Helly's theorem (see, e.g., [Billingsley, 1968]) there exists a further subsequence $(H_{k(m)})$ of $(H_{n(k)})$ and a distribution function Φ_x with support on $\mathbb{I}^{d-1}$ such that $(H_{k(m)})$ converges weakly to Φ_x. Since $\mathbf{y} \mapsto H(x, \mathbf{y})$ is increasing and right-continuous in each place one concludes that $H(x, \mathbf{y}) = \Phi_x(\mathbf{y})$ for almost every $\mathbf{y} \in \mathbb{I}^{d-1}$. Define $G : \mathbb{I}^d \to \mathbb{I}$ via

$$G(x, \mathbf{y}) := \Phi_x(\mathbf{y})\,\mathbf{1}_{F_1}(x) + \mathbf{1}_{\mathbb{I}^{d-1}}(\mathbf{y})\,\mathbf{1}_{F_1^c}(x).$$

Now $x \mapsto G(x, \mathbf{y})$ is measurable for every $\mathbf{y} \in \mathbb{I}^{d-1}$, while, by construction, $\mathbf{y} \mapsto G(x, \mathbf{y})$ is a distribution function with support on $\mathbb{I}^{d-1}$, for every $x \in \mathbb{I}$. Set $K(x, [\mathbf{0}, \mathbf{y}]) := G(x, \mathbf{y})$, and, by the usual measure-theoretical technique, extend its definition to the family $\mathscr{B}(\mathbb{I}^{d-1})$ of the Borel sets of $\mathbb{I}^{d-1}$, for every $x \in \mathbb{I}$.

By construction G coincides with H λ_d-a.e. One can now claim that $(x, \mathbf{y}) \mapsto K(x, \mathbf{y})$ is a regular conditional distribution of a copula $C \in \mathscr{C}_d$. To this end define $C : \mathbb{I}^d \to \mathbb{I}$ via

$$C(x, \mathbf{y}) := \int_{[0,x]} G(t, \mathbf{y})\,d\lambda(t).$$

Now C as a function of x is absolutely continuous, while, as a function of $\mathbf{y}$, it is increasing for every $x \in \mathbb{I}$. Moreover, one has

$$C_{k(j)}(x, \mathbf{y}) \xrightarrow[j \to +\infty]{} C(x, \mathbf{y})$$

on a dense set of $\mathbb{I}^{d-1}$. But all the $C_{k(j)}$'s are copulas, so that their limit is also a copula by Theorem 1.7.5; therefore C is a copula. The dominated convergence theorem then yields $D_1(C_{k(j)}, C) \to 0$ as j tends to $+\infty$. The entire sequence (C_n) converges to C as a consequence of the triangle inequality

$$D_1(C_n, C) \le D_1(C_n, C_{k(j)}) + D_1(C_{k(j)}, C).$$

This proves that $(\mathscr{C}_d, D_1)$ is complete.

In order to prove the separability of $(\mathscr{C}_d, D_1)$ consider the family $\mathscr{S}_n$ of n-th order checkerboard approximations of $\mathscr{C}_d$; according to Theorem 4.5.8 $\mathscr{S}_n$ is dense in $\mathscr{C}_d$ with respect to ∂-convergence; consider also the subset $\widetilde{\mathscr{S}_n}$ of $\mathscr{S}_n$ composed by those checkerboard approximations in $\mathscr{S}_n$ with the property that the stochastic measure of every rectangle $R_{n,j} = \times_{k=1}^d [(j_k - 1)/n, j_k/n]$ induced by a checkerboard copula C_n equals a rational number. Thus the set $\widetilde{\mathscr{S}_n}$ is countable as is also

$$\widetilde{\mathscr{S}} := \bigcup_{n \geq 2} \widetilde{\mathscr{S}_n}.$$

By Theorem 4.3.8, $\mathscr{S}_n$ is dense in $(\mathscr{C}_d, D_1)$. Consider a copula $B \in \mathscr{S}_n$ and let $\varepsilon > 0$. Order the rectangles $R_{n,j}$ lexicographically and then consider the embedding of the B-volumes of these rectangles into $\mathbb{I}^{n^d}$; thus $\mathscr{S}_n$ is isomorphic to a box S_n in $\mathbb{I}^{n^d}$. Since the set E_n of extreme points of S_n is finite and every point α in S_n is a convex combination of $m \leq n^d + 1$ elements of E_n (see [Gruber, 2007]), and since $\mathbb{Q}$ is dense in $\mathbb{I}$, one can find a vector $(\beta_1, \ldots, \beta_m)$ such that both

$$\max_{j \in \{1, \ldots, n\}^d} |\alpha_j - \beta_j| < \frac{\varepsilon}{n^d + 1} \qquad \sum_{j=1}^m \beta_j = 1$$

hold. This implies the existence of a checkerboard copula $\widetilde{B} \in \widetilde{\mathscr{S}_n}$ such that

$$\max_{j \in \{1, \ldots, n\}^d} \left| \mu_B(R_{n,j}) - \mu_{\widetilde{B}}(R_{n,j}) \right| < \frac{\varepsilon}{n^d + 1}.$$

It follows that $D_1(B, \widetilde{B}) < \varepsilon$; this shows that $\widetilde{\mathscr{S}_n}$, which is countable, is dense in $\mathscr{S}_n$ and proves the assertion. $\qquad \square$

As a consequence, one may also prove the following result.

Theorem 4.3.11. *The metric space $(\mathscr{C}_d, D_\partial)$ is complete and separable.*

Proof. Let (C_n) be a Cauchy sequence in $(\mathscr{C}_d, D_\partial)$; then, for every $j \in \{1, \ldots, d\}$, $(C_n^{\tau_j})$ (defined as in eq. (4.3.7)) is a Cauchy sequence in $(\mathscr{C}_d, D_1)$, which has been proved to be complete in the previous theorem. Therefore, there exist d copulas $C^{(j)}$ in $\mathscr{C}_d$ such that

$$\lim_{n \to +\infty} D_1\left(C_n^{\tau_j}, C^{(j)}\right) = 0 \qquad (j = 1, \ldots, d).$$

By Theorem 4.3.4, one has also

$$\lim_{n \to +\infty} d_\infty\left(C_n^{\tau_j}, C^{(j)}\right) = 0 \qquad (j = 1, \ldots, d).$$

This, in its turn, implies $\lim_{n \to +\infty} d_\infty\left(C_n, (C^{(j)})^{\tau_j}\right) = 0$ $(j = 1, \ldots, d)$. But a limit with respect to the metric d_∞ is unique, so that there exists a copula $C \in \mathscr{C}_d$ with $C = C^{\tau_j}$ for every $j \in \{1, \ldots, d\}$. Then

$$\lim_{n \to +\infty} D_\partial(C_n, C) = \lim_{n \to +\infty} \sum_{j=1}^{d} D_1\left(C_n^{\tau_j}, C^{\tau_j}\right) = 0.$$

The separability of $(\mathscr{C}_d, D_\partial)$ is proved in the same way as in Theorem 4.3.10. □

Further readings 4.3.12. The Markov kernel representation of copulas has been introduced by Trutschnig [2011] and Fernández-Sánchez and Trutschnig [2015], where also the related distances are considered. The ∂-convergence is presented in [Mikusiński and Taylor, 2009]. ∎

4.4 Copulas and Markov operators

In this section, we explore the correspondence between copulas and a special class of operators, known as Markov operators. The usefulness of this correspondence for the approximation of copulas will be illustrated in Section 4.5.

Let two probability spaces $(\Omega_1, \mathscr{F}_1)$ and $(\Omega_2, \mathscr{F}_2)$ be given and consider the measurable space $(\Omega_1 \times \Omega_2, \mathscr{F}_1 \otimes \mathscr{F}_2)$, where $\mathscr{F}_1 \otimes \mathscr{F}_2$ is the product σ-field. If $\mathbb{P}$ is a probability measure on $(\Omega_1 \times \Omega_2, \mathscr{F}_1 \otimes \mathscr{F}_2)$, the marginals of $\mathbb{P}$ are defined as the probability measures $\mathbb{P}_1$ and $\mathbb{P}_2$ on $(\Omega_1, \mathscr{F}_1)$ and $(\Omega_2, \mathscr{F}_2)$, respectively, such that for every set $A \in \mathscr{F}_1$, $\mathbb{P}_1(A) = \mathbb{P}(A \times \Omega_2)$ and, for every set $B \in \mathscr{F}_2$, $\mathbb{P}_2(B) = \mathbb{P}(\Omega_1 \times B)$.

If a function $f : \Omega_1 \to \mathbb{R}$ is measurable, then it may be considered as a function $\tilde{f} : \Omega_1 \times \Omega_2 \to \mathbb{R}$ by setting $f(x) = \tilde{f}(x, y)$ for all $x \in \Omega_1$ and $y \in \Omega_2$. Once this has been stipulated, the notation f will be used also for $\tilde{f}$. With this notation, one has, if f belongs to $L^p(\Omega_1)$, for $1 \leq p \leq \infty$,

$$\int_{\Omega_1 \times \Omega_2} f \, d\mathbb{P} = \int_{\Omega_1} f \, d\mathbb{P}_1.$$

As a consequence, one has the trivial inclusion $L^p(\Omega_1) \subset L^p(\Omega_1 \times \Omega_2)$. Analogously, one has $L^p(\Omega_2) \subset L^p(\Omega_1 \times \Omega_2)$.

Taking into account the previous convention, the following definition may be given.

Definition 4.4.1. A linear operator $T : L^\infty(\Omega_1) \to L^\infty(\Omega_2)$ is said to be a *Markov operator* if

(M1) T is positive, viz. $Tf \geq 0$ whenever $f \geq 0$;

(M2) $T1 = 1$ (here 1 is the constant function $f \equiv 1$);

(M3) T preserves the expectation of every function $f \in L^\infty(\Omega_1)$, namely $\mathbb{E}_2(Tf) = \mathbb{E}_1(f)$, where $\mathbb{E}_j$ denotes the expectation in the probability space $(\Omega_j, \mathscr{F}_j, \mathbb{P}_j)$ $(j = 1, 2)$. ◇

Properties (M1) and (M2) ensure that a Markov operator is bounded. Although a Markov operator has been defined on L^∞ its definition may be extended to L^p with $p \in [1, +\infty]$.

Theorem 4.4.2. *Every Markov operator $T : L^\infty(\Omega_1) \to L^\infty(\Omega_2)$ has an extension to a bounded operator $T : L^p(\Omega_1) \to L^p(\Omega_2)$ for every $p \geq 1$.*

Proof. Let f be in $L^\infty(\Omega_1)$. In view of [Lasota and Mackey, 1994, Proposition 3.1.1], one has

$$\int_{\Omega_2} |Tf| \, d\mathbb{P}_2 \leq \int_{\Omega_2} T |f| \, d\mathbb{P}_2 = \int_{\Omega_1} |f| \, d\mathbb{P}_1 \,,$$

and since $L^\infty(\Omega_1)$ is dense in $L^1(\Omega_1)$, T has a unique extension to a bounded operator $T : L^1(\Omega_1) \to L^1(\Omega_2)$. Therefore, by the Riesz-Thorin theorem (see, e.g., [Garling, 2007]), T has a unique extension to a bounded operator $T : L^p(\Omega_1) \to L^p(\Omega_2)$ for every $p \in [1, +\infty[$. $\qquad\square$

It is not hard to check that the operator norm of T, $\|T\|$, is equal to 1. Also its extension to L^1 has norm equal to 1. We shall exploit this possibility by proving results in L^∞ rather than in L^1, whenever this is convenient.

Example 4.4.3. Let $\varphi : \Omega_2 \to \Omega_1$ be measure-preserving, namely $\mathbb{P}_1(B) = \mathbb{P}_2\left(\varphi^{-1}(B)\right)$ for every set $B \in \mathscr{F}_1$; then the mapping $T : L^\infty(\Omega_1) \to L^\infty(\Omega_2)$ defined by $Tf := f \circ \varphi$ is a Markov operator. To see this, it is enough to check the expectation invariance property (M3). The "change of variables" formula yields

$$\mathbb{E}_2(Tf) = \int_{\Omega_2} Tf \, d\mathbb{P}_2 = \int_{\Omega_2} f \circ \varphi \, d\mathbb{P}_2$$

$$= \int_{\Omega_1} f \, d(\mathbb{P}_2 \circ \varphi^{-1}) = \int_{\Omega_1} f \, d\mathbb{P}_1 = \mathbb{E}_1(f) \,,$$

whence the assertion. $\qquad\blacksquare$

The next result establishes a one-to-one correspondence between Markov operators from $L^\infty(\Omega_1)$ into $L^\infty(\Omega_2)$ and probability measures on the measurable space $(\Omega_1 \times \Omega_2, \mathscr{F}_1 \otimes \mathscr{F}_2)$.

Theorem 4.4.4. *Let $(\Omega_1, \mathscr{F}_1, \mathbb{P}_1)$ and $(\Omega_2, \mathscr{F}_2, \mathbb{P}_2)$ be probability spaces and consider a Markov operator $T : L^\infty(\Omega_1) \to L^\infty(\Omega_2)$. The set function $\mathbb{P} : \mathscr{F}_1 \times \mathscr{F}_2$ defined through*

$$\mathbb{P}(A \times B) := \int_{\Omega_2} (T \, \mathbf{1}_A) \, \mathbf{1}_B \, d\mathbb{P}_2 \tag{4.4.1}$$

can be extended to a probability measure on $\mathscr{F}_1 \otimes \mathscr{F}_2$ that has $\mathbb{P}_1$ and $\mathbb{P}_2$ as its marginals.

Conversely, if $\mathbb{P}$ is a probability measure on $(\Omega_1 \times \Omega_2, \mathscr{F}_1 \otimes \mathscr{F}_2)$ having $\mathbb{P}_1$

and $\mathbb{P}_2$ as marginals, then there exists a unique Markov operator $T : L^\infty(\Omega_1) \to L^\infty(\Omega_2)$ such that, for every $f \in L^\infty(\Omega_1)$ and for every $g \in L^\infty(\Omega_2)$,

$$\int_{\Omega_2} (Tf)\,g\,\mathrm{d}\mathbb{P}_2 = \int_{\Omega_1 \times \Omega_2} f\,g\,\mathrm{d}\mathbb{P}. \tag{4.4.2}$$

Proof. The extension of $\mathbb{P}$ from the family $\mathscr{F}_1 \times \mathscr{F}_2$ of measurable rectangles to the product σ-field $\mathscr{F}_1 \otimes \mathscr{F}_2$ can be made by a standard measure-theoretical argument as presented, e.g., in [Ash, 2000, Theorem 1.2.8]. Thus, let $(A_n \times B_n)_{n \in \mathbb{N}}$ be an increasing sequence of measurable rectangles and set

$$A \times B = \bigcup_{n \in \mathbb{N}} (A_n \times B_n).$$

The linearity and the boundedness of T together with property (M1) imply

$$T\mathbf{1}_{A_n} \xrightarrow[n \to +\infty]{} T\mathbf{1}_A \quad \mathbb{P}_2\text{-a.e.},$$

$$(T\mathbf{1}_{A_n})\,\mathbf{1}_{B_n} \xrightarrow[n \to +\infty]{} (T\mathbf{1}_A)\,\mathbf{1}_B \quad \mathbb{P}_2\text{-a.e.}$$

Now the monotone convergence theorem yields

$$\mathbb{P}(A_n \times B_n) = \int_{\Omega_2} (T\mathbf{1}_{A_n})\,\mathbf{1}_{B_n}\,\mathrm{d}\mathbb{P}_2 \xrightarrow[n \to +\infty]{} \int_{\Omega_2} (T\mathbf{1}_A)\,\mathbf{1}_B\,\mathrm{d}\mathbb{P}_2 = \mathbb{P}(A \times B).$$

This proves that $\mathbb{P}$ can be extended to be a σ-additive measure. Moreover, because of (M2), it is a probability measure and its marginals are $\mathbb{P}_1$ and $\mathbb{P}_2$.

Now let $\mathbb{P}$ be a probability measure on $\mathscr{F}_1 \times \mathscr{F}_2$ having marginals $\mathbb{P}_1$ and $\mathbb{P}_2$.

Take two functions f and g in $L^2(\Omega_1)$ and $L^2(\Omega_2)$, respectively. By Schwarz inequality

$$\left| \int_{\Omega_1 \times \Omega_2} f\,g\,\mathrm{d}\mathbb{P} \right| \le \left(\int_{\Omega_1 \times \Omega_2} f^2\,\mathrm{d}\mathbb{P} \right)^{1/2} \left(\int_{\Omega_1 \times \Omega_2} g^2\,\mathrm{d}\mathbb{P} \right)^{1/2}$$

$$= \|f\|_{L^2(\Omega_1)} \|g\|_{L^2(\Omega_2)}. \tag{4.4.3}$$

Thus, for every $f \in L^2(\Omega_1)$,

$$\varphi_f(g) := \int_{\Omega_1 \times \Omega_2} f\,g\,\mathrm{d}\mathbb{P}$$

defines a bounded linear functional on $L^2(\Omega_2)$. By the Riesz–Fréchet representation theorem [Riesz, 1907; Fréchet, 1907] there exists a unique Tf in $L^2(\Omega_2)$ such that, for every g in $L^2(\Omega_2)$,

$$\int_{\Omega_1 \times \Omega_2} f\,g\,\mathrm{d}\mathbb{P} = \int_{\Omega_2} (Tf)\,g\,\mathrm{d}\mathbb{P}_2. \tag{4.4.4}$$

The map $T : L^2(\Omega_1) \to L^2(\Omega_2)$ thus defined is linear by definition and bounded by (4.4.3). It is immediately seen that property (M1) is satisfied for every f in $L^2(\Omega_1)$. Since

$$\int_{\Omega_2} (T1)\, g\, d\,\mathbb{P}_2 = \int_{\Omega_1 \times \Omega_2} g\, d\,\mathbb{P} = \int_{\Omega_2} g\, d\,\mathbb{P}_2\,,$$

for every g in $L^2(\Omega_2)$, one has $T1 = 1$. Finally, setting $g = 1$ in (4.4.4), one has

$$\mathbb{E}_2(Tf) = \int_{\Omega_2} (Tf)\, d\,\mathbb{P}_2 = \int_{\Omega_1 \times \Omega_2} f\, d\,\mathbb{P} = \int_{\Omega_1} f\, d\,\mathbb{P}_1 = \mathbb{E}_1(f)\,,$$

which proves that T is a Markov operator. $\qquad\square$

The probability measure defined by (4.4.1) will be denoted by $\mathbb{P}_T$, while the Markov operator defined by (4.4.2) will be denoted by $T_{\mathbb{P}}$.

A simple method of constructing Markov operators is given in the following theorem.

Theorem 4.4.5. *Let $(\Omega_1, \mathscr{F}_1, \mathbb{P}_1)$ and $(\Omega_2, \mathscr{F}_2, \mathbb{P}_2)$ be probability spaces. If the function $K : \Omega_1 \times \Omega_2 \to \mathbb{R}$ is positive, measurable with respect to the product σ-field $\mathscr{F}_1 \otimes \mathscr{F}_2$, and such that*

$$\int_{\Omega_1} K(x,y)\, d\,\mathbb{P}_1(x) = 1 \quad \mathbb{P}_2\text{-}a.e.\,, \tag{4.4.5}$$

$$\int_{\Omega_2} K(x,y)\, d\,\mathbb{P}_2(y) = 1 \quad \mathbb{P}_1\text{-}a.e.\,, \tag{4.4.6}$$

then

(a) *a Markov operator from $L^\infty(\Omega_1)$ into $L^\infty(\Omega_2)$ is defined via*

$$(Tf)(y) := \int_{\Omega_1} K(x,y)\, f(x)\, d\,\mathbb{P}_1(x)\,;$$

(b) *the unique probability measure $\mathbb{P}_T$ associated with T is given, for $S \in \mathscr{F}_1 \otimes \mathscr{F}_2$, by*

$$\mathbb{P}_T(S) = \int_S K(x,y)\, d(\mathbb{P}_1 \otimes \mathbb{P}_2)(x,y)\,, \tag{4.4.7}$$

where $\mathbb{P}_1 \otimes \mathbb{P}_2$ is the product probability on $\Omega_1 \times \Omega_2$.

Proof. (a) Let f be in $L^\infty(\Omega_1)$; then, because of (4.4.5),

$$|(Tf)(y)| = \left| \int_{\Omega_1} K(x,y)\, f(x)\, d\,\mathbb{P}_1(x) \right| \le \|f\|_{L^\infty(\mathbb{P}_1)}\,,$$

which shows that Tf is in $L^\infty(\Omega_2)$. Since properties (M1) and (M2) are obvious, only (M3) remains to be checked. By Fubini's theorem and by (4.4.6)

$$
\begin{aligned}
\mathbb{E}_2(Tf) &= \int_{\Omega_2} (Tf)(y)\,\mathrm{d}\,\mathbb{P}_2(y) = \int_{\Omega_2} \mathrm{d}\,\mathbb{P}_2(y) \int_{\Omega_1} K(x,y)\,f(x)\,\mathrm{d}\,\mathbb{P}_1(x) \\
&= \int_{\Omega_1} f(x)\,\mathrm{d}\,\mathbb{P}_1(x) \int_{\Omega_2} K(x,y)\,\mathrm{d}\,\mathbb{P}_2(y) = \int_{\Omega_1} f(x)\,\mathrm{d}\,\mathbb{P}_1(x) = \mathbb{E}_1(f)\,,
\end{aligned}
$$

which establishes (a).

(b) It suffices to find the expression of $\mathbb{P}_T$ on the family $\mathscr{F}_1 \times \mathscr{F}_2$ of measurable rectangles. Thus, if $S = A \times B$, (4.4.1) yields

$$
\begin{aligned}
\mathbb{P}_T(A \times B) &= \int_{\Omega_2} (T\,\mathbf{1}_A)(y)\,\mathbf{1}_B(y)\,\mathrm{d}\,\mathbb{P}_2(y) \\
&= \int_{\Omega_2} \mathbf{1}_B(y)\,\mathrm{d}\,\mathbb{P}_2(y) \int_{\Omega_1} K(x,y)\mathbf{1}_A(x)\,\mathrm{d}\,\mathbb{P}_1(x) \\
&= \int_B \mathrm{d}\,\mathbb{P}_2(y) \int_A K(x,y)\,\mathrm{d}\,\mathbb{P}_1(x) = \int_{A\times B} K(x,y)\,\mathrm{d}\,\mathbb{P}_1(x)\,\mathrm{d}\,\mathbb{P}_2(y) \\
&= \int_{A\times B} K(x,y)\,\mathrm{d}(\mathbb{P}_1 \otimes \mathbb{P}_2)(x,y)\,,
\end{aligned}
$$

which establishes (b) and concludes the proof. $\square$

The function K of Theorem 4.4.5 will be called the *kernel* of the Markov operator T or of the probability measure $\mathbb{P}_T$ with respect to the marginals $\mathbb{P}_1$ and $\mathbb{P}_2$.

Corollary 4.4.6. *Let $\mathbb{P}_1$ and $\mathbb{P}_2$ be the marginals of the probability measure $\mathbb{P}$ on the measurable space $(\Omega_1 \times \Omega_2, \mathscr{F}_1 \otimes \mathscr{F}_2)$. Then the following statements are equivalent:*

(a) *the Markov operator $T_\mathbb{P} : L^\infty(\Omega_1) \to L^\infty(\Omega_2)$ can be represented in the form (4.4.7) for a kernel K;*

(b) *$\mathbb{P}$ is absolutely continuous with respect to the product measure $\mathbb{P}_1 \otimes \mathbb{P}_2$ and the kernel K is given by the Radon–Nikodym derivative*

$$
K = \frac{\mathrm{d}\mathbb{P}}{\mathrm{d}(\mathbb{P}_1 \otimes \mathbb{P}_2)}\,.
$$

Proof. Since (a)$\Longrightarrow$(b) can be easily derived, the converse implication is considered. To this end, assume that $\mathbb{P}$ is absolutely continuous with respect to $\mathbb{P}_1 \otimes \mathbb{P}_2$. Then, by the Radon–Nikodym theorem, set K equal to the Radon–Nikodym derivative of $\mathbb{P}$ with respect to $\mathbb{P}_1 \otimes \mathbb{P}_2$. From this and $\mathbb{P}(A \times B) = \int_B T\mathbf{1}_A\,\mathrm{d}\mathbb{P}_2$ one can deduce

$$
\int_B Tf\,\mathrm{d}\mathbb{P}_2 = \int_B \int_{\Omega_1} K(x,y)f(x)\,\mathrm{d}\mathbb{P}_1\,\mathrm{d}\mathbb{P}_2\,,
$$

whenever $f \in L^\infty(\Omega_1)$ and B is a measurable subset of Ω_2. It is then easily seen that

$$\int_{\Omega_1} K(x,y) \, d\mathbb{P}_2 = 1 \quad \text{and} \quad \int_{\Omega_2} K(x,y) \, d\mathbb{P}_1 = 1,$$

which concludes the proof. $\square$

It is remarkable that a Markov operator can be expressed as a conditional expectation. To see this, one introduces the canonical projection $\pi : X \times Y \to Y$.

Corollary 4.4.7. *Let $\mathbb{P}$ be a probability measure on $(\Omega_1 \times \Omega_2, \mathscr{F}_1 \otimes \mathscr{F}_2)$ with marginals $\mathbb{P}_1$ and $\mathbb{P}_2$. Consider the associated Markov operator $T : L^\infty(\Omega_1) \to L^\infty(\Omega_2)$. If $f \in L^\infty(\Omega_1, \mathbb{P}_1)$ is regarded as a random variable on $(\Omega_1 \times \Omega_2, \mathscr{F}_1 \otimes \mathscr{F}_2, \mathbb{P})$, then*

$$Tf(y) = E(f \mid \pi = y) \qquad \mathbb{P}_2\text{-a.e.} \tag{4.4.8}$$

Proof. For every $B \in \mathscr{F}_2$, in view of (4.4.2) with $g = 1_B \in L^\infty(\Omega_2)$ (or, analogously, $g = 1_{\pi \in B} \in L^\infty(\Omega_1 \times \Omega_2)$), one has

$$\int_B (Tf) \, d\mathbb{P}_2 = \int_{\pi^{-1}(B)} f \, d\mathbb{P}.$$

By definition of conditional expectation (see, e.g., eq. 3.4.2), one also has

$$\int_{\pi^{-1}(B)} f \, d\mathbb{P} = \int_B E(f \mid \pi = y) \, d\mathbb{P}_2(y),$$

from which the desired assertion follows. $\square$

Theorem 4.4.4 can be reformulated in terms of copulas and associated probability measures in the following way.

Factor $\mathbb{I}^d$ as $\mathbb{I}^d = \mathbb{I} \times \mathbb{I}^{d-1}$, and use the convention $(x, \mathbf{y}) \in \mathbb{I}^d$, where $x \in \mathbb{I}$. For a copula $C \in \mathscr{C}_d$ let μ_C be the d-fold stochastic measure associated with it and let μ_C^1 be its marginal on the Borel sets of $\mathbb{I}^{d-1}$, so that $\mu_C^1(S) = \mu_C(\mathbb{I} \times S)$ for every Borel set S of $\mathbb{I}^{d-1}$. Consider the associated Markov operator

$$T : L^\infty(\mathbb{I}^{d-1}, \mu_C^1) \to L^\infty(\mathbb{I}, \lambda). \tag{4.4.9}$$

If π_1 is the canonical projection $\pi_1(\mathbf{u}) = u_1$ on $\mathbb{I}^d$, then one has

$$Tf(x) = E(f \mid \pi_1 = x) \qquad \text{for } \lambda \text{ almost every } x \in \mathbb{I}.$$

In particular, if $f = 1_R$, where $R = [\mathbf{0}, \mathbf{y}] \subseteq \mathbb{I}^{d-1}$, one has

$$T1_R(x) = E(1_R \mid \pi_1 = x) = \partial_1 C(x, \mathbf{y}) \qquad \text{for } \lambda\text{-almost every } x \in \mathbb{I},$$

where the last equality follows by Theorem 3.4.1. Moreover, in view of Theorem 3.4.4, it follows that

$$T1_R(x) = K_C(x, [\mathbf{0}, \mathbf{y}]) \qquad \text{for } \lambda\text{-almost every } x \in \mathbb{I}.$$

The previous procedure can be repeated when we factorise $\mathbb{I}^d$ with respect to a different coordinate. Specifically, factor $\mathbb{I}^d$ in the following way: $\mathbb{I}^d = \mathbb{I}^{j-1} \times \mathbb{I} \times \mathbb{I}^{d-j}$ and write

$$\mathbb{I}^{(d,j)} := \{(u_1, \ldots, u_{j-1}, u_{j+1}, \ldots, u_d) : u_i \in \mathbb{I} \ (i = 1, \ldots, j-1, j+1, \ldots, d)\} \ .$$

For a copula $C \in \mathscr{C}_d$ let μ_C be the d-fold stochastic measure associated with it and let $\mu^{d,j}$ be its marginal on the Borel sets of $\mathbb{I}^{(d,j)}$, so that $\mu^{d,j}(S) = \mu_C(S \times \mathbb{I})$ for every Borel set S of $\mathbb{I}^{(d,j)}$. For every $k = 1, \ldots, d$, one has a Markov operator

$$T^{(j)} : L^\infty(\mathbb{I}^{(d,j)}) \to L^\infty(\mathbb{I}, \lambda) \ . \tag{4.4.10}$$

Now, analogously to the previous case, the following result can be formulated.

Theorem 4.4.8. *Let C be a copula in $\mathscr{C}_d$ and μ_C its associated measure. For $k = 1, \ldots, d$, let $T^{(j)}$ be the Markov operator of (4.4.10) and let $\pi_j : \mathbb{I}^d \to \mathbb{I}$ be the canonical j-th projection, i.e., $\pi_j(\mathbf{u}) = u_j$. For $(u_1, \ldots, u_{j-1}, u_{j+1}, \ldots, u_d)$, let*

$$R := [0, u_1] \times \cdots \times [0, u_{j-1}] \times [0, u_{j+1}] \times \ldots [0, u_d] \ .$$

Then, for $\mathbf{u}_j(t) \in \mathbb{I}^d$, one has

(a) $T^{(j)} f(t) = E(f \mid \pi_j = t)$ *for λ-almost all $t \in \mathbb{I}$.*

(b) $T^{(j)} 1_R(t) = \partial_j C(u_1, \ldots, u_{j-1}, t, u_{j+1}, \ldots, u_d)$ *for λ-a.e. $t \in \mathbb{I}$.*

(c) $T^{(j)} 1_R(t) = K_{C^{\tau_j}}(t, R)$ *for λ-a.e. $t \in \mathbb{I}$, where $= K_{C^{\tau_j}}$ is the Markov kernel associated with the copula C^{τ_j}, where τ_j is the permutation of $\mathbb{I}^d$ defined in eq. (4.3.7).*

4.4.1   Bivariate copulas and Markov operators

Since 2-copulas are in one-to-one correspondence with doubly stochastic measures (which are probability measures), Theorem 4.4.4 can be applied to this situation as well. Here, the construction of the Markov operator associated with a 2-copula can be made explicit, as the following result shows.

Theorem 4.4.9. *For every Markov operator T on $L^\infty(\mathbb{I})$ the function C_T defined on $\mathbb{I}^2$ via*

$$C_T(u, v) := \int_0^u \left(T1_{[0,v]}\right)(s) \, ds \tag{4.4.11}$$

is a 2-copula.
Conversely, for every 2-copula C, the operator T_C defined on $L^\infty(\mathbb{I})$ via

$$(T_C f)(x) := \frac{d}{dx} \int_0^1 \partial_2 C(x, t) f(t) \, dt \tag{4.4.12}$$

is a Markov operator on $L^\infty(\mathbb{I})$. Furthermore, for every $f \in L^\infty(\mathbb{I})$ and for every $g \in L^\infty(\mathbb{I})$,

$$\int_{\mathbb{I}} (Tf)\, g \, \mathrm{d}\lambda = \int_{\mathbb{I}^2} f\, g \, \mathrm{d}C \,. \tag{4.4.13}$$

Proof. The first part is a direct consequence of Theorem 4.4.4 in view of eq. (4.4.1). For the second part, consider that, for every function $f \in L^\infty(\mathbb{I})$, the function

$$\mathbb{I} \ni x \mapsto \varphi(x) := \int_0^1 \partial_2 C(x,t)\, f(t)\, \mathrm{d}t \tag{4.4.14}$$

satisfies a Lipschitz condition with constant $\|f\|_\infty$. For, if $x_1 < x_2$, then, because of (1.6.3),

$$|\varphi(x_2) - \varphi(x_1)| = \left| \int_0^1 (\partial_2 C(x_2,t) - \partial_2 C(x_1,t))\, f(t)\, \mathrm{d}t \right|$$

$$\leq \|f\|_\infty \int_0^1 |\partial_2 C(x_2,t) - \partial_2 C(x_1,t)| \, \mathrm{d}t$$

$$\leq \|f\|_\infty (C(x_2,1) - C(x_1,1)) = \|f\|_\infty (x_2 - x_1) \,.$$

Since the function $x \mapsto \partial_2 C(x,t)$ is increasing for almost all $t \in \mathbb{I}$, the functions

$$x \mapsto \int_0^1 \partial_2 C(x,t)\, (|f(t)| - f(t))\, \mathrm{d}t, \quad x \mapsto \int_0^1 \partial_2 C(x,t)\, (|f(t)| + f(t))\, \mathrm{d}t$$

are both increasing, and, as a consequence, are differentiable almost everywhere. Since the function (4.4.14) is a linear combination of the above functions, it is also differentiable almost everywhere. Therefore, the derivative in eq. (4.4.12) exists a.e.

It follows from what has just been proved that the derivative in eq. (4.4.12) is bounded above by $\|f\|_\infty$, so that $T_C f$ belongs to L^∞. It also follows from the above considerations and from eq. (1.6.2) that $\partial_2 C(x,t)\, f(t)$ is positive whenever f is positive, so that T_C is also positive. Property $T_C 1 = 1$ follows from a direct calculation. In order to show that T_C is expectation invariant, one has, since both f and $\partial_2 C$ are bounded a.e.,

$$\mathbb{E}(T_C f) = \int_0^1 (T_C f)(x)\, \mathrm{d}x = \int_0^1 \left(\frac{d}{dx} \int_0^1 \partial_2 C(x,t)\, f(t)\, \mathrm{d}t \right) \mathrm{d}x$$

$$= \int_0^1 (\partial_2 C(1,t) - \partial_2 C(0,t))\, f(t)\, \mathrm{d}t = \int_0^1 f(t)\, \mathrm{d}t = \mathbb{E}(f) \,.$$

Here use has been made of the fact that the function (4.4.14) satisfies a Lipschitz condition and is, therefore, absolutely continuous.

Finally, for every rectangle A and B in $\mathbb{I}$, it can be easily proved that

$$\int_0^1 (T 1_B) 1_A = \int_{\mathbb{I}^2} 1_B\, 1_A \, \mathrm{d}\mathbb{P} = \mathbb{P}(A \times B) \,.$$

As a consequence of the properties of a Markov operator, it follows that T_C also satisfies (4.4.13).   □

Example 4.4.10. The Markov operators corresponding to the three 2-copulas W_2, M_2 and Π_2 are easily computed. For every $f \in L^\infty(\mathbb{I})$ and for every $x \in \mathbb{I}$, one has

$$(T_{W_2} f)(x) = f(1-x), \tag{4.4.15}$$

$$(T_{M_2} f)(x) = f(x), \tag{4.4.16}$$

$$(T_{\Pi_2} f)(x) = \int_0^1 f \, d\lambda. \tag{4.4.17}$$

■

If $\mathscr{T}$ denotes the class of Markov operators in $L^\infty(\mathbb{I}, \mathscr{B}(\mathbb{I}), \lambda)$, Theorem 4.4.9 introduces the following two mappings:

- $\Phi : \mathscr{C}_2 \to \mathscr{T}$ defined by $\Phi(C) := T_C$, where T_C is given by eq. (4.4.12),
- $\Psi : \mathscr{T} \to \mathscr{C}_2$ defined by $\Psi(T) := C_T$, where C_T is given by eq. (4.4.11).

The next theorem is about the mappings Φ and Ψ.

Theorem 4.4.11. *The composition mappings $\Phi \circ \Psi$ and $\Psi \circ \Phi$ are the identity operators on $\mathscr{T}$ and $\mathscr{C}_2$, respectively.*

Proof. Consider, first, the composition $\Psi \circ \Phi$. Let a copula $C \in \mathscr{C}_2$ be given.

$$(\Psi \circ \Phi(C))(u,v) = C_{\Phi(C)}(u,v) = \int_0^u \left(T_C \mathbf{1}_{[0,v]}\right)(s)\,ds$$

$$= \int_0^u ds \frac{d}{ds} \int_0^1 \partial_2 C(s,t)\,\mathbf{1}_{[0,v]}(t)\,dt$$

$$= \int_0^1 \partial_2 C(u,t)\,\mathbf{1}_{[0,v]}(t)\,dt = \int_0^v \partial_2 C(u,t)\,dt = C(u,v).$$

Therefore $\Psi \circ \Phi = \mathrm{id}_{\mathscr{C}_2}$.

In order to show that $\Phi \circ \Psi$ is the identity operator $\mathrm{id}_{\mathscr{T}}$ on $\mathscr{T}$ it suffices to show that, for every Markov operator T and for every indicator function $\mathbf{1}_{[0,v]}$, one has

$$\Phi \circ \Psi(T)\mathbf{1}_{[0,v]} = T\mathbf{1}_{[0,v]};$$

in fact, then $\Phi \circ \Psi$ and $\mathrm{id}_{\mathscr{T}}$ agree on all the linear combinations of indicators of intervals of the type $]v, v'] \subseteq \mathbb{I}$. These linear combinations are dense in L^∞, so that $\Phi \circ \Psi$ and $\mathrm{id}_{\mathscr{T}}$ coincide in L^∞. For every $v \in \mathbb{I}$,

$$\left(\Phi \circ \Psi(T)\,\mathbf{1}_{[0,v]}\right)(x) = \left(T_{\Psi(T)}\mathbf{1}_{[0,v]}\right)(x)$$

$$= \frac{d}{dx} \int_0^1 \mathbf{1}_{[0,v]}\,dt \frac{\partial}{\partial t} \int_0^x \left(T\mathbf{1}_{[0,t]}\right)(s)\,ds$$

$$= \frac{d}{dx} \int_0^v dt \frac{\partial}{\partial t} \int_0^x \left(T\mathbf{1}_{[0,t]}\right)(s)\,ds$$

$$= \frac{d}{dx} \int_0^x \left(T\mathbf{1}_{[0,v]}\right)(s)\,ds = T\mathbf{1}_{[0,v]}(x),$$

which proves the assertion. □

We examine next the case of Markov operators corresponding to copulas represented through measure-preserving transformations (see Section 3.5). The Markov operator corresponding to the copula $C_{f,g}$ will be denoted by $T_{f,g}$ rather than by $T_{C_{f,g}}$.

Theorem 4.4.12. *Let $f : \mathbb{I} \to \mathbb{I}$ be measure-preserving and let i denote the identity function on $\mathbb{I}$, $i = \mathrm{id}_{\mathbb{I}}$. Then, for every $\varphi \in L^1(\mathbb{I})$,*

$$T_{i,f}\,\varphi = \varphi \circ f \qquad a.e.$$

Moreover, if there exists a function $\psi \in L^1(\mathbb{I})$ such that $\varphi = \psi \circ f$ a.e., then

$$T_{f,i}\,\varphi = \psi \qquad a.e.$$

Proof. Assume $\varphi \in C^\infty(\mathbb{I})$. Then, an integration by parts and (3.5.6) yield, for every $x \in \mathbb{I}$,

$$
\begin{aligned}
(T_{i,f}\,\varphi)\,(x) &= \frac{d}{dx}\int_{\mathbb{I}} \partial_2 C_{i,f}(x,t)\,\varphi(t)\,\mathrm{d}t \\
&= \frac{d}{dx}\,C_{i,f}(x,1)\,\varphi(1) - \int_{\mathbb{I}} \partial_1 C_{i,f}(x,t)\,\varphi'(t)\,\mathrm{d}t \\
&= \varphi(1) - \int_0^1 \mathbf{1}_{[f(x),1]}(t)\,\varphi'(t)\,\mathrm{d}t = \varphi(1) - \int_{f(x)}^1 \varphi'(t)\,\mathrm{d}t \\
&= \varphi(1) - \varphi(1) + \varphi(f(x)) = \varphi(f(x))\,.
\end{aligned}
$$

But $C^\infty(\mathbb{I})$ is dense in $L^1(\mathbb{I})$, and f is measure-preserving so that, if φ is in $L^1(\mathbb{I})$ and (φ_n) is a sequence of functions of $C^\infty(\mathbb{I})$ converging to φ in $L^1(\mathbb{I})$,

$$
\begin{aligned}
\|\varphi_n \circ f - \varphi \circ f\|_1 &= \int_{\mathbb{I}} |\varphi_n \circ f - \varphi \circ f|\,\mathrm{d}\lambda = \int_{\mathbb{I}} |\varphi_n - \varphi|\,\mathrm{d}(\lambda f^{-1}) \\
&= \int_{\mathbb{I}} |\varphi_n - \varphi|\,\mathrm{d}\lambda = \|\varphi_n - \varphi\|_1 \xrightarrow[n\to+\infty]{} 0\,.
\end{aligned}
$$

Thus

$$
\begin{aligned}
\|T_{i,f}\,\varphi - \varphi \circ f\|_1 &\le \|T_{i,f}\,\varphi - T_{i,f}\varphi_n\|_1 + \|T_{i,f}\,\varphi_n - \varphi_n \circ f\|_1 \\
&\quad + \|\varphi_n \circ f - \varphi \circ f\|_1 \le 2\,\|\varphi_n \circ f - \varphi \circ f\|_1 \xrightarrow[n\to\infty]{} 0\,.
\end{aligned}
$$

Therefore $T_{i,f}\,\varphi = \varphi \circ f$ a.e., which proves the first assertion.

Now let $\varphi = \psi \circ f$ for some function $\psi \in L^1(\mathbb{I})$; then, because of (3.5.5) and since f is measure-preserving, one has a.e.

$$
\begin{aligned}
(T_{f,i}\,\varphi)\,(x) &= \frac{d}{dx}\int_{\mathbb{I}} \partial_2 C_{f,i}(x,t)\,\varphi(t)\,\mathrm{d}t = \frac{d}{dx}\int_{\mathbb{I}} \mathbf{1}_{[0,x]}(f(t))\,\psi(f(t))\,\mathrm{d}t \\
&= \frac{d}{dx}\int_{\mathbb{I}} \mathbf{1}_{[0,x]}(t)\,\psi(t)\,\mathrm{d}t = \frac{d}{dx}\int_0^x \psi(t)\,\mathrm{d}t = \psi(x)\,,
\end{aligned}
$$

which concludes the proof. $\square$

Historical remark 4.4.13. The study of Markov operators and copulas (doubly stochastic measures) was started by Brown [1965, 1966] and continued by Darsow et al. [1992], and, more recently, by Mikusiński and Taylor [2009, 2010]. The relationship between 2-copulas and Markov operators has been considered by Olsen et al. [1996]. The representation (4.4.12) improves previous results by Ryff [1963], who, using an older characterisation by Kantorovich and Vulich [1937], proved that every Markov operator T on $L^\infty(\mathbb{I})$ may be represented in the form

$$(Tf)(x) = \frac{d}{dx} \int_0^1 K(x,t)\,f(t)\,\mathrm{d}t,$$

where the kernel K is measurable and satisfies a suitable set of conditions. It is easily seen that the partial derivative $\partial_2 C$ of a copula C satisfies those conditions.

4.5   Convergence in the sense of Markov operators

The representation of a copula in terms of Markov operators suggests a natural way to introduce another type of convergence, as illustrated below.

Definition 4.5.1. Given a sequence (C_n) of copulas in $\mathscr{C}_d$ one says that it converges in the $\mathfrak{M}$ sense to the copula $C \in \mathscr{C}_d$, and writes

$$C_n \xrightarrow[n\to\infty]{\mathfrak{M}} C \qquad \text{or} \qquad \mathfrak{M} - \lim_{n\to\infty} C_n = C\,,$$

if, for every $j = 1, \ldots, d$ and for every bounded measurable function $f : \mathbb{I}^{d-1} \to \mathbb{R}$, one has

$$\lim_{n\to\infty} \int_0^1 \left| T_n^{(j)} f - T^{(j)} f \right|\, \mathrm{d}\lambda = 0\,,$$

or, equivalently,

$$\lim_{n\to\infty} \left\| T_n^{(j)} f - T^{(j)} f \right\|_{L^1(\mathbb{I})} = 0$$

where $T_n^{(j)}$ and $T^{(j)}$ are the Markov operators associated with C_n and C, respectively, from Theorem 4.4.8. The topology of $\mathfrak{M}$-convergence will be called the *Markov topology*. $\diamond$

The following result holds.

Theorem 4.5.2. *Given a sequence (C_n) of copulas in $\mathscr{C}_d$, if $C_n \xrightarrow[n\to\infty]{\mathfrak{M}} C$, then*
$C_n \xrightarrow[n\to\infty]{\partial} C.$

Proof. It suffices to set, for $k \in \{1, \ldots, d\}$,

$$f = \mathbf{1}_{[0,u_1] \times \cdots \times [0,u_{k-1}] \times [0,u_{k+1}] \times \cdots \times [0,u_d]}\,.$$

Then, because of Theorem 4.4.8 and with the same notation,

$$\int_0^1 |\partial_k C_n(u_k(t)) - \partial_k C(u_k(t))| \ dt = \int_0^1 \left| (T_n^{(k)} f)(t) - (T^{(k)} f)(t) \right| \ dt \,,$$

which tends to zero as n goes to $+\infty$. $\qquad\square$

In the remainder of this section it will be seen that the implication of Theorem 4.5.2 cannot be reversed.

Remark 4.5.3. Both ∂- and $\mathfrak{M}$-convergences amount to L^1 convergence of conditional expectations in the sense that $C_n \to C$ they be equivalently written as

$$\mathbb{E}_n(f \mid \pi_j = t) \to \mathbb{E}(f \mid \pi_j = t) \,,$$

where the expectation is calculated with respect to the measure induced by C_n and C, respectively, while f is any function belonging to a specified class. Specifically, in the case of $\mathfrak{M}$-convergence the function f is in $L^\infty(\mathbb{I}^{d-1})$, while for ∂-convergence, $f = 1_R$ for a suitable rectangle $R \subseteq \mathbb{I}^{d-1}$. $\qquad\blacksquare$

Below we present sufficient conditions that ensure $\mathfrak{M}$-convergence.

Theorem 4.5.4. *Let* $(C_n)_{n\in\mathbb{Z}_+}$ *be a sequence of copulas in* $\mathscr{C}_d$ *and let* $(\mu_n)_{n\in\mathbb{Z}_+}$ *be the corresponding sequence of measures. If*

(a) *each of the copulas* C_n *has continuous density* c_n $(n \in \mathbb{Z}_+)$,

(b) $\|c_n - c_0\|_1 \xrightarrow[n\to\infty]{L^1(\mathbb{I}^d, \lambda_d)} 0,$

then

$$\mathfrak{M} - \lim_{n\to\infty} C_n = C_0 \,.$$

Proof. Consider the factorisation $\mathbb{I}^d = \mathbb{I} \times \mathbb{I}^{d-1}$. Since C_n admits a continuous density, the Markov operator $T_n : L^\infty(\mathbb{I}^{d-1}) \to L^\infty(\mathbb{I})$ associated with C_n is given, for $f \in L^\infty(\mathbb{I}^{d-1})$, by

$$(T_n f)(u) = \int_{\mathbb{I}^{d-1}} c_n(u, \mathbf{v}) \, f(\mathbf{v}) \ d\mathbf{v} \,,$$

where $u \in \mathbb{I}$ and $\mathbf{v} \in \mathbb{I}^{d-1}$ (see [Mikusiński and Taylor, 2009, Remark 2]).

It follows that, for every bounded measurable $f : \mathbb{I}^{d-1} \to \mathbb{R}$, one has

$$0 \le \int_{\mathbb{I}} |T_n f - T_0 f| \ d\lambda = \int_{\mathbb{I}} \left| \int_{\mathbb{I}^{d-1}} (c_n(u, \mathbf{v}) - c_0(u, \mathbf{v})) \, f(\mathbf{v}) \, d\mathbf{v} \right| \ du$$

$$\le \|f\|_\infty \int_{\mathbb{I}^d} |c_n(u, \mathbf{v}) - c_0(u, \mathbf{v})| \ d\mathbf{v} \ du \xrightarrow[n\to\infty]{} 0 \,,$$

whence the assertion. $\qquad\square$

The preceding theorem has the following consequence for the checkerboard and Bernstein approximations.

Theorem 4.5.5. *Let the copula $C \in \mathscr{C}_d$ be absolutely continuous with continuous density c and let (C_n) and (B_n) be the sequences of checkerboard and Bernstein approximations, respectively. Then*

$$\mathfrak{M} - \lim_{n\to\infty} C_n = C \qquad and \qquad \mathfrak{M} - \lim_{n\to\infty} B_n = C\,.$$

Proof. We shall only prove the result for the checkerboard approximations, the proof for the Bernstein approximation being similar, but notationally more involved. It will be proved that the sequence of densities (c_n) converges uniformly λ_d-a.e. to c, the density of C. But uniform convergence λ_d-a.e. implies convergence in $L^1(\mathbb{I}^d, \lambda_d)$; this yields the assertion by virtue of Theorem 4.5.4.

Let (C_n) be the sequence of checkerboard approximations. Because of the uniform continuity of c on $\mathbb{I}^d$, for $\varepsilon > 0$, one can choose n large enough to have

$$|c(\mathbf{u}) - c(\mathbf{v})| < \varepsilon\,,$$

for all $\mathbf{u}$ and $\mathbf{v}$ in $B_{\mathbf{j}}$ of eq. (4.1.1). Also because of the continuity of c, for every $\mathbf{j} = (j_1, \ldots, j_d)$, there exists $\mathbf{u_j} \in B_{\mathbf{j}}$ such that

$$\mu_C(B_{\mathbf{j}}) = \int_{B_{\mathbf{j}}} c\, \mathrm{d}\lambda_d = c(\mathbf{u_j})\,\frac{1}{n^d}\,.$$

Choose $\mathbf{u}$ in the interior of some box $B_{\mathbf{j}}$; the set of such $\mathbf{u}$'s has λ_d-measure 1. The density c_n of C_n may be written in the form

$$c_n = n^d \sum_{\mathbf{j}} \mu_C(B_{\mathbf{j}}) \mathbf{1}_{I_{d,n,\mathbf{j}}}\,;$$

then

$$.\,|c_n(\mathbf{u}) - c(\mathbf{u})| = \left| n^d \sum_{\mathbf{j}} \mu_C(B_{\mathbf{j}}) \mathbf{1}_{B_{\mathbf{j}}}(\mathbf{u}) - c(\mathbf{u}) \right| = |c(\mathbf{u_j}) - c(\mathbf{u})| < \varepsilon\,,$$

which establishes the assertion for the checkerboard approximations. $\square$

Example 4.5.6. Let M_3 be the comonotonicity 3-copula. Then μ_{M_3} concentrates the probability mass on the main diagonal $\{(u_1, u_2, u_3) : u_1 = u_2 = u_3\}$ of $\mathbb{I}^3$. Consider the associated Markov operator $T_{M_3} : L^\infty(\mathbb{I}^2) \to L^\infty(\mathbb{I})$; for a measurable bounded function $f : \mathbb{I}^2 \to \mathbb{R}$ and $\varphi \in L^\infty(\mathbb{I})$, one has

$$\int_{\mathbb{I}} (T_{M_3}f)(t)\, \varphi(t)\, \mathrm{d}t = \int_{\mathbb{I}^3} f(u_1, u_2)\, \varphi(t)\, \mathrm{d}\mu_{M_3}(u_1, u_2, t) = \int_0^1 f(t, t)\, \varphi(t)\, \mathrm{d}t\,.$$

Therefore, if $\widetilde{\mu}$ denotes the marginal measure of μ_{M_3} on $\mathbb{I}^2$, one has

$$(T_{M_3}f)(t) = f(t,t) \quad \widetilde{\mu}\text{-a.e.}$$

Let C_n be the checkerboard approximation of M_3; Li et al. [1998] show that the corresponding Markov operator $T_n : L^\infty(\mathbb{I}^2) \to L^\infty(\mathbb{I})$ is given by

$$(T_n f)(t) = n^2 \sum_{j=0}^{n-1} I(j,n) \, \mathbf{1}_{]\frac{j}{n}, \frac{j+1}{n}]}(t) \,,$$

where

$$I(j,n) := \int_{]\frac{j}{n}, \frac{j+1}{n}]^2} f \, d\lambda_2 \,.$$

For the function

$$f(u_1, u_2) = \begin{cases} 1, & u_1 = u_2 \,, \\ 0, & \text{elsewhere,} \end{cases}$$

one has $T_{M_3}f = 1$.

On the other hand, let $\widetilde{\gamma}_n$ be the marginal on $\mathbb{I}^2$ of the measure induced by C_n. Since f has a support of λ_2-measure equal to zero, $f = 0$ $\widetilde{\gamma}_n$-a.e. Thus $T_n f = 0$ and, for every $n \in \mathbb{N}$,

$$\int_{\mathbb{I}} |T_n f - T_{M_3}f| \, d\lambda = 1 \,,$$

which, obviously, does not converge to zero. One concludes that (C_n) does not converge to M_3 in the $\mathfrak{M}$ sense.   ∎

While the sequence (C_n) of checkerboard approximations converges to C if the d-copula C and all the copulas involved are absolutely continuous with continuous densities, as was seen above (Theorem 4.5.5), a weaker result holds in general. In order to prove it the following preliminary result will be needed.

Lemma 4.5.7. *Let $(T_n)_{n \in \mathbb{Z}_+}$ be a sequence of Markov operators and let D be a dense subset of $\mathbb{I}^{d-1}$. Consider the rectangle*

$$R := \mathbf{1}_{[0,u_1] \times \cdots \times [0,u_{k-1}] \times [0,u_{k+1}] \times \cdots \times [0,u_d]}$$

for $k = 1, \ldots, d$ and for $\mathbf{u}_k = (u_1, \ldots, u_{k-1}, u_{k+1}, \ldots, u_d) \in D$. If for every $\mathbf{u}_k \in D$,

$$\lim_{n \to +\infty} T_n \mathbf{1}_R = T_0 \mathbf{1}_R \qquad \text{a.e. on } \mathbb{I} \,,$$

then, for every $\mathbf{u}_k \in D$,

$$\lim_{n \to +\infty} \int_{\mathbb{I}^d} |T_n \mathbf{1}_R - T_0 \mathbf{1}_R| \, d\lambda_d = 0 \,.$$

Proof. Since Markov operators have L^∞-norm 1, one has $\|T_n f\|_\infty \le \|f\|_\infty$ for every function $f \in L^\infty(\mathbb{I}^d)$ and for every $n \in \mathbb{Z}_+$. The assertion is now an immediate consequence of the dominated convergence theorem.   □

Theorem 4.5.8. *For every copula $C \in \mathscr{C}_d$ the sequences $(C_n)_{n\in\mathbb{N}}$ and $(B_n)_{n\in\mathbb{N}}$ of its checkerboard and Bernstein approximations ∂-converge to C.*

Proof. Also in this case, as in the proof of Theorem 4.5.5, and for the same reasons, we shall present the proof only for the checkerboard approximations.

In view of Lemma 4.5.7 one need only show that

$$\lim_{n\to+\infty} T_{C_n}^{(j)} \mathbf{1}_R = T_C^{(j)} \mathbf{1}_R \qquad a.e. \text{ on } \mathbb{I},$$

for every $\mathbf{u}_k(t) = (u_1, \ldots, u_{k-1}, t, u_{k+1}, \ldots, u_d)$ belonging to a sufficiently fine mesh D_n. But, since

$$(T_C^{(k)} \mathbf{1}_R)(t) = \partial_k C(t, \mathbf{u}_k(t)),$$

and if t lies in the interval $\left] \frac{j_0-1}{n}, \frac{j_0}{n} \right[$,

$$(T_{C_n}^{(k)} \mathbf{1}_R)(t) = \partial_k C_n(t, \mathbf{u}_k(t)) = \frac{1}{n} \left\{ C\left(\frac{j_0}{n}, \mathbf{u}_k(t)\right) - C\left(\frac{j_0-1}{n}, \mathbf{u}_k(t)\right) \right\}.$$

Since the partial derivatives of C exist a.e., the assertion is now proved.   □

As a consequence of this latter result it follows that the sequence (C_n) of checkerboard approximations of M_3, recall Example 4.5.6, converge to M_3 in the ∂-sense, but not in the $\mathfrak{M}$-sense which proves that the implication (a) $\implies$ (b) in Theorem 4.5.2 cannot be reversed.

Historical remark 4.5.9. The modes of convergence of this section were introduced by Li et al. [1998] in the case $d = 2$; later Mikusiński and Taylor [2009, 2010] extended their study to the general case $d > 2$.

Chapter 5

The Markov product of copulas

This chapter is devoted to the study of a remarkable binary operation on the set of bivariate copulas $\mathscr{C}_2$, usually called $*$-*product*, which was introduced by Darsow et al. [1992]. We shall also refer to this operation as the *Markov product* for 2-copulas, since it exhibits interesting connexions with Markov operators and processes.

5.1 The Markov product

The basic definition of the Markov product is given below.

Theorem 5.1.1. *For all copulas A and B in $\mathscr{C}_2$, consider the function $A * B : \mathbb{I}^2 \to \mathbb{I}$ defined by*

$$(A * B)(u, v) := \int_0^1 \partial_2 A(u, t)\, \partial_1 B(t, v)\, dt\,. \tag{5.1.1}$$

Then $A * B$ is a 2-copula.

Proof. First, notice that, since the derivatives of both A and B exist a.e., are measurable and bounded below by 0 and above by 1, where they exist, the integral in eq. (5.1.1) exists and is finite.

Set $C = A * B$. Then C satisfies the boundary conditions for a copula. Moreover, for all x, u, y and v in $\mathbb{I}$ with $x \leq u$ and $y \leq v$, one has

$$C(u, v) + C(x, y) - C(x, v) - C(u, y) = \int_0^1 \partial_2 A(u, t)\, \partial_1 B(t, v)\, dt$$
$$+ \int_0^1 \partial_2 A(x, t)\, \partial_1 B(t, y)\, dt - \int_0^1 \partial_2 A(x, t)\, \partial_1 B(t, v)\, dt$$
$$- \int_0^1 \partial_2 A(u, t)\, \partial_1 B(t, y)\, dt$$
$$= \int_0^1 \partial_2 \left(A(u, t) - A(x, t)\right) \partial_1 \left(B(t, v) - B(t, y)\right)\, dt \geq 0\,,$$

because of eq. (1.6.3). $\qquad\square$

Now direct and simple computations yield, for every copula $C \in \mathscr{C}_2$ and for all u and v in $\mathbb{I}$,

$$\Pi_2 * C = C * \Pi_2 = \Pi_2, \tag{5.1.2}$$

$$M_2 * C = C * M_2 = C, \tag{5.1.3}$$

$$(W_2 * C)(u, v) = C^{\sigma_1}(u, v), \tag{5.1.4}$$

$$(C * W_2)(u, v) = C^{\sigma_2}(u, v), \tag{5.1.5}$$

where C^{σ_1} and C^{σ_2} are obtained via symmetry of $\mathbb{I}^2$ and also considered in Corollary 2.4.4. Notice that, in particular, one has

$$W_2 * W_2 = M_2, \tag{5.1.6}$$

while $(W_2 * C) * W_2 = \widehat{C}$, the survival copula associated to C.

In algebraic notation, $*$ is a binary operation on $\mathscr{C}_2$. Moreover, Π_2 is a (right and left) annihilator (i.e., a null element) of $(\mathscr{C}_2, *)$, while M_2 is a unit element of $(\mathscr{C}_2, *)$.

Remark 5.1.2. The space $(\mathscr{C}_2, *)$ has various similarities with the space of all $(n \times n)$ doubly stochastic matrices endowed with the matrix multiplication. In particular, the latter space contains a unit element, which corresponds to the identity matrix, and a null matrix that corresponds to the $n \times n$ matrix each of whose entries is $1/n$. For more details, see Darsow et al. [1992]. ∎

Example 5.1.3. Let $A \in \mathscr{C}_2$ be any bivariate copula and consider the tent copula C_θ of Example 3.2.5. Then a little algebra shows that

$$(C_\theta * A)(u, v) = \begin{cases} \theta A\left(\frac{u}{\theta}, v\right), & u \leq \theta, \\ v - (1 - \theta) A\left(\frac{1-u}{1-\theta}, v\right), & u > \theta, \end{cases}$$

and that

$$(A * C_\theta)(u, v) = A(u, \theta v) + u - A(u, 1 - (1 - \theta)v);$$

by the previous theorem both are copulas in $\mathscr{C}_2$. ∎

The following property is of immediate proof.

Theorem 5.1.4. *As a binary operation on $\mathscr{C}_2$ the $*$-product is both right and left distributive over convex combinations.*

One of the basic properties of the $*$-product is associativity. In order to prove it, a preliminary result will be needed, which is of general interest.

Theorem 5.1.5. *Consider a sequence $(A_n)_{n \in \mathbb{N}}$ of 2-copulas and a copula $B \in \mathscr{C}_2$. If A_n converges (uniformly) to $A \in \mathscr{C}_2$ as $n \to +\infty$, then both the statements*

$$A_n * B \xrightarrow[n \to +\infty]{} A * B \quad and \quad B * A_n \xrightarrow[n \to +\infty]{} B * A$$

hold. In other words the Markov product is continuous in each place with respect to the uniform convergence of copulas.

Proof. Only the first statement will be proved, the other one having a completely analogous proof. For fixed u and v in $\mathbb{I}$, set

$$g_v(t) := B(t, v) \qquad \text{and} \qquad f_{n,u}(t) := A(u, t) - A_n(u, t).$$

Because of Theorem 1.6.1, the derivatives g_v' and $f_{n,u}'$ exist a.e. on $\mathbb{I}$ and belong to $L^\infty(\mathbb{I})$; in particular $\|f_{n,u}'\|_\infty \leq 2$. There exists a non-zero simple function

$$\varphi = \sum_{j=1}^{k} \alpha_j \, 1_{[u_{j-1}, u_j]},$$

such that

$$\|\varphi - g_v'\|_\infty \leq \frac{\varepsilon}{3};$$

here $0 = u_0 < u_1 < \cdots < u_k = 1$. Since $(f_{n,u}(u_j))_{n \in \mathbb{N}}$ tends to zero for every index $j \in \{1, \ldots, k\}$, there is a natural number $N = N(\varepsilon)$ such that, for every $n \geq N$ and for every $j \in \{1, \ldots, k\}$,

$$|f_{n,u}(u_j) - f_{n,u}(u_{j-1})| < \frac{\varepsilon}{3 \sum_{j=1}^{k} |\alpha_j|}.$$

Then

$$\left| \int_0^1 g_v'(t) f_{n,u}'(t) \, dt \right| \leq \left| \int_0^1 (g_v'(t) - \varphi(t)) f_{n,u}'(t) \, dt \right| + \left| \int_0^1 \varphi(t) f_{n,u}'(t) \, dt \right|$$

$$\leq \frac{2\varepsilon}{3} + \sum_{j=1}^{k} |\alpha_j| \left| \int_{u_{j-1}}^{u_j} f_{n,u}'(t) \, dt \right|$$

$$= \frac{2\varepsilon}{3} + \sum_{j=1}^{k} |\alpha_j| \, |f_{n,u}(u_j) - f_{n,u}(u_{j-1})|$$

$$< \frac{2\varepsilon}{3} + \frac{\varepsilon}{3} = \varepsilon.$$

This concludes the proof. ⊔

Theorem 5.1.6. *The Markov product is associative, viz. $A * (B * C) = (A * B) * C$, for all 2-copulas A, B and C.*

Proof. Let A, B and C be copulas. In view of Theorem 4.1.11 one may limit one's attention to absolutely continuous, but otherwise generic, copulas B such that the derivatives $\partial_2 \partial_1 B$ and $\partial_1 \partial_2 B$ exist a.e., are integrable and are equal almost everywhere. Set $\varphi(t) := A(u, t)$ and $\psi(s) := C(s, v)$ for fixed u and v in $\mathbb{I}$. Then, by

Fubini's theorem

$$
\begin{aligned}
(A * (B * C))(u,v) &= \int_0^1 \varphi'(t) \frac{\mathrm{d}}{\mathrm{d}t} \left(\int_0^1 \partial_2 B(t,s)\, \psi'(s)\, \mathrm{d}s \right)\, \mathrm{d}t \\
&= \int_0^1 \varphi'(t)\, \mathrm{d}t \int_0^1 \partial_1 \partial_2 B(t,s)\, \psi'(s)\, \mathrm{d}s \\
&= \int_0^1 \psi'(s)\, \mathrm{d}s \int_0^1 \partial_2 \partial_1 B(t,s)\, \varphi'(t)\, \mathrm{d}t \\
&= \int_0^1 \psi'(s)\, \mathrm{d}s \frac{\mathrm{d}}{\mathrm{d}s} \left(\int_0^1 \partial_2 \partial_1 B(t,s)\, \varphi'(t)\, \mathrm{d}t \right) \\
&= ((A * B) * C)(u,v),
\end{aligned}
$$

which concludes the proof. $\square$

As a consequence of Theorem 5.1.6, one has

Corollary 5.1.7. $(\mathscr{C}_2, *)$ *is a semigroup with identity.*

However, as shown in Example 5.1.3, the $*$-product is not commutative, so that the semigroup $(\mathscr{C}_2, *)$ is not Abelian.

Further readings 5.1.8. The $*$-product was introduced by Darsow et al. [1992]; for some of its properties see also [Darsow and Olsen, 2010; Trutschnig and Fernández Sánchez, 2012; Trutschnig, 2013]. Lagerås [2010] calls it the *Markov product*. The continuity of the $*$-product with respect to different distances in $\mathscr{C}_2$ has been considered by Darsow and Olsen [1995].   ∎

5.2   Invertible and extremal elements in $\mathscr{C}_2$

In this section, we are interested in the investigations of invertible elements of $\mathscr{C}_2$ with respect to the $*$-product. As will be seen, these studies are related to the determination of extremal copulas (in the sense of Definition 1.7.9).

Definition 5.2.1. Given a copula $C \in \mathscr{C}_2$, a copula $A \in \mathscr{C}_2$ will be said to be a *left inverse* of C if $A * C = M_2$, while a copula $B \in \mathscr{C}_2$ will be said to be a *right inverse* of C if $C * B = M_2$. A copula that possesses both a left and a right inverse is said to be *invertible*.   ◇

Left and right inverses are characterised in the next two theorems.

Theorem 5.2.2. *For a copula $C \in \mathscr{C}_2$ the following statements are equivalent:*

(a) *for every $u \in \mathbb{I}$ there exists $a = a(u) \in \,]0,1[$ such that, for almost every $v \in \mathbb{I}$,*
   $\partial_1 C(u,v) = \mathbf{1}_{[a(u),1]}(v)$;

(b) *C has a left inverse.*

In either case the transpose C^T of C is a left inverse of C.

Proof. (a) $\Longrightarrow$ (b) Since for almost every u the map $v \mapsto \partial_1 C(u, v)$ is increasing, by Corollary 1.6.4, and bounded above by 1, it follows that for $s \leq t$ and for almost every $u \in \mathbb{I}$,

$$\partial_1 C(u, s) \, \partial_1 C(u, t) = \partial_1 C(u, s) \,.$$

Therefore

$$(C^T * C)(u, v) = \int_0^1 \partial_2 C^T(u, t) \, \partial_1 C(t, v) \, dt = \int_0^1 \partial_1 C(t, u) \, \partial_1 C(t, v) \, dt$$

$$= \int_0^1 \partial_1 C(t, \min\{u, v\}) \, dt = \min\{u, v\}.$$

Thus C has a left inverse and this equals C^T.

(b) $\Longrightarrow$ (a) Assume $A * C = M_2$; then, by Schwartz's inequality and by the fact that the derivatives of a copula, where they exist, take values in $\mathbb{I}$, one has, for every $v \in \mathbb{I}$,

$$v = \int_0^1 \partial_2 A(v, t) \, \partial_1 C(t, v) \, dt$$

$$\leq \left(\int_0^1 (\partial_2 A(v, t))^2 \, dt \right)^{1/2} \left(\int_0^1 (\partial_1 C(t, v))^2 \, dt \right)^{1/2}$$

$$\leq v^{1/2} \left(\int_0^1 (\partial_1 C(t, v))^2 \, dt \right)^{1/2} \leq v^{1/2} \left(\int_0^1 \partial_1 C(t, v) \, dt \right)^{1/2} = v \,.$$

Thus all the above inequalities are, in fact, equalities, so that, if $v > 0$, then

$$\int_0^1 \left\{ \partial_1 C(t, v) - (\partial_1 C(t, v))^2 \right\} \, dt = 0 \,.$$

Since the integrand is positive one has

$$\partial_1 C(t, v) - (\partial_1 C(t, v))^2 = 0 \,,$$

or, equivalently, either $\partial_1 C(t, v) = 0$ or $\partial_1 C(t, v) = 1$ for almost every t. Since $v \mapsto \partial_1 C(t, v)$ is increasing, there exists $a(u)$ in $\mathbb{I}$ such that $\partial_1 C(u, v) = 1_{[a(u),1]}(v)$. For $v = 0$, one has $\partial_1 C(u, 0) = 0$, by the boundary conditions. This completes the proof. $\qquad\square$

An analogous result holds in characterising copulas with a right inverse. Its proof is based on the consideration of transpose copulas.

Theorem 5.2.3. *For a copula $C \in \mathscr{C}_2$ the following statements are equivalent:*

(a) *for every $v \in \mathbb{I}$ there exists $b = b(v) \in \,]0, 1[$ such that $\partial_2 C(u, v) = 1_{[b(v),1]}(u)$, for almost every $u \in \mathbb{I}$;*

(b) *C has a right inverse.*

In either case the transpose C^T of C is a right inverse of C.

As a consequence of the two previous theorems one immediately sees that the product copula Π_2 has neither a left nor a right inverse.

Notice that a copula $C \in \mathscr{C}_2$ may have a right but not a left inverse or, conversely, a left but not a right inverse. The following example exhibits a family of copulas with a left inverse but without a right inverse.

Example 5.2.4. For $\theta \in]0,1[$, consider the one-parameter family C_θ of Example 3.2.5. Then, while one has $\partial_1 C_\theta = 0$ or $\partial_1 C_\theta = 1$ a.e., one also has $\partial_2 C_\theta(u,v) = \theta$ whenever u is in $]\theta\, v, 1 - (1 - \theta)\, v[$. Therefore Theorems 5.2.2 and 5.2.3 imply that, for every $\theta \in]0,1[$, C_θ has a left but not a right inverse. ∎

The existence of inverses in the semigroup with identity $(\mathscr{C}_2, *)$ is related to the concept of extremal copula, as can be seen below.

Theorem 5.2.5. *If a copula $C \in \mathscr{C}_2$ possesses either a left or right inverse, then it is extremal.*

Proof. Let $C \in \mathscr{C}_2$ have a left inverse; then, by Theorem 5.2.2, the unit square $\mathbb{I}^2$ may be decomposed into the disjoint union of two sets I_1 and I_2, $\mathbb{I}^2 = I_1 \cup I_2$, such that $\partial_1 C = 0$ a.e. in I_1 and $\partial_1 C = 1$ a.e. in I_2. If there exists $\alpha \in]0,1[$ such that $C = \alpha A + (1-\alpha) B$, then one has $\partial_1 C = \alpha\, \partial_1 A + (1-\alpha)\, \partial_1 B$ almost everywhere. Because of the properties of the derivatives of copulas, one has $\partial_1 A = \partial_1 B = 0$ a.e. in I_1 and $\partial_1 A = \partial_1 B = 1$ a.e. in I_2. Therefore, $\partial_1 A = \partial_1 B = \partial_1 C$ a.e. in $\mathbb{I}^2$. Since all copulas share the same boundary conditions, integrating yields $A = B = C$, so that C is extreme. □

The following result establishes the uniqueness of left or right inverses when they exist.

Theorem 5.2.6. *When they exist, left and right inverses of copulas in $(\mathscr{C}_2, *)$ are unique.*

Proof. It suffices to prove the result only in the case of left inverses. Let A and B be two left inverses of the copula $C \in \mathscr{C}_2$, namely, $A * C = B * C = M_2$. Then also $(A+B)/2$ is a 2-copula and is a left inverse of C, or, equivalently, C is a right inverse of $(A + B)/2$, which, then, by Theorem 5.2.5 is extreme. Therefore $A = B$. □

The next result considers the $*$-product for copulas represented by means of measure-preserving transformations (see Section 3.5).

Theorem 5.2.7. *Let f and g be measure-preserving transformations and let i be the identity map on $\mathbb{I}$, $i = id_\mathbb{I}$. Then*

(a) $C_{f,g} = C_{f,i} * C_{i,g}$;

(b) $C_{f,i} = C_{g,i}$, *if, and only if, $f = g$ a.e.;*

(c) $C_{f,i} * C_{g,i} = C_{f \circ g, i}$ *and* $C_{i,g} * C_{i,f} = C_{i, f \circ g}$.

Proof. (a) Because of equations (3.5.5) and (3.5.6), one has

$$(C_{f,i} * C_{i,g})(s,t) = \int_0^1 1_{f^{-1}[0,s]}(u) \, 1_{g^{-1}[0,t]}(u) \, du$$

$$= \int_0^1 1_{f^{-1}[0,s] \cap g^{-1}[0,t]}(u) \, du$$

$$= \lambda \left(f^{-1}[0,s] \cap g^{-1}[0,t] \right) = C_{f,g}(s,t).$$

(b) If $C_{f,i} = C_{g,i}$, then also $C_{f,i}^T = C_{g,i}^T$ so that $C_{i,f} = C_{i,g}$, on account of Theorem 3.5.3 (d). By (a) and by Theorem 3.5.3 (b),

$$M_2 = C_{f,f} = C_{f,i} * C_{i,f} = C_{f,i} * C_{i,g} = C_{f,g};$$

thus $f = g$ almost everywhere. If $f = g$ a.e., then

$$M_2 = C_{f,g} = C_{f,i} * C_{i,g}.$$

(c) Using the obvious relation $i \circ g = g$, one has

$$C_{g,i} * C_{i,f \circ g} = C_{g,f \circ g} = C_{i \circ g, f \circ g} = C_{i,f},$$

where use has been made of (3.5.4). □

Theorem 5.2.8. *If the copula $C \in \mathscr{C}_2$ has a left inverse, the function $f : \mathbb{I} \to \mathbb{I}$ defined by*

$$f(u) := \inf\{v \in \mathbb{I} : \partial_1 C(u,v) = 1\}$$

is measure-preserving and $C = C_{i,f}$.

Proof. Because of Theorem 5.2.2, $\partial_1 C(u,v)$ equals either 0 or 1 for every $v \in \mathbb{I}$ and for almost every $u \in \mathbb{I}$. Notice that if v_0 is such that $\partial_1 C(u,v_0) = 1$ then $\partial_1 C(u,v)$ exists and is equal to 1 for every $v > v_0$. In fact, for $h > 0$, the 2-increasing property applied first to the rectangle $[u, u+h] \times [v_0, v]$ and, then, to $[u, u+h] \times [v, 1]$ gives

$$\frac{C(u+h, v_0) - C(u, v_0)}{h} \leq \frac{C(u+h, v) - C(u, v)}{h}$$

$$\leq \frac{C(u+h, 1) - C(u, 1)}{h} = 1,$$

hence the assertion as h goes to zero.

Next, consider the set $S(a) := \{u \in \mathbb{I} : \partial_1 C(u,a) = 1\}$. We claim that $S(a)$ is a Borel set for every $a \in \mathbb{I}$. To see this, let (h_n) be an infinitesimal sequence of strictly positive numbers; then the function $u \mapsto \partial_1 C(u,a)$ is the pointwise limit, where it exists, of the continuous and, hence, also measurable, functions

$$u \mapsto \frac{C(u+h_n, a) - C(u,a)}{h_n}, \quad \text{and} \quad u \mapsto \frac{C(u,a) - C(u-h_n, a)}{h_n}.$$

The derivative $\partial_1 C(u, a)$ does not exist at those points where the maximum and minimum limits of the above functions are not all equal; these points form a null Borel set, since the partial derivative $\partial_1 C(u, a)$ exists for almost every $u \in \mathbb{I}$. One may redefine $\partial_1 C(u, a)$ to be zero where the derivative does not exist so as to have a Borel function defined on the whole interval $\mathbb{I}$. Since the sets $S(a)$ are the level sets of the Borel functions thus defined, they are Borel sets; moreover, one has $S(a) \subseteq S(b)$ whenever $a < b$.

Now let $\{q_n\}$ be an enumeration of the rationals of $\mathbb{I}$

$$\mathbb{Q} \bigcap \mathbb{I} = \bigcup_{n \in \mathscr{B}} \{q_n\};$$

then the function $f : \mathbb{I} \to \mathbb{R}$ defined by

$$f(x) := \inf \left\{ q_n \in \mathbb{Q} \bigcap \mathbb{I} : x \in S(q_n) \right\}$$

is Borel-measurable (see, e.g., [Royden, 1988]).

If $f(x) < a$, then there exists a rational number q_n in $]f(x), a[$ such that $\partial_1 C(x, q_n) = 1$; then, necessarily $\partial_1 C(x, a) = 1$, and, as a consequence, $\partial_1 C(x, q_n) = 1$ for every rational number $q_n > a$ and $f(x) \leq a$. Therefore, for every $a \in \mathbb{I}$ one has

$$\{f < a\} \subseteq \{x \in \mathbb{I} : \partial_1 C(x, a) = 1\} \subseteq \{f \leq a\}.$$

For $\varepsilon > 0$ one has

$$\{x \in \mathbb{I} : \partial_1 C(x, a) = 1\} \subseteq \{f \leq a\} \subseteq \{f \leq a + \varepsilon\} \subseteq \{x \in \mathbb{I} : \partial_1 C(x, a + \varepsilon) = 1\},$$

and hence

$$\lambda\left(\{x \in \mathbb{I} : \partial_1 C(x, a) = 1\}\right) \leq \lambda(\{f \leq a\}) \leq \lambda\left(\{x \in \mathbb{I} : \partial_1 C(x, a + \varepsilon) = 1\}\right).$$

Since $\partial_1 C$ takes only the values 0 and 1, one has

$$\lambda\left(\{x \in \mathbb{I} : \partial_1 C(x, a) = 1\}\right) = \int_0^1 \partial_1 C(x, a)\, \mathrm{d}t = C(1, a) - C(0, a) = a.$$

Similarly, $\lambda\left(\{x \in \mathbb{I} : \partial_1 C(x, a + \varepsilon) = 1\}\right) = a + \varepsilon$. Therefore $a \leq \lambda(\{f \leq a\}) \leq a + \varepsilon$ for every $\varepsilon > 0$, and, because of the arbitrariness of ε, $\lambda(\{f \leq a\}) = a$ so that f is measure-preserving. Since $\lambda(\{f < a\}) = a$, one also has $\lambda(\{f = a\}) = 0$. It follows that for every $a \in \mathbb{I}$ and for almost every $t \in \mathbb{I}$ one has

$$\partial_1 C(t, a) = \mathbf{1}_{f^{-1}[0,a]}(t).$$

Now integrating between 0 and u yields

$$C(u, a) = \int_0^u \mathbf{1}_{f^{-1}[0,a]}(t)\, \mathrm{d}t = \lambda\left([0, u] \cap f^{-1}[0, a]\right) = C_{i,f}(u, a),$$

which concludes the proof.   □

A measure-preserving transformation f will be said to possess an *essential inverse* g if g is also measure-preserving and $f \circ g = g \circ f = i$ a.e.

Theorem 5.2.9. *For a copula $C \in \mathscr{C}_2$ the following are equivalent:*

(a) *it has a two-sided inverse;*

(b) *there exists a measure-preserving function f with an essential inverse such that C can be represented in the form $C = C_{f,i}$.*

Proof. (a) $\Longrightarrow$ (b) Because of Theorems 5.2.2 and 5.2.3 one has $C * C^T = C^T * C = M_2$; therefore C^T has a left inverse, which means, by Theorem 5.2.8, that there exists a measure-preserving transformation f such that $C^T = C_{i,f}$ or, equivalently, $C = C_{f,i}$. But also C has a left inverse so that, by the same theorem, $C = C_{i,g}$ for a suitable measure-preserving g. Then $C_{f,i} = C_{i,g}$ and, as a consequence of Theorem 3.5.3 (b) and 5.2.7 (a) and (c),

$$C_{i,i} = M_2 = C_{g,g} = C_{g,i} * C_{i,g} = C_{g,i} * C_{f,i} = C_{g \circ f,i} \,;$$

it follows from Theorem 5.2.7 (b) that $g \circ f = i$ almost everywhere. Similarly

$$C_{i,i} = M_2 = C_{f,f} = C_{f,i} * C_{i,f} = C_{f,i} * C_{i,g}^T = C_{f,i} * C_{g,i} = C_{f \circ g,i} \,,$$

whence $f \circ g = i$ almost everywhere.

(b) $\Longrightarrow$ (a) Let $C = C_{f,i}$, where f has an essential inverse g; then $C_{g,i} * C_{f,i} = C_{i,i} = M_2$ by Theorem 5.2.7 (c), and similarly $C_{f,i} * C_{g,i} = C_{f \circ g,i} = C_{i,i} = M_2$, so that $C = C_{f,i}$ has both a left and a right inverse. $\qquad\square$

In particular, any shuffle of Min is invertible.

We conclude this section with an application of the previous results. It was seen above (Theorem 5.1.5) that the $*$-product is continuous in each place with respect to the topology of uniform convergence of copulas. However, it is not jointly continuous in this topology.

Theorem 5.2.10. *The $*$-product is not jointly continuous with respect to the topology of uniform convergence.*

Proof. By Corollary 4.1.16 the shuffles of Min are dense in $\mathscr{C}_2$. Thus, there is a sequence (C_n) of shuffles of Min that converges uniformly to Π_2. Since the transpose operation is continuous, also the sequence (C_n^T) of transposes converges uniformly to $\Pi_2^t = \Pi_2$. If the $*$-product were jointly continuous one would have, by eq. (5.1.2),

$$C_n^T * C_n \xrightarrow[n \to +\infty]{} \Pi_2 * \Pi_2 = \Pi_2 \,.$$

But this is impossible since $C_n^T * C_n = M_2$ for every $n \in \mathbb{N}$. $\qquad\square$

Further readings 5.2.11. The study of inverses of a copula was introduced by Darsow et al. [1992]. More recently, these studies have also been connected with a special kind of Sobolev-type norm introduced in the space of copulas; see, for instance, [Siburg and Stoimenov, 2008b; Ruankong et al., 2013].

5.3   Idempotent copulas, Markov operators and conditional expectations

In view of the one-to-one correspondence between copulas and Markov operators of Section 4.4.1, it is natural to investigate the Markov operator T_{A*B} corresponding to the $*$-product of the two copulas A and B. At the same time, the introduction of the $*$-product allows one to characterise the composition of two Markov operators.

Theorem 5.3.1. *For all copulas A and B in $\mathscr{C}_2$, one has*

$$T_{A*B} = T_A \circ T_B \,, \tag{5.3.1}$$

namely, the Markov operator corresponding to the $*$-product $A*B$ is the composition of the Markov operators T_A and T_B corresponding to the copulas A and B.

Proof. Let f be in $C^1(\mathbb{I})$. Then

$$
\begin{aligned}
(T_A \circ T_B)(f)(x) &= \frac{d}{dx} \int_0^1 \partial_2 A(x,t)\,(T_B f)(t)\,dt \\
&= \frac{d}{dx} \int_0^1 \partial_2 A(x,t)\,dt\,\frac{d}{dt} \int_0^1 \partial_2 B(t,s)\,f(s)\,ds \\
&= \frac{d}{dx} \int_0^1 \partial_2 A(x,t)\,\frac{d}{dt}\left(t\,f(1) - \int_0^1 B(t,s)\,f'(s)\,ds \right) dt \\
&= \frac{d}{dx}\left(x\,f(1) - \int_0^1 f'(s)\,ds \int_0^1 \partial_2 A(x,t)\,\partial_1 B(t,s)\,dt \right) \\
&= \frac{d}{dx}\left(x\,f(1) - \int_0^1 (A*B)(x,s)\,f'(s)\,ds \right) \\
&= \frac{d}{dx}\left(x\,f(1) - x\,f(1) + \int_0^1 \partial_2(A*B)(x,s)\,f(s)\,ds \right) \\
&= (T_{A*B} f)(x)\,.
\end{aligned}
$$

Thus the assertion holds in $C^1(\mathbb{I})$, which is dense in L^1; therefore, it holds for every f in L^1.   □

It is now possible to collect the results of Theorems 5.3.1 and 4.4.9 in the following Isomorphism Theorem.

Theorem 5.3.2. *The mapping $\Phi : \mathscr{C}_2 \to \mathscr{T}_M$ given by (4.4.12) is an isomorphism between the set $(\mathscr{C}_2, *)$ and the set of Markov operators on $L^1(\mathbb{I})$ under composition. Specifically, if $\Phi(C) := T_C$, then Φ is a bijection and*

(a) $\Phi(C_1 * C_2) = \Phi(C_1) \circ \Phi(C_2)$;

(b) $\Phi(\alpha\,C_1 + (1 - \alpha)\,C_2) = \alpha\,\Phi(C_1) + (1 - \alpha)\,\Phi(C_2)$ *for every $\alpha \in [0,1]$.*

Theorem 5.3.2 might help to show that the Markov product $A * B$ can be also interpreted in terms of standard composition of the Markov kernels of A and B, as follows.

Theorem 5.3.3. *Let $A, B \in \mathscr{C}_2$ and let K_A and K_B denote the corresponding Markov kernels. Then, for every Borel set $F \subseteq \mathbb{I}^d$,*

$$K_A \circ K_B(x, F) := \int_0^1 K_B(y, F) K_A(x, \mathrm{d}y) \qquad (5.3.2)$$

is the Markov kernel of $A * B$.

Proof. It is well known that the right-hand side of (5.3.2) is a Markov kernel (see, e.g., [Klenke, 2008, section 14]). Now, let $f = \sum_{i=1}^n \alpha_i \mathbf{1}_{E_i}$ be a positive simple function with $(E_i)_{i=1}^n$ being a measurable partition of $\mathbb{I}$. Then

$$\int_{\mathbb{I}^2} f(y) K_A(x, \mathrm{d}y) \, \mathrm{d}\lambda(x) = \int_0^1 \sum_{i=1}^n \alpha_i K_A(x, E_i) \, \mathrm{d}\lambda(x) \qquad (5.3.3)$$

$$= \sum_{i=1}^n \alpha_i \lambda(E_i) = \int_0^1 f \, \mathrm{d}\lambda. \qquad (5.3.4)$$

Since the class of simple functions is dense in $L^1(\mathbb{I}, \mathscr{B}(\mathbb{I}), \lambda)$, one has

$$\int_{\mathbb{I}^2} K_B(y, F) K_A(x, \mathrm{d}y) = \int_0^1 K_B(x, F) \, \mathrm{d}\lambda(x) = \lambda(F),$$

which shows that $K_A \circ K_B$ satisfies eq. (3.4.8). Finally, if $E \in \mathscr{B}(\mathbb{I})$, then it follows from Theorem 5.3.2 that, for λ-almost every $x \in \mathbb{I}$,

$$K_{A*B}(x, E) = (T_A \circ T_B)(\mathbf{1}_E)(x) = T_A(K_B(\cdot, E)(x)) = \int_{\mathbb{I}} K_B(y, E) K_A(x, \mathrm{d}y),$$

which completes the proof. $\qquad\qquad\qquad\qquad\qquad\qquad\qquad\qquad\qquad\qquad\square$

The relationship between copulas and Markov operators is used below in order to establish the deep connexion between 2-copulas and conditional expectations (CEs). An ample literature exists on characterisations of CEs. Here, we shall quote only the characterisation given by Pfanzagl [1967]. Pfanzagl's characterisation allows to identify those Markov operators that are also CEs, when the probability space under consideration is $(\mathbb{I}, \mathscr{B}(\mathbb{I}), \lambda)$; it is given in the following theorem.

Theorem 5.3.4. *Let $\mathscr{H}$ be a subset of $L^1(\Omega, \mathscr{F}, \mathbb{P})$ endowed with the following properties:*

(1) αf *belongs to $\mathscr{H}$ for all $f \in \mathscr{H}$ and $\alpha \in \mathbb{R}$;*

(2) $1 + f$ *belongs to $\mathscr{H}$ whenever f is in $\mathscr{H}$;*

(3) *the pointwise minimum $f \wedge g$ of two functions f and g in $\mathscr{H}$ belongs to $\mathscr{H}$;*

(4) *if $(f_n)_{n \in \mathbb{N}}$ is a decreasing sequence of elements of $\mathscr{H}$ that tends to a function f in L^1, then f belongs to $\mathscr{H}$.*

Then an operator $T : \mathscr{H} \to \mathscr{H}$ is the restriction to $\mathscr{H}$ of a conditional expectation if, and only if, it satisfies the following conditions:

(a) T is increasing: $Tf \leq Tg$ whenever $f \leq g$ $(f, g \in \mathscr{H})$;

(b) T is homogeneous: $T(\alpha f) = \alpha T f$ for all $\alpha \in \mathbb{R}$ and $f \in \mathscr{H}$;

(c) T is translation invariant: $T(1 + f) = 1 + Tf$ for every $f \in \mathscr{H}$;

(d) T is expectation invariant: $\mathbb{E}(Tf) = \mathbb{E}(f)$ for every $f \in \mathscr{H}$;

(e) T is idempotent: $T^2 := T \circ T = T$.

When these conditions are satisfied, then $T = \mathbb{E}_{\mathscr{G}}$, where

$$\mathscr{G} = \{A \in \mathscr{F} : T \mathbf{1}_A = \mathbf{1}_A\} \, .$$

When the probability space is $(\mathbb{I}, \mathscr{B}(\mathbb{I}), \lambda)$, a comparison of the Definition 4.4.1 of a Markov operator with Theorem 5.3.4 yields the following result.

Theorem 5.3.5. *For a Markov operator* $T : L^\infty(\mathbb{I}) \to L^\infty(\mathbb{I})$ *the following conditions are equivalent:*

(a) T is the restriction to $L^\infty(\mathbb{I})$ of a CE;

(b) T is idempotent, viz. $T^2 = T$.

When either condition is satisfied, then $T = \mathbb{E}_{\mathscr{G}}$, where

$$\mathscr{G} := \{A \in \mathscr{B}(\mathbb{I}) : T \mathbf{1}_A = \mathbf{1}_A\} \, .$$

Proof. Take as $\mathscr{H}$ the entire space $L^\infty(\mathbb{I})$, which obviously satisfies all the conditions (1) through (4). It is immediate from the definition that a CE is an idempotent Markov operator.

Conversely, let T be an idempotent Markov operator. Since T, as a Markov operator, is linear, it is both homogeneous and translation invariant (the constant function 1 is in both L^1 and L^∞); but T is also positive, and, as a consequence, it is increasing. Thus, all the conditions of Theorem 5.3.4 are satisfied; now, the assertion follows from that theorem.   □

In view of Theorem 4.4.9 one can state

Theorem 5.3.6. *A Markov operator* T *is idempotent with respect to composition* $T^2 = T$, *if, and only if, the unique copula* $C_T \in \mathscr{C}_2$ *that corresponds to it satisfies* $C_T = C_T * C_T$.

Proof. Let T be a Markov operator and let $C_T \in \mathscr{C}_2$ be the copula corresponding to it. Then T is idempotent if, and only if,

$$T_{C_T} = T = T \circ T = T_{C_T} \circ T_{C_T} = T_{C_T * C_T} \, .$$

The result now follows from the Isomorphism Theorem 5.3.2.   □

In analogy to the case of Markov operators, a copula $C \in \mathscr{C}_2$ will be said to be *idempotent* if $C * C = C$.

Both the copulas Π_2 and M_2 are idempotent. Another idempotent copula is given, for every $\alpha \in]0, 1[$, by

$$C_\alpha(u, v) := \frac{\Pi_2(u, v)}{\alpha} \mathbf{1}_{[0,\alpha] \times [0,\alpha]}(u, v) + M_2(u, v) \mathbf{1}_{\mathbb{I}^2 \setminus [0,\alpha] \times [0,\alpha]}(u, v). \quad (5.3.5)$$

The copula of eq. (5.3.5) is an ordinal sum. An example of a copula that is *not* idempotent is given below.

Example 5.3.7. Consider the tent copula of Example 3.2.5. Then it follows from Example 5.1.3 that

$$(C_\theta * C_\theta)(u, v) = \begin{cases} \theta\, C_\theta\left(\frac{u}{\theta}, v\right), & u \le \theta, \\ v - (1 - \theta)\, C_\theta\left(\frac{1-u}{1-\theta}, v\right), & u > \theta, \end{cases}$$

so that C_θ is not idempotent. ∎

We can now collect the previous results of this section.

Theorem 5.3.8. *For a copula $C \in \mathscr{C}_2$, the following statements are equivalent:*

(a) *the corresponding Markov operator T_C is a CE restricted to $L^\infty(\mathbb{I}, \mathscr{B}(\mathbb{I}), \lambda)$;*

(b) *the corresponding Markov operator T_C is idempotent;*

(c) *C is idempotent.*

Since the copulas Π_2 and M_2 are idempotent, the corresponding Markov operators T_{Π_2} and T_{M_2} are CEs. One easily computes

$$T_{\Pi_2} f = \mathbb{E}(f) = \int_0^1 f(t)\, dt \qquad \text{and} \qquad T_{M_2} f = f$$

for every f in $L^1(\mathbb{I})$. Therefore $T_{\Pi_2} = \mathbb{E}_\mathscr{N}$, where $\mathscr{N}$ is the trivial σ-field $\{\emptyset, \mathbb{I}\}$, and $T_{M_2} = \mathbb{E}_\mathscr{B}$.

As a consequence of the above considerations a one-to-one correspondence has been established between CEs and idempotent Markov operators, or, equivalently, between CEs and idempotent copulas. It is now possible to state, as a final result,

Theorem 5.3.9. *To every sub-σ-field $\mathscr{G}$ of $\mathscr{B}$, the Borel σ-field of $\mathbb{I}$, there corresponds a unique idempotent copula $C(\mathscr{G}) \in \mathscr{C}_2$ such that $\mathbb{E}_\mathscr{G} = T_{C(\mathscr{G})}$. Conversely, to every idempotent copula $C \in \mathscr{C}_2$ there corresponds a unique sub-σ-field $\mathscr{G}(C)$ of $\mathscr{B}$ such that $T_C = \mathbb{E}_{\mathscr{G}(C)}$.*

The problems of determining, for a given sub-σ-field $\mathscr{G}$ of $\mathscr{B}$, the explicit expression of the copula $C(\mathscr{G})$ that corresponds to it, and, conversely, of determining, for a given idempotent copula C, the corresponding sub-σ-field $\mathscr{G}(C)$ of $\mathscr{B}$ are still open.

Let $\mathscr{I}_2$ be the the subset of idempotents of $\mathscr{C}_2$. A relation $\le_*$ may be defined on $\mathscr{I}_2$ in the following way: if A and B are idempotent copulas one writes $A \le_* B$ if $A * B = B * A = A$.

Theorem 5.3.10. *The relation $\leq_*$ is a partial order on $\mathscr{I}_2$, so that $(\mathscr{I}_2, \leq_*)$ is a poset.*

Proof. Assume $A \leq_* B$ and $B \leq_* C$, where A, B and C are idempotents; then

$$A * C = (A * B) * C = A * (B * C) = A * B = A,$$
$$C * A = C * (B * A) = (C * B) * A = B * A = A.$$

Therefore $\leq_*$ is transitive. Since A is an idempotent $A = A * A$, which means that $\leq_*$ is reflexive on $\mathscr{I}_2$. Moreover the relations $A \leq_* B$ and $B \leq_* A$ imply

$$A = A * B = B * A \qquad \text{and} \qquad B = B * A = A * B,$$

so that $A = B$; thus $\leq_*$ is antisymmetric, which proves that $(\mathscr{I}_2, \leq_*)$ is a partial order. $\qquad\square$

The partial order $\leq_*$ on $\mathscr{I}_2$ allows one to show the reflex of the so-called "tower property" (see, e.g., [Williams, 1991]) of conditional expectations on their corresponding idempotent copulas.

Theorem 5.3.11. *Let $\mathscr{G}_1$ and $\mathscr{G}_2$ be sub-σ-fields of $\mathscr{B}$ and let C_1 and C_2 be the (uniquely determined) idempotent copulas corresponding to them. Then the following statements are equivalent:*

(a) $\mathscr{G}_1 \subseteq \mathscr{G}_2$;

(b) $C_1 \leq_* C_2$.

Proof. (a) $\Longrightarrow$ (b) The conditional expectations $\mathbb{E}_1 := \mathbb{E}_{\mathscr{G}_1}$ and $\mathbb{E}_2 := \mathbb{E}_{\mathscr{G}_2}$ are idempotent Markov operators; as is well known $\mathscr{G}_1 \subseteq \mathscr{G}_2$ implies $\mathbb{E}_1 \mathbb{E}_2 = \mathbb{E}_2 \mathbb{E}_1 = \mathbb{E}_1$ so that one has $C_1 * C_2 = C_2 * C_1 = C_1$, namely $C_1 \leq_* C_2$.

(b) $\Longrightarrow$ (a) By the one-to-one correspondence between copulas and conditional expectations it follows that $\mathbb{E}_1 \mathbb{E}_2 = \mathbb{E}_2 \mathbb{E}_1 = \mathbb{E}_1$, so that, if B belongs to $\mathscr{G}_1$, then $\mathbb{E}_1 \mathbf{1}_B = \mathbf{1}_B$ and

$$\mathbb{E}_2 \mathbf{1}_B = \mathbb{E}_2 (\mathbb{E}_1 \mathbf{1}_B) = (\mathbb{E}_2 \mathbb{E}_1) \mathbf{1}_B = \mathbb{E}_1 \mathbf{1}_B = \mathbf{1}_B,$$

which means that B is in $\mathscr{G}_2$; therefore $\mathscr{G}_1 \subseteq \mathscr{G}_2$. $\qquad\square$

Theorem 5.3.12. *Let $A \in \mathscr{C}_2$ be a copula with a left inverse, but no right inverse; then the copula $A * A^T$ is idempotent.*

Proof. It follows from Theorem 5.2.2 that the left inverse of A is its transpose A^T, $A^T * A = M_2$. Then, on account of the associativity of the $*$-product (Theorem 5.1.6),

$$(A * A^T) * (A * A^T) = A * (A^T * A) * A^T = A * M_2 * A^T = A * A^T.$$

Notice that $A * A^T \neq M_2$, since, otherwise, A^T would be a right inverse of A, contrary to the assumption. $\qquad\square$

Not all idempotent copulas can be expressed in this way; for instance, there is no left invertible copula A such that the idempotent copula Π_2 can be written in the form $\Pi_2 = A * A^T$, for, if this were possible, then

$$\Pi_2 = A^T * \Pi_2 * A = A^T * (A * A^T) * A = (A^T * A) * (A^T * A) = M_2 * M_2 = M_2\,,$$

a contradiction, since $\Pi_2 \neq M_2$.

When a copula is represented through measure-preserving transformations, the copula A of the previous theorem can be represented as $A = C_{i,f}$; then $A * A^T = C_{i,f} * C_{f,i}$ is idempotent.

Remark 5.3.13. In looking for the idempotents of $*$-operation, one should ideally try to solve the integro-differential equation that defines an idempotent, namely for all $(u, v) \in \mathbb{I}^2$,

$$C(u, v) = \int_0^1 \partial_2 C(u, t)\, \partial_1 C(t, v)\, \mathrm{d}t\,. \tag{5.3.6}$$

While it is not hard to find solutions of this equation that are different from the idempotents listed above, these usually do not satisfy the boundary conditions of a copula. However, an indirect approach is possible in the search for idempotent copulas: one can exploit the connexion between idempotent copulas and conditional expectations that has just been established. Albanese and Sempi [2004] study the idempotents corresponding to the conditional expectations with respect to the sub-σ-fields of $\mathscr{B}$ generated by a countable, or possibly finite, (measurable) partition of the unit interval $\mathbb{I}$. ∎

Further readings 5.3.14. In this section we follow the second author's presentation [Sempi, 2002]. Darsow and Olsen [2010] prove that an idempotent 2-copula is necessarily symmetric, that the relation $E * F = F * E = E$ introduces a partial order, $E \leq_* F$, in the set of idempotent copulas and that this latter set is a lattice with respect to the partial order $\leq_*$. ∎

5.4 The Markov product and Markov processes

Here we show how to use copulas in order to characterise Markov processes. We recall some basic definitions.

A (continuous time) stochastic process $(X_t)_{t>0}$ is called a *Markov process* if, for every choice of $n \in \mathbb{N}$, of $t_1, t_2, \ldots, t_n$ in $[0, +\infty[$ with $t_1 < t_2 < \cdots < t_n < t$ and for every choice of a Borel set A, the conditional independence property

$$\mathbb{P}(X_t \in A \mid X_{t_1}, X_{t_2}, \ldots, X_{t_n}) = \mathbb{P}(X_t \in A \mid X_{t_n})$$

is satisfied or, equivalently, in view of the fact that the σ-field of Borel sets is generated by the sets $]-\infty, \alpha]$ $(\alpha \in \mathbb{R})$,

$$\mathbb{P}(X_t \leq \alpha \mid X_{t_1}, X_{t_2}, \ldots, X_{t_n}) = \mathbb{P}(X_t \leq \alpha \mid X_{t_n})\,. \tag{5.4.1}$$

Moreover, the following characterisation of a Markov process is used.

Theorem 5.4.1. *For a stochastic process $(X_t)_{t \in T}$ the following statements are equivalent:*

(a) *(X_t) satisfies the conditional independence property (5.4.1) for $n = 2$;*

(b) *for all t_1, t_2, t_3 in T with $t_1 < t_2 < t_3$,*

$$\mathbb{P}\left(X_{t_1} \leq \alpha_1, X_{t_3} \leq \alpha_3 \mid X_{t_2}\right)$$
$$= \mathbb{P}\left(X_{t_1} \leq \alpha_1 \mid X_{t_2}\right) \mathbb{P}\left(X_{t_3} \leq \alpha_3 \mid X_{t_2}\right). \tag{5.4.2}$$

Proof. For convenience of notation we write X_j rather than X_{t_j}. Let B be any Borel set; then the proof follows from reading the following sequence of equalities in the two directions.

$$\int_{X_2^{-1}(B)} \mathbb{E}\left(\mathbf{1}_{\{X_1 \leq \alpha_1\}}, \mathbf{1}_{\{X_3 \leq \alpha_3\}} \mid X_2\right) \, d\mathbb{P} = \int_{X_2^{-1}(B)} \mathbf{1}_{\{X_1 \leq \alpha_1\}} \mathbf{1}_{\{X_3 \leq \alpha_3\}} \, d\mathbb{P}$$

$$= \int_{X_2^{-1}(B) \cap X_1^{-1}(]-\infty, \alpha_1])} \mathbf{1}_{\{X_3 \leq \alpha_3\}} \, d\mathbb{P}$$

$$= \int_{X_2^{-1}(B) \cap X_1^{-1}(]-\infty, \alpha_1])} \mathbb{E}\left(\mathbf{1}_{\{X_3 \leq \alpha_3\}} \mid X_1, X_2\right) \, d\mathbb{P}$$

$$= \int_{X_2^{-1}(B) \cap X_1^{-1}(]-\infty, \alpha_1])} \mathbb{E}\left(\mathbf{1}_{\{X_3 \leq \alpha_3\}} \mid X_2\right) \, d\mathbb{P}$$

$$= \int_{X_2^{-1}(B)} \mathbf{1}_{\{X_1 \leq \alpha_1\}} \mathbb{E}\left(\mathbf{1}_{\{X_3 \leq \alpha_3\}} \mid X_2\right) \, d\mathbb{P}$$

$$= \int_{X_2^{-1}(B)} \mathbb{E}\left(\mathbf{1}_{\{X_1 \leq \alpha_1\}} \mathbb{E}\left(\mathbf{1}_{\{X_3 \leq \alpha_3\}} \mid X_2\right) \mid X_2\right) \, d\mathbb{P}$$

$$= \int_{X_2^{-1}(B)} \mathbb{E}\left(\mathbf{1}_{\{X_1 \leq \alpha_1\}} \mid X_2\right) \mathbb{E}\left(\mathbf{1}_{\{X_3 \leq \alpha_3\}} \mid X_2\right) \, d\mathbb{P},$$

which proves the assertion since B is arbitrary. $\qquad \square$

The starting point of the link between Markov product and Markov processes is the interpretation of the $*$-product in terms of conditional independence.

Theorem 5.4.2. *Let X, Y and Z be continuous random variables on the probability space $(\Omega, \mathscr{F}, \mathbb{P})$. If X and Z are conditionally independent given Y, then*

$$C_{XZ} = C_{XY} * C_{YZ}. \tag{5.4.3}$$

Proof. As a consequence of eq. (3.4.4), one has, for all s and v in $\mathbb{I}$,

$$
\begin{aligned}
(C_{XY} * C_{YZ})(s, v) &= \int_0^1 \partial_2 C_{XY}(s, t) \, \partial_1 C_{YZ}(t, v) \, \mathrm{d}t \\
&= \int_{-\infty}^{+\infty} \partial_2 C_{XY}(F_X(x), F_Y(y)) \, \partial_1 C_{YZ}(F_Y(y), F_Z(z)) \, \mathrm{d}F_Y(y) \\
&= \int_\Omega \mathbb{E}\left(1_{\{X \le x\}} \mid Y\right) \mathbb{E}\left(1_{\{Z \le z\}} \mid Y\right) \, \mathrm{d}\mathbb{P} \\
&= \int_\Omega \mathbb{E}\left(1_{\{X \le x\}} 1_{\{Z \le z\}} \mid Y\right) \, \mathrm{d}\mathbb{P} = \mathbb{P}(X \le x, Z \le z) = C_{XZ}(s, v),
\end{aligned}
$$

which proves the assertion. $\square$

For simplicity's sake, consider a (continuous) Markov process $(X_n)_{n \in \mathbb{Z}}$ in discrete time. Then, the content of the preceding theorem is that if $C_{j-1,j}$ is the copula of (X_{j-1}, X_j) and $C_{j,j+1}$ is the copula of (X_j, X_{j+1}), then $C_{j-1,j} * C_{j,j+1}$ is the copula of (X_{j-1}, X_{j+1}) with $j \ge 1$. If, moreover, $(X_n)_{n \in \mathbb{Z}}$ is homogeneous, the copula of (X_0, X_n) is

$$
C^{*n} := C * C * \cdots * C \qquad (n \text{ times}).
$$

In particular, because of (5.1.3) and (5.1.6), one has

$$
W_2^{*n} = \begin{cases} M_2, & \text{for } n \text{ even,} \\ W_2, & \text{for } n \text{ odd,} \end{cases}
$$

while $M_2^{*n} = M_2$ for every $n \in \mathbb{N}$.

Lemma 5.4.3. *For every function f in $L^1(\mathbb{I})$ and for every copula A, one has*

$$
\int_0^x \mathrm{d}u \int_0^1 f(t) \, \partial_1 A(u, \mathrm{d}t) = \int_0^1 f(t) \, A(x, \mathrm{d}t).
$$

Proof. By the usual argument of measure theory, since simple functions are dense in $L^1(\mathbb{I})$, it suffices to prove the assertion for indicators of sets of Borel σ-field $\mathscr{B}$; then, for E in $\mathscr{B}$ and for every $x \in \mathbb{I}$, Fubini's theorem yields

$$
\begin{aligned}
\int_0^x \mathrm{d}u \int_0^1 1_E(t) \, \partial_1 A(u, \mathrm{d}t) &= \int_0^1 1_E(t) \int_0^x \partial_1 A(u, \mathrm{d}t) \, \mathrm{d}u \\
&= \int_0^1 1_E(t) \, A(x, \mathrm{d}t),
\end{aligned}
$$

which yields the assertion. $\square$

Lemma 5.4.4. *Let A and B be copulas; then, for every $y \in \mathbb{I}$ and for almost every $x \in \mathbb{I}$,*

$$\partial_1(A * B)(x, y) = \frac{\partial}{\partial x} \int_0^1 \partial_2 A(x, t) \, \partial_1 B(t, y) \, \mathrm{d}t$$

$$= \int_0^1 \partial_1 B(t, y) \, \partial_1 A(x, \mathrm{d}t) . \tag{5.4.4}$$

Proof. Let φ be any function in $C^\infty(\mathbb{I})$ such that $\varphi(0) = \varphi(1) = 1$. Then, because of Lemma 5.4.3, two integrations by parts lead to

$$\int_0^1 \varphi(x) \, \mathrm{d}x \int_0^1 \partial_1 B(t, y) \, \partial_1 A(x, \mathrm{d}t)$$

$$= - \int_0^1 \varphi'(x) \, \mathrm{d}x \int_0^x \mathrm{d}u \int_0^1 \partial_1 B(t, y) \, \partial_1 A(u, \mathrm{d}t)$$

$$= - \int_0^1 \varphi'(x) \, \mathrm{d}x \int_0^1 \partial_1 B(t, y) \, A(x, \mathrm{d}t)$$

$$= - \int_0^1 \varphi'(x) \, \mathrm{d}x \int_0^1 \partial_2 A(x, t) \, \partial_1 B(t, y) \, \mathrm{d}t$$

$$= \int_0^1 \varphi(x) \, \mathrm{d}x \frac{\partial}{\partial x} \int_0^1 \partial_2 A(x, t) \, \partial_1 B(t, y) \, \mathrm{d}t .$$

Therefore (5.4.4) holds. $\qquad\qquad\Box$

The probabilistic interpretation of the $*$-product is given in the following theorem.

Theorem 5.4.5. *Let $(X_t)_{t \in T}$ be a real stochastic process, let each random variable X_t be continuous for every $t \in T$ and let C_{st} denote the (unique) copula of the random variables X_s and X_t $(s, t \in T)$. Then the following statements are equivalent:*

(a) *for all s, t, u in T,*

$$C_{st} = C_{su} * C_{ut} ; \tag{5.4.5}$$

(b) *the transition probabilities $\mathbb{P}(s, x, t, A) := \mathbb{P}(X_t \in A \mid X_s = x)$ satisfy the Chapman-Kolmogorov equations*

$$\mathbb{P}(s, x, t, A) = \int_{\mathbb{R}} \mathbb{P}(u, \xi, t, A) \, \mathbb{P}(s, x, u, \mathrm{d}\xi) \tag{5.4.6}$$

for every Borel set A, for all s and t in T with $s < t$, for every $u \in \,]s, t[\,\cap\, T$ and for almost every $x \in \mathbb{R}$.

Proof. (a) $\Longrightarrow$ (b) The map $A \mapsto \mathbb{P}(s, x, t, A)$ is a probability measure for all s and

t in T and for almost every $x \in \mathbb{R}$. Therefore, it suffices to verify the Chapman-Kolmogorov equations (5.4.6) for Borel sets of the form $A =]-\infty, a]$. By Theorem 3.4.1, the transition probabilities for sets of this form can be written as

$$\mathbb{P}(s, x, t, A) = \partial_1 C_{st}\left(F_s(x), F_t(a)\right) \qquad \text{a.e.} \qquad (5.4.7)$$

Then, by Lemma 5.4.4,

$$\int_{\mathbb{R}} \mathbb{P}(u, \xi, t, A)\, \mathbb{P}(s, x, u, \mathrm{d}\xi) = \int_{\mathbb{R}} \partial_1 C_{ut}\left(F_u(\xi), F_t(a)\right) \partial_1 C_{su}\left(F_s(x), F_u(\mathrm{d}\xi)\right)$$

$$= \int_0^1 \partial_1 C_{ut}\left(\eta, F_t(a)\right) \partial_1 C_{su}\left(F_s(x), \mathrm{d}\eta\right)$$

$$= \frac{\partial}{\partial \zeta} \int_0^1 \partial_1 C_{ut}\left(\eta, F_t(a)\right) \partial_2 C_{su}(\zeta, \eta)\, \mathrm{d}\eta \Big|_{\zeta = F_s(x)}$$

$$= \frac{\partial}{\partial \zeta}\left(C_{su} * C_{ut}\right)\left(\zeta, F_t(a)\right)\Big|_{\zeta = F_s(x)}$$

$$= \partial_1\left(C_{su} * C_{ut}\right)\left(F_s(x), F_t(a)\right) = \partial_1 C_{st}\left(F_s(x), F_t(a)\right)$$

$$= \mathbb{P}(s, x, t,]-\infty, a]).$$

(b) $\Longrightarrow$ (a) If the Chapman-Kolmogorov equation holds and if $A =]-\infty, a]$, the last equation yields in view of (5.4.7)

$$\partial_1 C_{st}\left(F_s(x), F_t(a)\right) = \mathbb{P}(s, x, t, A) = \int_{\mathbb{R}} \mathbb{P}(u, \xi, t, A)\, \mathbb{P}(s, x, u, \mathrm{d}\xi)$$

$$= \partial_1\left(C_{su} * C_{ut}\right)\left(F_s(x), F_t(a)\right),$$

so that (5.4.5) holds. $\qquad\square$

It is well known that satisfaction of the Chapman–Kolmogorov equation is a necessary but not a sufficient condition for a Markov process; see, e.g., [Grimmett and Stirzaker, 2001, Example 6.1.14]. In order to state a sufficient condition, we need to introduce another operation on copulas, known as the $\star$-product.

Definition 5.4.6. For an m-copula A, $A \in \mathscr{C}_m$ and an n-copula B, $B \in \mathscr{C}_n$, define a function $A \star B : \mathbb{I}^{m+n-1} \to \mathbb{I}$ via

$$(A \star B)(x_1, x_2, \ldots, x_{m+n-1})$$

$$:= \int_0^{x_m} \partial_m A(x_1, \ldots, x_{m-1}, t)\, \partial_1 B(t, x_{m+1}, \ldots, x_{m+n-1})\, \mathrm{d}t. \qquad (5.4.8)$$

$\diamondsuit$

By a technique very similar to that of Theorem 5.1.1 one proves the following result.

Theorem 5.4.7. *For every m-copula A, $A \in \mathscr{C}_m$ and for every n-copula B, $B \in \mathscr{C}_n$, the mapping $A \star B$ defined by (5.4.8) yields an $(m + n - 1)$-copula, namely $\star : \mathscr{C}_m \times \mathscr{C}_n \to \mathscr{C}_{m+n-1}$.*

Notice that the $\star$-product (5.4.8) allows one to construct a copula in $\mathscr{C}_{m+n-1}$ starting from a copula A in $\mathscr{C}_m$ and one B in $\mathscr{C}_n$. In particular, if both A and B are in $\mathscr{C}_2$, then $A \star B$ is in $\mathscr{C}_3$.

When $m = n = 2$, then the $*$- and $\star$-products are related by

$$(A * B)(x, y) = (A \star B)(x, 1, y).$$

The next theorem is shown in a manner similar to that of the other results of this section (Theorems 5.1.4, 5.1.5 and 5.1.6).

Theorem 5.4.8. *The following results hold.*

(a) *the $\star$-product is distributive over convex combinations;*

(b) *if $(A_j)_{j \in \mathbb{N}}$ is a sequence in $\mathscr{C}_m$ that converges in $\mathscr{C}_m$ to A, and B is in $\mathscr{C}_n$, then*

$$A_j \star B \xrightarrow[j \to +\infty]{} A \star B \,;$$

similarly, A is in $\mathscr{C}_m$ and $(B_j)_{j \in \mathbb{N}}$ is a sequence in $\mathscr{C}_n$ that converges to B in $\mathscr{C}_n$, then

$$A \star B_j \xrightarrow[j \to +\infty]{} A \star B \,;$$

(c) *the $\star$-product is associative, viz., for all $A \in \mathscr{C}_m$, $B \in \mathscr{C}_n$ and $C \in \mathscr{C}_r$,*

$$A \star (B \star C) = (A \star B) \star C \,.$$

The definition of $\star$-product is used in the following result.

Theorem 5.4.9. *For a stochastic process $(X_t)_{t \in T}$ such that the random variable X_t has a continuous distribution for every $t \in T$, the following statements are equivalent:*

(a) *(X_t) is a Markov process;*

(b) *for every choice of $n \geq 2$ and of $t_1, t_2, \ldots, t_n$ in T such that $t_1 < t_2 < \cdots < t_n$, one has*

$$C_{t_1, t_2, \ldots, t_n} = C_{t_1 t_2} \star C_{t_2 t_3} \star \cdots \star C_{t_{n-1} t_n} \,, \tag{5.4.9}$$

where $C_{t_1, t_2, \ldots, t_n}$ is the copula of the random vector $(X_{t_1}, X_{t_2}, \ldots, X_{t_n})$ and $C_{t_j t_{j+1}}$ is the copula of the random variables X_{t_j} and $X_{t_{j+1}}$.

Proof. For convenience of notation we write X_j rather than X_{t_j}, C_{ij} rather than C_{t_i, t_j} and the like.

(a) $\implies$ (b) Integrating both sides of (5.4.2) over the set $X_2^{-1}(]-\infty, \alpha_2])$ yields, in view of Theorem 3.4.1,

$$
F_{123}(\alpha_1, \alpha_2, \alpha_3) = \int_{-\infty}^{\alpha_2} \partial_2 C_{12}\left(F_1(\alpha_1), F_2(\xi)\right) \partial_1 C_{23}\left(F_2(\xi), F_3(\alpha_3)\right) \, dF_2(\xi)
$$

$$
= \int_0^{F_2(\alpha_2)} \partial_2 C_{12}\left(F_{(\alpha_1)}, \eta\right) \partial_1 C_{23}\left(\eta, F_3(\alpha_3)\right) \, d\eta
$$

$$
= (C_{12} \star C_{23})\left(F_1(\alpha_1), F_2(\alpha_2), F_3(\alpha_3)\right) .
$$

This is equation (5.4.9) for the case $n = 3$.

In the general case, $n > 3$, one proceeds by induction. Consider the times

$$
t_1 < t_2 < \cdots < t_n < t_{n+1} .
$$

The conditional independence condition (5.4.1) may be written in the form

$$
\mathbb{E}\left(\mathbf{1}_{\{X_t \le \alpha\}} \mid X_1, X_2, \ldots, X_n\right) = \mathbb{E}\left(\mathbf{1}_{\{X_t \le \alpha\}} \mid X_n\right) \qquad a.e.
$$

Set $C_n := C_{t_1 t_2} \star C_{t_2 t_3} \star \cdots \star C_{t_{n-1} t_n}$ and $A_j := X_j^{-1}(]-\infty, \alpha_j])$; moreover we shall write t instead of t_{n+1}, whenever this simplifies the notation. Then, integrating the last equation over the set $\cap_{j=1}^n A_j$, one has, because of Theorem 3.4.1,

$$
F_{1,\ldots,n,t}(\alpha_1, \ldots, \alpha_n, \alpha) = C(F_1(\alpha_1), \ldots, F_n(\alpha_1), F_t(\alpha))
$$

$$
= \int_{A_1 \cap \cdots \cap A_n} \partial_1 C_{n,t}\left(F_1(X_1(\omega)), \ldots, F_n(X_n(\omega)), F_t(\alpha)\right) \, d\mathbb{P}(\omega)
$$

$$
= \int_{-\infty}^{\alpha_1} \cdots \int_{-\infty}^{\alpha_n} \partial_1 C_{n,t}\left(F_1(\xi_1), \ldots, F_n(\xi_n), F_t(\alpha)\right) \, dF_{1,\ldots,n}(\xi_1, \ldots, \xi_n)
$$

$$
= \int_0^{F_1(\alpha_1)} \cdots \int_0^{F_n(\alpha_n)} \partial_1 C_{n,t}\left(\eta_1, \ldots, \eta_n, F_t(\alpha)\right) C_{1,\ldots,n}(\, d\eta_1, \ldots, d\eta_n)
$$

$$
= \int_0^{F_n(\alpha_n)} \partial_1 C_{n,t}\left(\eta_1, \ldots, \eta_n, F_t(\alpha)\right) C_{1,\ldots,n}\left(F_1(\alpha_1), \ldots, F_{n-1}(\alpha_{n-1}), d\eta_n\right)
$$

$$
= \int_0^{F_n(\alpha_n)} \partial_n C_{1,\ldots,n}\left(F_1(\alpha_1), \ldots, F_{n-1}(\alpha_{n-1}), \eta_n\right)
$$

$$
\times \partial_1 C_{n,t}\left(\eta_1, \ldots, \eta_n, F_t(\alpha)\right) \, d\eta_n
$$

$$
= (C_{1,\ldots,n} \star C_{nt})\left(F_1(\alpha_1), \ldots, F_n(\alpha_1), F_t(\alpha)\right) ,
$$

which concludes the induction argument.

(b) $\implies$ (a) Assume now that equation (5.4.9) holds for $n = 3$. We wish to show that, for every Borel set B, one has

$$
\int_B \mathbf{1}_{\{X_1 \le \alpha_1\}} \mathbf{1}_{\{X_3 \le \alpha_3\}} \, d\mathbb{P} = \int_B \mathbb{E}\left(\mathbf{1}_{\{X_1 \le \alpha_1\}} \mid X_2\right) \mathbb{E}\left(\mathbf{1}_{\{X_3 \le \alpha_3\}} \mid X_2\right) \, d\mathbb{P}.
$$

$$
\tag{5.4.10}
$$

Since they generate the Borel σ-field $\mathscr{B}$, it suffices to verify eq. (5.4.10) for sets of the form $]-\infty, \alpha_2]$.

$$
\int_{]-\infty,\alpha_2]} \mathbf{1}_{\{X_1 \leq \alpha_1\}} \mathbf{1}_{\{X_3 < \alpha_3\}} \, d\mathbb{P} = \int \mathbf{1}_{\{X_1 \leq \alpha_1\}} \mathbf{1}_{\{X_2 \leq \alpha_2\}} \mathbf{1}_{\{X_3 \leq \alpha_3\}} \, d\mathbb{P}
$$

$$
= \mathbb{P}\left(X_1 \leq \alpha_1, X_2 \leq \alpha_2, X_3 \leq \alpha_3\right) = C_{123}\left(F_1(\alpha_1), F_2(\alpha_2), F_3(\alpha_3)\right)
$$

$$
= \int_0^{F_2(\alpha_2)} \partial_2 C_{12}\left(F_1(\alpha_1), \xi\right) \partial_1 C_{23}\left(\xi, F_3(\alpha_3)\right) \, d\xi
$$

$$
= \int_{]-\infty,\alpha_2]} \mathbb{E}\left(\mathbf{1}_{\{X_1 < \alpha_1\}} \mid X_2\right) \mathbb{E}\left(\mathbf{1}_{\{X_3 < \alpha_3\}} \mid X_2\right) \, d\mathbb{P},
$$

where recourse has been made to equation (3.4.4). This proves eq. (5.4.10). $\qquad\square$

It is now possible to see from the standpoint of copulas why the Chapman-Kolmogorov equations (5.4.6) alone do not guarantee that a process is Markov. One can construct a family of n-copulas with the following two requirements:

- they do not satisfy the conditions of the equations (5.4.9);
- they do satisfy the conditions of the equations (5.4.5), and are, as a consequence of Theorem 5.4.5, compatible with the 2-copulas of a Markov process and, hence, with the Chapman-Kolmogorov equations.

In order to construct such a family, it is enough to consider a stochastic process (X_t) in which the random variables are pairwise independent. Thus the copula of every pair of random variables X_s and X_t is given by Π_2. Since, by equation (5.1.2), $\Pi_2 * \Pi_2 = \Pi_2$, the Chapman-Kolmogorov equations are satisfied. It is now an easy task to verify that for every $n > 2$, the n-fold $\star$-product of Π_2 yields

$$
(\Pi_2 \star \cdots \star \Pi_2)(u_1, \ldots, u_n) = \Pi_n(u_1, \ldots, u_n),
$$

so that it follows that the only Markov process with pairwise independent (continuous) random variables is one where all finite subsets of random variables in the process are independent.

On the other hand, there are many 3-copulas whose 2-marginals coincide with Π_2; such an instance is represented by the family of copulas

$$
C_\alpha(u_1, u_2, u_3) := u_1 u_2 u_3 + \alpha \, u_1 \left(1 - u_1\right) u_2 \left(1 - u_2\right) u_3 \left(1 - u_3\right), \qquad (5.4.11)
$$

for $\alpha \in \,]-1, 1[$.

Now consider a process (X_t) such that

- three of its random variables, call them X_1, X_2 and X_3, have C_α of eq. (5.4.11) as their copula;
- every finite set not containing all three of X_1, X_2 and X_3 is made of independent random variables;

- the n-copula $(n > 3)$ of a finite set containing all three of them is given by

$$C_{t_1,\ldots,t_n}(u_1,\ldots,u_n) = C_\alpha(u_1, u_2, u_3)\, \Pi_{n-3}(u_4,\ldots,u_n)\,,$$

where we set $\Pi_1(t) := t$.

Such a process exists since it is easily verified that the resulting joint distributions satisfy the compatibility of Kolmogorov's consistency theorem (see, e.g., [Doob, 1953]); this ensures the existence of a stochastic process with the specified joint distributions. Since any two random variables in this process are independent, the Chapman-Kolmogorov equations are satisfied. However, the copula of X_1, X_2 and X_3 is inconsistent with equation (5.4.9), so that the process is not a Markov process.

It is instructive to compare the traditional way of specifying a Markov process with the one presented in this section and due to Darsow et al. [1992]. In the traditional approach a Markov process is singled out by specifying the initial distribution F_0 a family of transition probabilities $\mathbb{P}(s, x, t, A)$ that satisfy the Chapman-Kolmogorov equations. Notice that in this, the classical approach, the transition probabilities are fixed, so that changing the initial distribution simultaneously varies all the marginal distributions. In the present approach, a Markov process is specified by giving all the marginal distributions and a family of 2-copulas that satisfies (5.4.5); as a consequence, holding the copulas of the process fixed and varying the initial distribution does not affect the other marginals.

Further readings 5.4.10. In this section, we follow the presentation of the paper [Darsow et al., 1992] with one important difference; we limit our presentation to the case of continuous random variables. If this is not the case, the results we state continue to hold by recourse to the method of bilinear interpolation presented in the proof of Lemma 2.3.4, which was used in establishing Sklar's theorem.

More has been written, and, hence more could be said, about copulas and Markov processes, but this would involve lengthy preliminaries. Thus we shall simply quote a few papers, where the reader may find more references. Beare [2010] and Longla and Peligrad [2012] study mixing conditions in terms of copulas, while Ibragimov [2009] shows that a Markov process of order k is fully determined by its $(k + 1)$-dimensional copulas and its marginals. A recent example of how to construct Markov processes is presented by Cherubini et al. [2011a,b]. ∎

5.5 A generalisation of the Markov product

The $*$-product of the copulas A and B may be written in the form

$$(A * B)(u, v) = \int_0^1 \Pi_2\left(\partial_2 A(u, t), \partial_1 B(t, v)\right)\, \mathrm{d}t\,.$$

This suggests the following generalisation. Let $\mathbf{C} = \{C_t\}_{t \in \mathbb{I}} \subseteq \mathscr{C}_2$ an arbitrary family of copulas in $\mathscr{C}_2$, and, for every pair A and B of 2-copulas, define a mapping

$A *_{\mathbf{C}} B : \mathbb{I}^2 \to \mathbb{I}$ by

$$(A *_{\mathbf{C}} B)(u, v) := \int_0^1 C_t \left(\partial_2 A(u, t), \partial_1 B(t, v)\right) \, dt. \tag{5.5.1}$$

A problem with this generalisation arises from the fact that the integrand in the last expression may not be measurable; notice that, if it is measurable, then it is also integrable since copulas are bounded above by 1.

Example 5.5.1. Let S be a subset of $\mathbb{I}$ that is not Lebesgue measurable and define the family $\{C_t\}_{t \in \mathbb{I}}$ by

$$C_t := \begin{cases} M_2, & t \in S, \\ W_2, & t \notin S. \end{cases}$$

Then, for instance, the mapping $t \mapsto C_t \left(\partial_1 \Pi_2(u, t), \partial_1 \Pi_2(t, v)\right)$ is not Lebesgue measurable for all u and v in $\mathbb{I}$. ∎

In the following $\mathcal{M}$ will denote the set of families $\mathbf{C} = \{C_t\}_{t \in \mathbb{I}}$ for which the mapping $t \mapsto C_t \left(\partial_2 A(u, t), \partial_1 B(t, v)\right)$ is Lebesgue measurable for every $(u, v) \in \mathbb{I}^2$.

Theorem 5.5.2. Let $\mathbf{C}$ be in $\mathcal{M}$; then the function $(A *_{\mathbf{C}} B)$ defined by eq. (5.5.1) is a copula in $\mathscr{C}_2$ for all copulas A and B in $\mathscr{C}_2$.

Proof. Since $\mathbf{C}$ is in $\mathcal{M}$, the operation of eq. (5.5.1) is well defined. The boundary conditions of a copula are obviously satisfied by $A *_{\mathbf{C}} B$. Then set $C = A *_{\mathbf{C}} B$; then, for all x, u, y and v in $\mathbb{I}$ with $x \leq u$ and $y \leq v$, one has

$$C(u, v) + C(x, y) - C(x, v) - C(u, y) = \int_0^1 C_t \left(\partial_2 A(u, t), \partial_1 B(t, v)\right) \, dt$$

$$+ \int_0^1 C_t \left(\partial_2 A(x, t), \partial_1 B(t, y)\right) \, dt - \int_0^1 C_t \left(\partial_2 A(x, t), \partial_1 B(t, v)\right) \, dt$$

$$- \int_0^1 C_t \left(\partial_2 A(u, t), \partial_1 B(t, y)\right) \, dt$$

$$= \int_0^1 \{ C_t \left(\partial_2 A(u, t), \partial_1 B(t, v)\right) + C_t \left(\partial_2 A(x, t), \partial_1 B(t, y)\right)$$

$$- C_t \left(\partial_2 A(x, t), \partial_1 B(t, v)\right) - C_t \left(\partial_2 A(u, t), \partial_1 B(t, y)\right) \} \, dt \geq 0,$$

since C_t is a copula, $\partial_2 A(x, t) \leq \partial_2 A(u, t)$ and $\partial_1 B(t, y) \leq \partial_1 B(t, v)$ for every $t \in \mathbb{I}$, by virtue of Corollary 1.6.4. □

For a family $\mathbf{C} = \{C_t\}_{t \in \mathbb{I}}$ of copulas in $\mathcal{M}$, and for a pair of copulas A and B of $\mathscr{C}_2$, the copula $A *_{\mathbf{C}} B$ is called the **C**-product of A and B.

Obviously, the $*$-product of Section 5.1 may recovered from the **C**-product when $C_t = \Pi_2$ for every $t \in \mathbb{I}$.

For every copula $A \in \mathscr{C}_2$, for every family $\mathbf{C} = \{C_t\}_{t\in\mathbb{I}}$ of copulas of $\mathscr{C}_2$ and for all u and v in $\mathbb{I}$, one has

$$A *_{\mathbf{C}} M_2 = M_2 *_{\mathbf{C}} A = A,$$
$$(A *_{\mathbf{C}} W_2)(s,t) = s - A(s, 1-t),$$
$$(W_2 *_{\mathbf{C}} A)(s,t) = t - A(1-s, t).$$

It is immediate to check that M_2 is the unique identity for the $*_{\mathbf{C}}$-product. No simple result exists in general for

$$(A *_{\mathbf{C}} \Pi_2)(u,v) = \int_0^1 C_t(\partial_2 A(u,t), v) \ \mathrm{d}t.$$

There is a special case in which the $\mathbf{C}$-product coincides with the $*$-product.

Theorem 5.5.3. *Let $\mathbf{C}$ be in $\mathscr{M}$. If either A is right invertible or B is left invertible with respect to $*$-product, then $A *_{\mathbf{C}} B = A * B$.*

Proof. It suffices to prove the statement when A is right invertible, as the proof of the other one is analogous. Because of Theorem 5.2.3, one has $\partial_2 A(u,v) = 1$ a.e. Let S be the set of points $(u,v) \in \mathbb{I}^2$ at which $\partial_2 A(u,v) = 1$; then

$$(A *_{\mathbf{C}} B)(u,v) = \int_0^1 C_t(\partial_2 A(u,t), \partial_1 B(t,v)) \ \mathrm{d}t = \int_S C_t(1, \partial_1 B(t,v)) \ \mathrm{d}t$$
$$= \int_S \partial_1 B(t,v) \ \mathrm{d}t = \int_0^1 \partial_2 A(u,t), \partial_1 B(t,v) \ \mathrm{d}t = (A * B)(u,v),$$

which concludes the proof.   □

Contrary to what holds for the $*$-product (see Theorem 5.1.6), the $*_{\mathbf{C}}$-product need not be associative. Consider, for instance, the copula C defined, for all u and v in $\mathbb{I}$, by

$$C(u,v) := uv + uv(1-u)(1-v).$$

Then a tedious calculation that presents no difficulty yields

$$C *_{\mathbf{C}} (C *_{\mathbf{C}} C) \neq (C *_{\mathbf{C}} C) *_{\mathbf{C}} C,$$

where $\mathbf{C}$ is the family formed by just one copula equal to C.

Some of the constructions of copulas presented in this book may be regarded as special cases of the $\mathbf{C}$-product. This is shown in the next example for the mixture of copulas.

Example 5.5.4. Let $\mathbf{C} = \{C_t\}_{t\in\mathbb{I}}$ be an arbitrary family in $\mathscr{M}$; then

$$(\Pi_2 *_{\mathbf{C}} \Pi_2)(u,v) = \int_0^1 C_t(u,v) \ \mathrm{d}t,$$

so that the $*_{\mathbf{C}}$-product $\Pi_2 *_{\mathbf{C}} \Pi_2$ is related to the construction of Corollary 1.4.7. ■

An ordinal sum of copulas may be derived from an $*_C$-product; the relationship between the two construction is given below.

Theorem 5.5.5. *Let $C = (\langle a_i, b_i, C_i \rangle)_{i \in I}$ be an ordinal sum from a finite set of indexes I. If $\mathbf{C} = (C_t)_{t \in \mathbb{I}}$ is any family of copulas in $\mathscr{C}_2$ such that $C_t = C_i$ whenever t is in $[a_i, b_i]$ for some $i \in I$, and if C_{Π_2} denotes the ordinal sum $C_{\Pi_2} := (\langle a_i, b_i, \Pi_2 \rangle)_{i \in I}$, then*

$$C = C_{\Pi_2} *_\mathbf{C} C_{\Pi_2}. \tag{5.5.2}$$

Proof. Let $[a, b]$ be a subinterval of $[0, 1]$ and let $U_{(a,b)}$ be the uniform distribution function on (a, b), namely

$$U_{(a,b)}(x) = \begin{cases} 0, & \text{if } x \le a, \\ \frac{x-a}{b-a}, & \text{if } x \in [a, b], \\ 1, & \text{if } x \ge b. \end{cases}$$

Notice that

$$\partial_2 C_{\Pi_2}(u_1, t) = \begin{cases} U_{(a_i, b_i)}(t), & \text{if } t \in [a_i, b_i], \\ \mathbf{1}_{[t,1]}(u_1), & \text{elsewhere.} \end{cases}$$

A similar expression holds for $\partial_1 C_{\Pi_2}(t, u_2)$.

Now assume that there is an index $i \in I$ such that both u_1 and u_2 are in $]a_i, b_i[$. Then

$$\begin{aligned} (C_{\Pi_2} *_\mathbf{C} C_{\Pi_2})(u_1, u_2) &= \int_0^{a_i} C_t(1, 1) \, dt \\ &\quad + \int_{a_i}^{b_i} C_t \left(U_{(a_i, b_i)}(u_1), U_{(a_i, b_i)}(u_2) \right) \, dt \\ &= a_i + (b_i - a_i) \, Ci \left(\frac{u_1 - a_i}{b_i - a_i}, \frac{u_2 - a_i}{b_i - a_i} \right) \\ &= C(u_1, u_2). \end{aligned}$$

If there is no index $i \in I$ such that (u_1, u_2) belongs to $]a_i, b_i[^2$, then one has

$$(C_{\Pi_2} *_\mathbf{C} C_{\Pi_2})(u_1, u_2) = \int_0^{u_1 \wedge u_2} C_t(1, 1) \, dt = M_2(u_1, u_2),$$

which concludes the proof. □

Just as in Section 5.1 for $*$-product, also the construction of the C-product may be modified in order to define a copula in $\mathscr{C}_3$ from two copulas in $\mathscr{C}_2$.

Theorem 5.5.6. *Let $\mathbf{C} = \{C_t\}_{t \in \mathbb{I}}$ be a family of copulas in $\mathscr{C}_2$. Then, for every pair A and B of copulas in $\mathscr{C}_2$, the mapping $A \star_\mathbf{C} B : \mathbb{I}^2 \to \mathbb{I}$ defined by*

$$(A \star_\mathbf{C} B)(u_1, u_2, u_3) := \int_0^{u_2} C_t(\partial_2 A(u_1, t), \partial_1 B(t, u_2)) \, dt \tag{5.5.3}$$

is a copula, namely $A \star_\mathbf{C} B \in \mathscr{C}_3$, provided that the above integral exists and is finite.

Proof. It is immediately seen that $A \star_{\mathbf{C}} B$ satisfies the boundary conditions of a copula of $\mathscr{C}_3$.

In order to prove that $A \star_{\mathbf{C}} B$ is 3-increasing, let, for $j = 1, 2, 3$, u_j and v_j be in $\mathbb{I}$ with $u_j \leq v_j$. Because of Corollary 1.6.4, one has $\partial_2 A(u_1, t) \leq \partial_2 A(v_1, t)$ and $\partial_1 B(t, u_3) \leq \partial_1 B(t, v_3)$ for every $t \in \mathbb{I}$; therefore, a simple calculation leads to

$$V_{A \star_{\mathbf{C}} B} \left([u_1, v_1] \times [u_2, v_2] \times [u_3, v_3] \right)$$
$$= \int_{u_2}^{v_2} V_{C_t} \left([\partial_2 A(u_1, t), \partial_2 A(v_1, t)] \times [\partial_1 B(t, u_3), \partial_1 B(t, v_3)] \right) \, \mathrm{d}t \geq 0 \,,$$

which concludes the proof. $\square$

The copula $A \star_{\mathbf{C}} B$ is called the **C**-*lifting* of A and B. When $C_t = \Pi_2$ for every $t \in [0, 1]$, the **C**-lifting of A and B coincides with the $\star$-operation of Section 5.1.

Example 5.5.7. Simple calculations show that, for every copula A in $\mathscr{C}_2$,

$$(A \star_{\mathbf{C}} M_2)(u_1, u_2, u_3) = A(u_1, u_2 \wedge u_3) \,,$$
$$(M_2 \star_{\mathbf{C}} A)(u_1, u_2, u_3) = A(u_1 \wedge u_2, u_3) \,,$$
$$(W_2 \star_{\mathbf{C}} W_2)(u_1, u_2, u_3) = W_2(u_1 \wedge u_3, u_2) \,.$$

■

It is also not hard to evaluate the three marginals of $A \star_{\mathbf{C}} B$ for every pair of copulas A and B in $\mathscr{C}_2$:

$$(A \star_{\mathbf{C}} B)(u_1, u_2, 1) = A(u_1, u_2) \,,$$
$$(A \star_{\mathbf{C}} B)(u_1, 1, u_3) = (A \ast_{\mathbf{C}} B)(u_1, u_3) \,,$$
$$(A \star_{\mathbf{C}} B)(1, u_2, u_3) = B(u_2, u_3) \,.$$

Thus, the $\star_{\mathbf{C}}$-product provides a general method for obtaining trivariate copulas when two of its bivariate marginals are assigned. This provides a link with the so-called compatibility problem, i.e. the problem of deriving a d-dimensional copula given some information about its k-dimensional marginals ($2 \leq k < d$). For a general overview see, e.g., [Dall'Aglio, 1991; Schweizer, 1991; Joe, 1997; Rüschendorf, 2013].

Further readings 5.5.8. The generalisations presented in this section appeared in [Durante et al., 2007b], but see also [Durante et al., 2008a] and [Ruankong and Sumetkijakan, 2011]. However, it should be noted that the $\star_{\mathbf{C}}$-product is related to the mixtures of conditional distributions as presented, for instance, by [Joe, 1997, section 4.5]. Nowadays, these methods have been largely considered in the general framework of *vine copulas* or *pair-copula constructions*; see, for instance, [Bedford and Cooke, 2002; Aas et al., 2009; Czado, 2010; Kurowicka and Joe, 2011; Czado and Brechmann, 2013; Joe, 2014].   ■

Chapter 6

A compendium of families of copulas

In Probability and Statistics of the past century, and this is still evident in many present day textbooks, the independence of the random variables involved in statistical models has often been one of the tenets of the theory; one of the exceptions has involved the extensive use of the multidimensional Gaussian distribution and its generalizations.

However, while the Gaussianity assumption made analytical results (and computations) somehow accessible, it was often not justified by the real situation that the model purported to describe. In this respect, de Finetti's perspective is still very valid [de Finetti, 1937]:

> [...] the unjustified and harmful habit of considering the Gaussian distribution in too exclusive a way, as if it represented the rule in almost all the cases arising in probability and in statistics, and as if each non-Gaussian distribution constituted an exceptional or irregular case (even the name of "normal distribution" may contribute to such an impression, and it would therefore perhaps be preferable to abandon it).[1]

In fact, there are two main features in the adoption of a multivariate Gaussian distribution that are not always encountered in the statistical practice (especially when working with financial data, as stressed by McNeil et al. [2005]).

- The tails of the univariate marginals are too "light", and, hence, they do not assign enough weight to extreme events. In the same manner, the joint tails of the distribution do not assign enough weight to the occurrence of several extreme outcomes at the same time.

- The distribution has a strong form of symmetry, known as elliptical symmetry.

Actually, the validity of the comfortable Gaussian assumption has been criticised (at least) since the early 1950s, as expressed, for example, in the following quotation by J. Tukey.

> As I am sure almost every geophysicist knows, distributions of actual errors and fluctuations have much more straggling extreme values than would correspond to the magic bell-shaped distribution of Gauss and Laplace.

---

[1] [...] e dell'abitudine ingiustificata e dannosa di considerare la distribuzione gaussiana in modo troppo esclusivo come se dovesse rappresentare la regola in quasi tutti i casi presentati dal calcolo della probabilità e dalla statistica, e se ogni diversa forma di distribuzione costituisse un caso eccezionale o irregolare (anche lo stesso nome di "legge normale" può contribuire a tale impressione, e sarebbe forse perciò preferibile abbandonarlo).

The introduction of copulas has greatly contributed to enlarge the field of possible stochastic models that can be used for the problem at hand. In fact, as a consequence of Sklar's theorem, having at one's disposal a variety of copulas turns out to be very useful for building stochastic models having different properties that are sometimes indispensable in practice. For this reason, in recent years, several investigations have been carried out about the construction of different families of copulas and the study of their properties.

The aim of this chapter is to collect selected families of copulas that have appeared in the literature and which, in our opinion, possess appealing features both from a theoretical and an applied point of view. It is worth mentioning that this is a partial selection that somehow reflects the authors' tastes: definitely, a number of other families that are also relevant are not listed here. For a more complete overview of families of copulas, we refer the reader to [Nelsen, 2006; Lai and Balakrishnan, 2009; Mai and Scherer, 2012b; Joe, 2014].

6.1 What is a family of copulas?

Before presenting these examples, we should like briefly to discuss the use of these families in statistical practice.

For several applied researchers and practitioners it seems enough to try to fit any stochastic model with the most convenient family and check that the fitting procedure is not so "bad". However, it should be stressed that any fitting procedure may be misleading if one were to describe any situation with families of copulas satisfying some assumptions (e.g., exchangeability, light tails, etc.) that are not necessary for the problem at hand. The usual quotation by G.E.P. Box, "All the models are wrong, some are useful", should be remembered!

In our opinion, before applying any statistical tool, one should not forget to analyse the main characteristics of the problem under consideration. At the same time, one should clearly bear in mind the final output of the investigation. Usually, in fact, fitting a copula (or a joint d.f.) to some data is just a tool for deriving some quantities of interest for the problem at hand (e.g., Value-at-Risk of a portfolio, return period of an extreme event, etc.). For such problems, a dramatic underestimation of the risk can be obtained when one tries to fit with copulas that do not exhibit any peculiar behaviour in the tails. This was exactly one of the main criticisms to the use of Li's model for credit risk using Gaussian copulas (see, for example, [Li, 2001; Whitehouse, 2005; Jones, 2009]).

Before proceeding, we should like to clarify the formal meaning of "family of copulas", a rather loose phrase. Then, inspired by considerations of Joe [1997], we discuss here some general properties that a "good" family of multivariate copulas should have.

Definition 6.1.1. Let Θ be a set. A mapping

$$\theta \in \Theta \mapsto C_\theta \in \mathscr{C}_d$$

is called a *family of copulas*. One writes: $(C_\theta)_{\theta \in \Theta}$. $\diamondsuit$

In other words, a family of copulas is any subset of $\mathscr{C}_d$ that can be indexed by a suitable set Θ. Such a Θ may be a subset of $\mathbb{R}^p$ ($p \geq 1$) or a set of functions with suitable properties. Statisticians usually refer to $\theta \in \Theta$ as a *parameter*. Depending on the properties of the mapping $\theta \mapsto C_\theta$ and on the structure of Θ, the following properties of a family of copulas may be introduced:

- *Identifiability.* A family $(C_\theta)_{\theta \in \Theta}$ is said to be *identifiable* if $\theta \mapsto C_\theta$ is injective. In other words, a copula in a given family cannot be parametrised in two different ways.

- *Monotonically ordered.* Suppose that there exists a partial order $\prec$ on Θ and a partial order $\leq$ on $\mathscr{C}_d$ (for instance, the PLOD order). A family $(C_\theta)_{\theta \in \Theta}$ is said to be *order-preserving* (respectively, *order-reversing*) if $\theta \prec \theta'$ implies $C_\theta \leq C_{\theta'}$ (respectively, $C_\theta \geq C_{\theta'}$). In other words, the order between the parameters is reflected by the same (respectively, opposite) $\leq$ order between the copulas.

Further properties that make a family of copulas appealing for practical purposes are the following.

- *Interpretability.* The members of the family may have a probabilistic interpretation suggesting "natural" situations where this family may be considered. This interpretation is often essential in order to simulate a random sample from the members of the family.

- *Flexible and wide range of dependence.* The members of the family should describe different types of dependence, including the independence copula Π_d and at least one of the Hoeffding-Fréchet bounds (possibly, as a limiting case with respect to the parameter). Having members with a variety of tail dependencies and asymmetries is a desirable property as well.

- *Easy-to-handle.* The members of the family should be expressed in a *closed form* or, at least, should be analytically tractable. Particular attention is devoted to the possibility of calculating closed forms for the density of the copulas and the measures of association related to them.

In the following we show several examples of copulas and clarify whether they satisfy some of the above properties. A further remark is needed here.

In the literature several classes of copulas are usually attributed to one or more researchers who, for the first time, have introduced and studied in a significant way its properties, or applied it to a broad field. However, since most of these families were actually presented in the language of distribution functions rather than that of copulas it has happened that they have been independently introduced by several authors in different years. In the following, we have tried, as much as possible, to give credit to the authors who actually proposed the above distributions. Therefore, we apologise in advance if we have overlooked earlier references or attributed names and notations in a way that may appear arbitrary to some readers.

6.2   Fréchet copulas

Following his studies about the upper and lower bounds in the class of distribution functions with fixed margins, Fréchet [1958] proposed a simple way to combine the upper and lower bound distributions in the Fréchet class in order to create a parametric family.

His two-parameter family may be represented in the form

$$C_{\alpha,\beta}^{\mathbf{Fre}}(u_1, u_2) = \alpha\, M_2(u_1, u_2) + (1 - \alpha - \beta)\, \Pi_2(u_1, u_2) + \beta\, W_2(u_1, u_2),$$

where α and β are in $\mathbb{I}$ with $\alpha + \beta \leq 1$. These copulas are obtained as a convex sum of the basic copulas M_2, Π_2 and W_2. The main idea of the construction consists of creating a family of copulas that can span all representative dependence properties (in such a case, independence, comonotonicity and countermonotonicity).

Given a bivariate Fréchet copula, the values of the Kendall's τ and Spearman's ρ associated with it are given, respectively, by

$$\tau(C_{\alpha,\beta}^{\mathbf{Fre}}) = \frac{1}{3}(\alpha - \beta)(\alpha + \beta + 2), \qquad \rho(C_{\alpha,\beta}^{\mathbf{Fre}}) = \alpha - \beta.$$

Moreover, the tail dependence coefficients associated with it are equal to

$$\lambda_L = \lambda_U = \alpha.$$

A subclass has been proposed by Mardia [1970]: a copula C_α belongs to the *Mardia family* if, for every $\alpha \in [-1, 1]$, it is defined by

$$C_\alpha(u_1, u_2) = \frac{1}{2}\, \alpha^2\, (1 - \alpha)\, W_2(u_1, u_2) + (1 - \alpha^2)\, \Pi_2(u_1, u_2)$$
$$+ \frac{1}{2}\, \alpha^2\, (1 + \alpha)\, M_2(u_1, u_2).$$

A closely related family has also been considered in [Gijbels et al., 2010].

Since the Fréchet lower bound is not a copula for $d \geq 3$, this family cannot be fully extended to the higher dimensional case. A possible d-dimensional extension of its subclass describing positive dependence is given by

$$C_\alpha^{\mathbf{Fre}}(\mathbf{u}) = \alpha\, M_d(\mathbf{u}) + (1 - \alpha)\, \Pi_d(\mathbf{u}), \tag{6.2.1}$$

for every $\alpha \in \mathbb{I}$. This family is order-preserving with respect to the PLOD order.

6.3   EFGM copulas

The so-called *Eyraud-Farlie-Gumbel-Morgenstern* (EFGM) distributions have been considered by Morgenstern [1956] and Gumbel [1958, 1960a] and further developed by Farlie [1960]. However, these distributions had already appeared in an earlier and, for many years, forgotten work by Eyraud [1936]. Here, we present the EFGM family of copulas derived from these distributions.

A bivariate EFGM copula has the following expression:

$$C_\alpha^{\mathbf{EFGM}}(u_1, u_2) = u_1 u_2 \left(1 + \alpha \left(1 - u_1\right) \left(1 - u_2\right)\right), \qquad (6.3.1)$$

with $\alpha \in [-1, 1]$. This family is order-preserving with respect to the PQD order. Given a bivariate EFGM copula, the values of the Kendall's τ and Spearman's ρ associated with it are given, respectively, by

$$\tau(C_\alpha^{\mathbf{EFGM}}) = \frac{2\alpha}{9}, \qquad \rho(C_\alpha^{\mathbf{EFGM}}) = \frac{\alpha}{3}.$$

Moreover, the tail dependence coefficients associated with it are equal to 0. It follows that $\tau(C_\alpha^{\mathbf{EFGM}}) \in [-2/9, 2/9]$ and $\rho(C_\alpha^{\mathbf{EFGM}}) \in [-1/3, 1/3]$. As noted several times in the literature, the limited range of dependence for these copulas restricts the usefulness of this family for modelling.

Now, consider the higher dimensional extension to $d \geq 3$. Let $\mathscr{S}$ be the class of all subsets of $\{1, \ldots, d\}$ having at least 2 elements, so that $\mathscr{S}$ contains $2^d - d - 1$ elements. To each $S \in \mathscr{S}$, we associate a real number α_S, with the convention that, when $S = \{i_1 \ldots, i_k\}$, $\alpha_S = \alpha_{i_1 \ldots i_k}$. An EFGM d-copula can be defined in the following form:

$$C_\alpha^{\mathbf{EFGM}}(\mathbf{u}) = \prod_{i=1}^d u_i \left(1 + \sum_{S \in \mathscr{S}} \alpha_S \prod_{j \in S} (1 - u_j)\right), \qquad (6.3.2)$$

for suitable values of the α_S's.

For instance, when $d = 3$ and for $\mathbf{u} \in \mathbb{I}^3$, EFGM copulas can be expressed in the following way

$$\begin{aligned} C_\alpha^{\mathbf{EFGM}}(\mathbf{u}) = u_1 u_2 u_3 \, [1 &+ \alpha_{12} \left(1 - u_1\right) \left(1 - u_2\right) + \alpha_{13} \left(1 - u_1\right) \left(1 - u_3\right) \\ &+ \alpha_{23} \left(1 - u_2\right) \left(1 - u_3\right) + \alpha_{123} \left(1 - u_1\right) \left(1 - u_2\right) \left(1 - u_3\right)] . \end{aligned} \qquad (6.3.3)$$

It is easily verified that the EFGM copulas are absolutely continuous with density given by

$$c_\alpha^{\mathbf{EFGM}}(\mathbf{u}) = 1 + \sum_{S \in \mathscr{S}} \alpha_S \prod_{j \in S} (1 - 2u_j). \qquad (6.3.4)$$

As a consequence, the parameters α_S's have to satisfy the following inequality

$$1 + \sum_{S \in \mathscr{S}} \alpha_S \prod_{j \in S} \xi_j \geq 0$$

for every $\xi_j \in \{-1, 1\}$. In particular, one has $|\alpha_S| \leq 1$.

Several extensions of EFGM copulas have been proposed in the literature, starting with the works by Farlie [1960] and Sarmanov [1966]. A complete survey of these generalised EFGM models of dependence is given by Drouet-Mari and Kotz [2001], where a list of several other references can be also found. More recent investigations are also provided in [Amblard and Girard, 2001, 2002, 2009; Fischer and

Klein, 2007; Rodríguez-Lallena and Úbeda Flores, 2004b]. Another possible extension of EFGM copulas is based on the construction of copulas that are quadratic in one variable [Quesada-Molina and Rodríguez-Lallena, 1995; Rodríguez-Lallena and Úbeda Flores, 2010].

Notice that EFGM copulas, among other ones, may be regarded as a "perturbation" of the product copula. As stressed by Durante et al. [2013c], all these copulas may be expressed in the following form

$$C(\mathbf{u}) = \Pi_d(\mathbf{u}) + P(\mathbf{u}), \tag{6.3.5}$$

where P is a suitable real function defined on $\mathbb{I}^d$.

Furthermore, it should be stressed that all absolutely continuous copulas can be expressed in the form (6.3.5) once a suitable representation of the density of P is given. This representation has been given by Rüschendorf [1985] (see also [de la Peña et al., 2006, eq. (3.1) and (3.2)]). Possible applications of EFGM copulas in copula-based stochastic processes are presented in [de la Peña et al., 2006, Section 5] and [Cherubini et al., 2011a].

6.4   Marshall-Olkin copulas

The construction of Marshall-Olkin copulas is related to one of the most popular distributions allowing for non-trivial singular components. For this reason, they have been applied extensively for years in fields like Reliability Theory and Credit Risk, where it is important to consider the possibility that two or more lifetimes coincide with a non-trivial probability (see, for instance, Cherubini et al. [2015] and references therein). Marshall-Olkin copulas have an intuitive representation in terms of stochastic models, as the following probabilistic interpretation shows.

Let $(\Omega, \mathscr{F}, \mathbb{P})$ be a probability space. For each non-empty subset $I \subseteq \{1, \ldots, d\}$, let E_I be a r.v. whose distribution function is the exponential distribution with mean $1/\lambda_I > 0$, where $\lambda_I \geq 0$. In case $\lambda_I = 0$, we set $\mathbb{P}(E_I = +\infty) = 1$. Assume that all the $2^d - 1$ r.v.'s E_I are independent. Moreover, for each $k \in \{1, \ldots, d\}$, assume $\sum_{I:k \in I} \lambda_I > 0$, namely each k belongs to at least one set I such that $\lambda_I > 0$. Under these conditions, one may define for every $k \in \{1, \ldots, d\}$ the r.v.'s

$$X_k := \min\{E_I : I \subseteq \{1, \ldots, d\}, k \in I\}.$$

The following results can be proved (see [Marshall and Olkin, 1967b,a] and [Mai and Scherer, 2012b, chapter 3]).

Theorem 6.4.1. *The survival function of the random vector* $(X_1, \ldots, X_d)$ *is given for all* $\mathbf{x} \in [0, +\infty[^d$ *by*

$$\overline{F}(\mathbf{x}) = \exp\left(- \sum_{\emptyset \neq I \subseteq \{1, \ldots, d\}} \lambda_I \max_{i \in I}\{x_i\} \right).$$

Moreover, the survival copula associated with $\overline{F}$ belongs to the family given by

$$C^{\mathbf{MO}}(\mathbf{u}) = \prod_{\emptyset \neq I \subseteq \{1,\ldots,d\}} \min \left\{ u_k^{\frac{\lambda_I}{\sum_{J:k \in J} \lambda_J}}, k \in I \right\}. \qquad (6.4.1)$$

Members of the family of copulas of type (6.4.1) are called *Marshall-Olkin copulas*.

For $d = 2$ Marshall-Olkin copulas can be expressed in the form

$$C^{\mathbf{MO}}(u_1, u_2) = u_1^{\frac{\lambda_{\{1\}}}{\lambda_{\{1\}}+\lambda_{\{1,2\}}}} u_2^{\frac{\lambda_{\{2\}}}{\lambda_{\{2\}}+\lambda_{\{1,2\}}}} \min \left\{ u_1^{\frac{\lambda_{\{1,2\}}}{\lambda_{\{1\}}+\lambda_{\{1,2\}}}}, u_2^{\frac{\lambda_{\{1,2\}}}{\lambda_{\{2\}}+\lambda_{\{1,2\}}}} \right\}$$

$$= \min \left\{ u_1^{1-\frac{\lambda_{\{1,2\}}}{\lambda_{\{1\}}+\lambda_{\{1,2\}}}} u_2, u_1 u_2^{1-\frac{\lambda_{\{1,2\}}}{\lambda_{\{2\}}+\lambda_{\{1,2\}}}} \right\}.$$

By taking into account the fact that both

$$\frac{\lambda_{\{1,2\}}}{\lambda_{\{1\}} + \lambda_{\{1,2\}}} \quad \text{and} \quad \frac{\lambda_{\{1,2\}}}{\lambda_{\{2\}} + \lambda_{\{1,2\}}}$$

belong to $\mathbb{I}$, Marshall-Olkin 2-copulas may be conveniently reparametrised as

$$C^{\mathbf{MO}}_{\alpha,\beta}(u_1, u_2) := \min \left\{ u_1^{1-\alpha} u_2, u_1 u_2^{1-\beta} \right\} = \begin{cases} u_1^{1-\alpha} u_2, & u_1^{\alpha} \geq u_2^{\beta}, \\ u_1 u_2^{1-\beta}, & u_1^{\alpha} < u_2^{\beta}, \end{cases}$$

for all α and β in $\mathbb{I}$.

Given a bivariate Marshall-Olkin copula, the values of the measures of association related to it are given by

$$\tau(C^{\mathbf{MO}}_{\alpha,\beta}) = \frac{\alpha\,\beta}{\alpha - \alpha\,\beta + \beta}, \qquad \rho(C^{\mathbf{MO}}_{\alpha,\beta}) = \frac{3\alpha\,\beta}{2\alpha - \alpha\,\beta + 2\beta}.$$

Moreover, the tail dependence coefficients associated with it are given by

$$\lambda_L(C^{\mathbf{MO}}_{\alpha,\beta}) = 0, \qquad \lambda_U(C^{\mathbf{MO}}_{\alpha,\beta}) = \min\{\alpha, \beta\}.$$

When $\min\{\alpha, \beta\} \neq 0$, $C^{\mathbf{MO}}_{\alpha,\beta}$ are not absolutely continuous, since they have a singular component along the curve $\{(u_1, u_2) \in \mathbb{I}^2 : u_1^{\alpha} = u_2^{\beta}\}$.

Interestingly, $C^{\mathbf{MO}}_{\alpha,\beta}$ is the copula associated with the bivariate distribution function (with exponential marginals) that is the unique solution of the functional equation

$$\mathbb{P}(X > s_1 + t, Y > s_2 + t \mid X > t, Y > t) = \mathbb{P}(X > s_1, Y > s_2)$$

for all $s_1 \geq 0$, $s_2 \geq 0$ and $t \geq 0$.

The exchangeable members of Marshall-Olkin 2-copulas can be expressed in the form

$$C_\alpha^{\mathbf{CA}}(u_1, u_2) := [\min\{u_1, u_2\}]^\alpha \, (u_1 u_2)^{1-\alpha} = \begin{cases} u_1 u_2^{1-\alpha}, & u_1 \leq u_2, \\ u_1^{1-\alpha} u_2, & u_1 > u_2, \end{cases} \quad (6.4.2)$$

with $\alpha \in \mathbb{I}$. They are usually called *Cuadras-Augé copulas* because they were also considered in [Cuadras and Augé, 1981]. Generalisations of the class of Cuadras-Augé 2-copulas can be found in [Durante, 2006] (see also [Durante et al., 2007c, 2008b; Durante and Salvadori, 2010]), [Cuadras, 2009] and in [Mai and Scherer, 2009; Mai et al., 2015].

6.5 Archimedean copulas

Archimedean copulas are widely used in applications and, as a consequence, have been extensively studied in the literature. Curiously, their introduction originated in connexion with the investigations on semigroups of the unit interval and probabilistic metric spaces (see, e.g., [Schweizer and Sklar, 1983] and chapter 8); only later have they been used in a statistical framework (see, e.g., [Genest and MacKay, 1986a]).

Archimedean copulas are parametrised via a one-dimensional function, which is defined below.

Definition 6.5.1. A function $\varphi : [0, +\infty[\to \mathbb{I}$ is said to be an *additive generator* if it is continuous, decreasing and $\varphi(0) = 1$, $\lim_{t \to +\infty} \varphi(t) = 0$ and is strictly decreasing on $[0, t_0]$, where $t_0 := \inf\{t > 0 : \varphi(t) = 0\}$. $\qquad \diamond$

Unless we say otherwise, when speaking of a generator we mean an additive generator. The *pseudo-inverse* of the generator φ is defined by

$$\varphi^{(-1)}(t) := \begin{cases} \varphi^{-1}(t), & t \in \,]0, 1]\,, \\ t_0, & t = 0. \end{cases} \quad (6.5.1)$$

Notice that $\varphi^{(-1)}(\varphi(t)) = \min\{t, t_0\}$ for every $t \geq 0$.

The following definition introduces d-dimensional *Archimedean copulas* for every $d \geq 2$.

Definition 6.5.2. A d-copula C with $d \geq 2$ is said to be *Archimedean* if a generator φ exists such that, for every $\mathbf{u} \in \mathbb{I}^d$,

$$C(\mathbf{u}) = \varphi\left(\varphi^{(-1)}(u_1) + \cdots + \varphi^{(-1)}(u_d)\right). \quad (6.5.2)$$

$\diamond$

With reference to the generator, the copula of eq. (6.5.2) is denoted by C_φ. When φ is strictly decreasing, its quasi-inverse $\varphi^{(-1)}$ equals its inverse, $\varphi^{(-1)} = \varphi^{-1}$, and the copula C_φ is said to be *strict*.

A d-dimensional Archimedean copula is, by construction, exchangeable. Moreover, if C_φ is generated by φ, then it is generated also by $\alpha\,\varphi$ for every $\alpha > 0$. Thus, the representation (6.5.2) is not unique.

Remark 6.5.3. The reader should also be alerted to the fact that it is common in the literature on copulas to represent the copulas of the present section through $\varphi^{(-1)}$ rather than by means of φ as we are now doing. ∎

Example 6.5.4. The copula Π_2 is Archimedean: take $\varphi(t) = e^{-t}$; since $\lim_{t\to+\infty} \varphi(t) = 0$ and $\varphi(t) > 0$ for every $t > 0$, φ is strict; then $\varphi^{-1}(t) = -\ln t$ and

$$\varphi\left(\varphi^{-1}(u) + \varphi^{-1}(v)\right) = \exp\left(-(-\ln u - \ln v)\right) = uv = \Pi_2(u, v).$$

Also the lower Hoeffding–Fréchet bound W_2 is Archimedean; in order to see this, it suffices to take $\varphi(t) := \max\{1 - t, 0\}$. Since $\varphi(1) = 0$, φ is not strict. Its quasi-inverse is $\varphi^{(-1)}(t) = 1 - t$.

On the contrary, the upper Hoeffding–Fréchet bound is not Archimedean, since no generator exists such that $M_2(t, t) = t = \varphi(2\varphi^{(-1)})(t))$ for every $t \in \mathbb{I}$. ∎

It is important to determine the properties that a generator has to enjoy in order that the function C_φ defined by (6.5.2) is a d-copula. This question will be addressed and answered in Theorem 6.5.7 below via the following preliminary definition.

Definition 6.5.5. A function $f : \,]a, b[\, \to \mathbb{R}$ is called d-*monotone* in $\,]a, b[$, where $-\infty \le a < b \le +\infty$ and $d \ge 2$ if

- it is differentiable up to order $d - 2$;
- for every $x \in \,]a, b[$, its derivatives satisfy

$$(-1)^k f^{(k)}(x) \ge 0$$

for $k = 0, 1, \ldots, d - 2$;
- $(-1)^{d-2} f^{(d-2)}$ is decreasing and convex in $\,]a, b[$. ◇

Moreover, if f has derivatives of every order in $\,]a, b[$ and if

$$(-1)^k f^{(k)}(x) \ge 0,$$

for every $x \in \,]a, b[$ and for every $k \in \mathbb{Z}_+$, f is said to be *completely monotone*. ◇

For $d = 1$, f is said to be (1-)monotone on $\,]a, b[$ if it is decreasing and positive on this interval. In the case $d = 2$, notice that a 2-monotone function f is simply decreasing and convex.

Definition 6.5.6. Let $I \subseteq \mathbb{R}$ be an interval. A function $f : I \to \mathbb{R}$ is said to be d-*monotone* (respectively, *completely monotone*) on I, with $d \in \mathbb{N}$, if it is continuous on I and if its restriction to the interior I° of I is d-monotone (respectively, completely monotone). ◇

Theorem 6.5.7. *Let $\varphi : [0, +\infty[\to \mathbb{I}$ be a generator. Then the following statements are equivalent:*

(a) *φ is d-monotone on $[0, +\infty[$;*

(b) *the function $C_\varphi : \mathbb{I}^d \to \mathbb{I}$ defined by (6.5.2) is a d-copula.*

Proof. See McNeil and Nešlehová [2009] and Malov [2001].   □

The following is an immediate consequence of Theorem 6.5.7.

Corollary 6.5.8. *Let φ be a generator. Then the following statements are equivalent:*

(a) *φ is convex;*

(b) *the function $C_\varphi : \mathbb{I}^2 \to \mathbb{I}$ defined by*

$$C_\varphi(u_1, u_2) = \varphi\left(\varphi^{(-1)}(u_1) + \varphi^{-1}(u_2)\right) \tag{6.5.3}$$

is a 2-copula.

Remark 6.5.9. A bivariate copula can be interpreted as a binary operation of $\mathbb{I}$. In particular, if it has the form (6.5.3), it is an associative operation on $\mathbb{I}$ since

$$C_\varphi\left(C_\varphi(u_1, u_2), u_3\right) = C_\varphi\left(u_1, C_\varphi(u_2, u_3)\right)$$

for all u_1, u_2, and u_3 in $\mathbb{I}$. Now, the adjective "Archimedean" assigned to copulas of type (6.5.3) refers to the fact that they satisfy a special property of the associative operations, as will be clarified in Definition 8.2.7.   ■

For $d = 2, 3, \ldots, \infty$, let $\mathscr{G}_d$ denote the set of generators of an Archimedean d-copula. It follows from Theorem 6.5.7 that the following inclusions hold

$$\mathscr{G}_2 \supseteq \mathscr{G}_3 \supseteq \cdots \supseteq \mathscr{G}_\infty .$$

All the inclusions are strict (see, e.g., [McNeil and Nešlehová, 2009]). For instance, it was seen in Example 6.5.4 that the Archimedean generator of the 2-copula W_2 is $\varphi_2(t) = \max\{1 - t, 0\}$; however, for $d \geq 3$,

$$W_d(\mathbf{u}) = \varphi_2\left(\varphi_2^{(-1)}(u_1) + \cdots + \varphi_2^{(-1)}(u_d)\right) = \max\{u_1 + \cdots + u_d - d + 1, 0\}$$

is not a d-copula as was proved in Section 1.7.

Remark 6.5.10. In some cases it may be expedient to represent an Archimedean copula by means of a multiplicative, rather than an additive, generator; this is achieved through the transformation $h(t) := \varphi(-\ln t)$. Then the copula is represented in the form

$$C(\mathbf{u}) = h\left(h^{(-1)}(u_1) \cdot h^{(-1)}(u_2) \cdots h^{(-1)}(u_d)\right) \tag{6.5.4}$$

for every $\mathbf{u} \in \mathbb{I}^d$.   ■

A stochastic representation of d-dimensional Archimedean copulas has been provided by McNeil and Nešlehová [2009]. To this end, two auxiliary definitions will be recalled.

Definition 6.5.11. A random vector $\mathbf{X}$ is said to have an ℓ_1-*norm symmetric distribution* if there exists a positive random variable R independent of the random vector $\mathbf{S}_d$, which is uniformly distributed on the simplex

$$\mathscr{S}_d := \left\{ \mathbf{x} \in [0, +\infty[^d : \sum_{j=1}^d x_j = 1 \right\}, \qquad (6.5.5)$$

so that $\mathbf{X}$ has the same probability distribution as $R\,\mathbf{S}_d$, $\mathbf{X} \overset{d}{=} R\,\mathbf{S}_d$. The r.v. R is called the *radial part* of $\mathbf{X}$ and its distribution as the *radial distribution*. ◇

Definition 6.5.12. Let F be the d.f. of a positive random variable X; the *Williamson d-transformation* of F is defined, for $d \geq 2$, by

$$\mathfrak{M}_d F(x) := \int_x^{+\infty} \left(1 - \frac{x}{t}\right)^{d-1} dF(t) = \begin{cases} E\left[\left(1 - \frac{x}{X}\right)_+^{d-1}\right], & \text{for } x > 0, \\ 1 - F(0), & \text{for } x = 0. \end{cases}$$

◇

It is now possible to collect the above results in the following characterisation.

Theorem 6.5.13. *The following results hold.*

(a) *If the random vector $\mathbf{X}$ has a d-dimensional ℓ_1-norm symmetric distribution with radial d.f. F_R satisfying $F_R(0) = 0$, then its survival copula is Archimedean with generator $\varphi = \mathfrak{M}_d F_R$.*

(b) *Let the random vector $\mathbf{U} = (U_1, \ldots, U_d)$ on $\mathbb{I}^d$ be distributed according to the d-dimensional Archimedean copula C with generator φ. Then $\left(\varphi^{(-1)}(U_1), \ldots, \varphi^{(-1)}(U_d)\right)$ has an ℓ_1-norm symmetric distribution with survival copula C and radial d.f. F_R given by $F_R = \mathfrak{M}_d^{-1}\varphi$.*

Thus, Archimedean copulas are naturally associated with ℓ_1-norm symmetric law $\mathbf{X} \overset{d}{=} R\,\mathbf{S}_d$. Furthermore, there exists a one-to-one correspondence between the generator φ of an Archimedean copula and the distribution of the radial part R, which is provided by the Williamson transformation and its inverse. The term "Williamson transformation" was adopted by McNeil and Nešlehová [2009] in recognition of the work by Williamson [1956].

Now, consider the subclass of Archimedean copulas that are generated by completely monotone functions. As a consequence of Theorem 6.5.7 one also obtains the following result.

Corollary 6.5.14. *Let $\varphi : [0, +\infty[\to \mathbb{I}$ be a generator. Then the following statements are equivalent:*

(a) φ is completely monotone on $[0, +\infty[$;

(b) the function $C_\varphi : \mathbb{I}^d \to \mathbb{I}$ defined by (6.5.2) is a d-copula for every $d \geq 2$.

Notice that if the Archimedean copula $C_\varphi \in \mathscr{C}_d$ is generated by a completely monotone function φ, then $C_\varphi \geq \Pi_d$ for every $d \geq 2$ (see, e.g., [Alsina et al., 2006, Theorem 4.4.6] and also [Müller and Scarsini, 2005]).

Historical remark 6.5.15. Corollary 6.5.14 was formulated by [Sklar, 1973, Theorem 9] and Kimberling [1974]. A proof is also given by [Alsina et al., 2006, Theorem 4.4.6] and Ressel [2011]. ∎

Copulas of Corollary 6.5.14 also admit an alternative stochastic representation. In fact, if $X_1, \ldots, X_d$ are independent and identically distributed r.v.'s with exponential distribution with mean 1, and M is an independent and positive random variable with Laplace transform φ, then the random vector

$$(U_1, \ldots, U_d) := \left(\varphi\left(\frac{E_1}{M}\right), \ldots, \varphi\frac{E_d}{M} \right)$$

has the Archimedean copula C_φ as its joint distribution function. Moreover, φ is completely monotone. For more details see [Marshall and Olkin, 1988] (and also [Genest et al., 2011; Mai and Scherer, 2012b]).

Here we list several families of Archimedean copulas. Alsina et al. [2006] (Table 2.6), Nelsen [2006] (Table 4.1) and Joe [1997, 2014] present longer lists of Archimedean copulas and give a synopsis of the properties of the reported families and the related measures of association. Notice that general formulas for the calculation of Kendall's tau and tail dependence coefficients for Archimedean copulas have been provided in [Genest and MacKay, 1986a; McNeil and Nešlehová, 2009] and [Charpentier and Segers, 2007, 2009; Larsson and Nešlehová, 2011], respectively.

Example 6.5.16 (Gumbel-Hougaard copulas). The standard expression for members of this family of d-copulas is

$$C_\alpha^{\mathbf{GH}}(\mathbf{u}) = \exp\left(-\left(\sum_{i=1}^d (-\ln(u_i))^\alpha \right)^{1/\alpha} \right), \qquad (6.5.6)$$

where $\alpha \geq 1$. For $\alpha = 1$ one obtains the independence copula as a special case, and the limit of $C_\alpha^{\mathbf{GH}}$ for $\alpha \to +\infty$ is the comonotonicity copula M_d. The Archimedean generator of this family is given by $\varphi(t) = \exp\left(-t^{1/\alpha}\right)$. Each member of this class is absolutely continuous.

Gumbel-Hougaard 2-copulas are positively quadrant dependent; moreover, their family is order-preserving with respect to pointwise order between copulas. The Kendall's tau associated with 2-copulas of this family is given by

$$\tau(C_\alpha^{\mathbf{GH}}) = \frac{\alpha - 1}{\alpha}.$$

Moreover, the tail dependence coefficients take the form

$$\lambda_L(C_\alpha^{\mathbf{GH}}) = 0, \qquad \lambda_U(C_\alpha^{\mathbf{GH}}) = 2 - 2^{1/\alpha}.$$

Copulas of this type can be derived from the work by Gumbel [1960b] and have been further considered by Hougaard [1986]. For this reason, this family is called the *Gumbel-Hougaard family* [Hutchinson and Lai, 1990]. ∎

Example 6.5.17 (Mardia-Takahasi-Clayton copulas). The standard expression for members of this family of d-copulas is

$$C_\alpha^{\mathbf{MTC}}(\mathbf{u}) = \max\left\{\left(\sum_{i=1}^{d} u_i^{-\alpha} - (d-1)\right)^{-1/\alpha}, 0\right\}, \qquad (6.5.7)$$

where $\alpha \geq -1/(d-1)$, $\alpha \neq 0$. The limiting case $\alpha \to 0$ corresponds to the independence copula.

The Archimedean generator of this family is given by

$$\varphi_\alpha(t) = (\max\{1 + \alpha t, 0\})^{-1/\alpha}.$$

It was proved by McNeil and Nešlehová [2009] that, for every d-dimensional Archimedean copula C and for every $\mathbf{u} \in \mathbb{I}^d$, $C_\alpha^{\mathbf{MTC}}(\mathbf{u}) \leq C(\mathbf{u})$ for $\alpha = -1/(d-1)$.

The Mardia-Takahasi-Clayton family of 2-copulas is order-preserving with respect to pointwise order between copulas. The Kendall's tau for the 2-copulas of this family is given by

$$\tau(C_\alpha^{\mathbf{MTC}}) = \frac{\alpha}{\alpha + 2}.$$

Moreover, the tail dependence coefficients take the form

$$\lambda_L(C_\alpha^{\mathbf{MTC}}) = \begin{cases} 2^{-1/\alpha}, & \alpha > 0, \\ 0, & \alpha \in [-1, 0], \end{cases} \qquad \lambda_U(C_\alpha^{\mathbf{MTC}}) = 0.$$

Copulas of this type can be derived from two types of multivariate distributions: the Pareto distribution by Mardia [1972] and the Burr distribution by Takahasi [1965]. They were mentioned as bivariate copulas in [Kimeldorf and Sampson, 1975] and as multivariate copulas in [Cook and Johnson, 1981], subsequently extended by Genest and MacKay [1986b]; they are sometimes referred to as the *Cook-Johnson* copulas. A special member of this family can also be derived from the bivariate logistic model of first type by Gumbel [1961]. As shown by Oakes [1982], bivariate copulas of type (6.5.7) are associated with the Clayton model [Clayton, 1978]. For this reason, many authors referred to this family as the *Clayton* family of copulas. Here, we prefer to name it the *Mardia-Takahasi-Clayton* family of copulas. We note that in the book by Klement et al. [2000] the 2-copulas of this family are also called the *Schweizer-Sklar* copulas. ∎

Example 6.5.18 (Frank copula). The standard expression for members of the Frank family of d-copulas is

$$C_\alpha^{\text{Frank}}(\mathbf{u}) = -\frac{1}{\alpha} \ln \left(1 + \frac{\prod_{i=1}^d \left(e^{-\alpha u_i} - 1 \right)}{\left(e^{-\alpha} - 1 \right)^{d-1}} \right), \qquad (6.5.8)$$

where $\alpha > 0$. The limiting case $\alpha = 0$ corresponds to Π_d. For $d = 2$, the parameter α can be extended also to the case $\alpha < 0$.

Copulas of this type have been introduced by Frank [1979] in relation to a problem about associative functions on $\mathbb{I}$. They are absolutely continuous. Moreover, the family of Frank 2-copulas is order-preserving with respect to pointwise order between copulas.

The Archimedean generator is given by $\varphi_\alpha(t) = \frac{1}{\alpha} \ln \left(1 - (1 - e^{-\alpha}) e^{-t} \right)$.

The Kendall's tau and Spearman's rho associated with bivariate Frank copulas are given, respectively, by

$$\tau(C_\alpha^{\text{Frank}}) = 1 - \frac{4}{\alpha} \left(1 - D_1(\alpha) \right),$$

$$\rho(C_\alpha^{\text{Frank}}) = 1 - \frac{12}{\alpha} \left(D_1(\alpha) - D_2(\alpha) \right),$$

where D_n is the Debye function given, for any integer n, by

$$D_n(x) = \frac{n}{x^n} \int_0^x \frac{t^n}{e^t - 1} \, dt. \qquad (6.5.9)$$

The tail dependence coefficients of the 2-copulas of this family take the form $\lambda_L(C_\alpha^{\text{Frank}}) = \lambda_U(C_\alpha^{\text{Frank}}) = 0$ (for more details, see [Nelsen, 1986; Genest, 1987]). Moreover, Frank 2-copulas are radially symmetric. In fact, they represent the solution of the functional equation translating the simultaneous associativity of $F(x, y)$ and $x + y - F(x, y)$ (see [Frank, 1979]). ∎

Example 6.5.19 (Ali-Mikhail-Haq copulas). The standard expression for members of the Ali-Mikhail-Haq (AMH) family of 2-copulas is

$$C_\alpha^{\text{AMH}}(u, v) = \frac{uv}{1 - \alpha (1 - u) (1 - v)} \qquad (\alpha \in [-1, 1[).$$

It is a strict Archimedean copula and has generator

$$\varphi_\alpha(t) = \frac{1 - \alpha}{e^t - \alpha}.$$

For $\alpha = 0$ one has $C_0 = \Pi_2$. The family of AMH 2-copulas is order-preserving with respect to the pointwise order between copulas. The expressions of measures of associations related to this family are available, for instance, in [Nelsen, 2006, page 172]. These copulas cannot model strong positive (or negative) association because of the limited range of the values assumed by the measures of association.

The tail dependence coefficients of the AMH 2-copulas are both equal to 0.

This family was introduced by Ali et al. [1978]. In the context of triangular norms, these copulas are also known as the *Hamacher copulas* (see, e.g., [Hamacher, 1978] and [Grabisch et al., 2009, p. 80]).

Since the generator of this family is completely monotone, this family of copulas can be extended to any dimension. ∎

Further readings 6.5.20. Various generalisations of the class of Archimedean copulas have been presented in the literature in order to remove the exchangeability property that may be, especially in high dimension, a restrictive assumption. One of these methods produces the so-called *hierarchical Archimedean copulas* or *nested Archimedean copulas*; they have been illustrated by Hofert [2010], Hering et al. [2010], Mai and Scherer [2012a], Okhrin et al. [2013] and Joe [2014] among others. Another possible approach consists of suitable modifications of the stochastic representation of Theorem 6.5.13; it has been, for instance, exploited by McNeil and Nešlehová [2010], who introduced the *Liouville copulas*. ∎

6.6 Extreme-value copulas

Extreme-value (EV) copulas arise from the dependence structures related to componentwise maxima of stationary stochastic processes. For more details see, for instance, Beirlant et al. [2004]; Gudendorf and Segers [2010].

The construction of EV copulas can be given by the following mechanism. Let $\{\mathbf{X}_i = (X_{i1}, \dots, X_{id})\}_{i=1,\dots,n}$ be a random sample from a d-dimensional random vector $\mathbf{X}_0$ on the probability space $(\Omega, \mathscr{F}, \mathbb{P})$ with a common (continuous) d.f. H. Let $F_1, \dots, F_d$ be the marginals of H and let C_H be the unique copula of each $\mathbf{X}_i$. Consider the vector of componentwise maxima

$$\mathbf{M}_n := (M_{n,1}, \dots, M_{n,d}), \tag{6.6.1}$$

where

$$M_{n,j} := \bigvee_{i=1}^{n} X_{ij} = \max\{X_{ij} : i = 1, \dots, n\}.$$

It is well known that the d.f. of $M_{n,j}$ is F_j^n. Now the joint and marginal d.f.'s of $\mathbf{M}_n$ are given, respectively, by H^n and $F_1^n, \dots, F_d^n$. As a consequence, if C_n is the copula of $\mathbf{M}_n$, it follows that, for $\mathbf{u} \in \mathbb{I}^d$,

$$C_n(u_1, \dots, u_d) = C_H^n\left(u_1^{1/n}, \dots, u_d^{1/n}\right).$$

An extreme-value copula arises as the limit of the previous expression when the sample size n goes to ∞.

Definition 6.6.1. A d-copula C is said to be an *extreme-value copula* if there exists a copula $C_H \in \mathscr{C}_d$ such that, for every $\mathbf{u} = (u_1, \dots, u_d) \in \mathbb{I}^d$,

$$C(\mathbf{u}) = \lim_{n \to +\infty} C_H^n\left(u_1^{1/n}, \dots, u_d^{1/n}\right). \tag{6.6.2}$$

The copula C_H is said to lie in the *domain of attraction* of C.   ◇

The notion of extreme-value copula is closely related to that of max-stable copula, which is defined below.

Definition 6.6.2. A d-copula $C \in \mathscr{C}_d$ is said to be *max-stable*, if,

$$C(\mathbf{u}) = C^k \left(u_1^{1/k}, \ldots, u_d^{1/k} \right), \tag{6.6.3}$$

for every $k \in \mathbb{N}$ and for every $\mathbf{u} = (u_1, \ldots, u_d) \in \mathbb{I}^d$.   ◇

For the previous definitions, it is obvious that a max-stable copula is in its own domain of attraction and, hence, is an extreme-value copula. The converse is also true, as the following result shows (see, e.g., [Gudendorf and Segers, 2010]).

Theorem 6.6.3. *For a d-copula C the following statements are equivalent:*

(a) *C is max-stable;*

(b) *C is an extreme-value copula.*

By definition, the family of extreme-value copulas coincides with the set of copulas of *extreme-value distributions*, that is, the class of limit distributions with non-degenerate margins of

$$\left(\frac{M_{n,1} - b_{n,1}}{a_{n,1}}, \ldots, \frac{M_{n,d} - b_{n,d}}{a_{n,d}} \right)$$

with $M_{n,j}$ as in (6.6.1), centring constants $b_{n,j} \in \mathbb{R}$ and scaling constants $a_{n,j} > 0$. Representations of extreme-value distributions then yield representations of extreme-value copulas by using the results of de Haan and Resnick [1977] and Pickands [1981] (see also [Gudendorf and Segers, 2010]). First, we denote by Γ_{d-1} the set of discrete d-ary probability distributions, or, equivalently, the unit simplex in $\mathbb{R}^d$:

$$\Gamma_{d-1} := \left\{ \mathbf{w} = (w_1, \ldots, w_d) \in \mathbb{I}^d : \sum_{j=1}^d w_j = 1 \right\}.$$

Theorem 6.6.4. *For a d-copula C the following statements are equivalent:*

(a) *C is an extreme-value copula;*

(b) *there exists a finite measure ν_C on the Borel sets of Γ_{d-1}, called* spectral measure, *such that, for every $\mathbf{u} \in \,]0,1]^d$,*

$$C(\mathbf{u}) = \exp\left(-\ell(\ln u_1, \ldots, \ln u_d) \right) \tag{6.6.4}$$

where the stable tail dependence function $\ell : [0, +\infty[^d \to [0, +\infty[$ *is defined by*

$$\ell(x_1, \ldots, x_d) = \int_{\Gamma_{d-1}} \bigvee_{j=1}^n (w_j x_j) \, d\nu_C(\mathbf{w}). \tag{6.6.5}$$

The spectral measure ν_c is only required to satisfy the moment constraints

$$\int_{\Gamma_{d-1}} w_j \, d\nu_C(\mathbf{w}) = 1, \qquad (j = 1, \ldots, d). \tag{6.6.6}$$

Notice, first of all, the rôle played by the constraints (6.6.6). One has, for every $u \in \mathbb{I}$,

$$C(1, \ldots, 1, u, 1, \ldots, 1) = \exp\{-\ell(0, \ldots, 0, -\ln u, 0, \ldots, 0)\}$$

$$= \exp\left\{\int_{\Gamma_{d-1}} w_j\,(-\ln u)\,d\nu_C(\mathbf{w})\right\}$$

$$= \exp\left\{\ln u \int_{\Gamma_{d-1}} w_j\,d\nu_C(\mathbf{w})\right\} = u.$$

Thus the constraints (6.6.6) ensure that C represented by (6.6.4) has uniform margins. Notice also that

$$\nu_C(\Gamma_{d-1}) = \int_{\Gamma_{d-1}} d\nu_C(\mathbf{w}) = \int_{\Gamma_{d-1}} \sum_{j=1}^{n} w_j \, d\nu_C(\mathbf{w}) = \sum_{j=1}^{n} \int_{\Gamma_{d-1}} w_j \, d\nu_C(\mathbf{w}) = d.$$

The function ℓ of eq. (6.6.5) is convex, positively homogeneous of degree one, i.e.,

$$\ell(c\,\mathbf{x}) = c\,\ell(\mathbf{x}) \qquad (c > 0),$$

and, for all $\mathbf{x} = (x_1, \ldots, x_d) \in [0, +\infty[^d$, satisfies the inequalities

$$\max\{x_1, \ldots, x_d\} \le \ell(\mathbf{x}) \le \sum_{j=1}^{d} x_j.$$

A characterisation of extreme-value copulas via the properties of the tail dependence function ℓ has been recently provided by Ressel [2013] and is reproduced below.

Theorem 6.6.5. *A function $C : \mathbb{I}^d \to \mathbb{I}$ is an extreme-value copula if, and only if, the function $\ell : [0, +\infty[^d \to [0, +\infty[$ defined by*

$$\ell(\mathbf{x}) = -\ln\left(C\left(e^{-x_1}, \ldots, e^{-x_d}\right)\right)$$

satisfies the following properties:

(a1) *ℓ is positively homogeneous of degree 1;*

(a2) *$\ell(\mathbf{e}_i) = 1$ for every $i = 1, 2, \ldots, d$, and $\mathbf{e}_i$ equal to the d-dimensional vector having all the components equal to 0 with the exception of the i-th one that is equal to 1;*

(a3) ℓ *is fully d-max-decreasing, namely for all* $\mathbf{x}$ *and* $\mathbf{h} \in [0, +\infty[^d$ *and for every subset* $J \subseteq \{1, \ldots, d\}$ *with* $\mathrm{card}(J) = k$,

$$\sum_{i_1, \ldots, i_k \in \{0,1\}} (-1)^{i_1 + \cdots + i_k} \, \ell \left(x_1 + i_1 \, h_1 \, \mathbf{1}_{\{1 \in J\}}, \ldots, x_d + i_d \, h_d \, \mathbf{1}_{\{d \in J\}} \right) \leq 0.$$

Example 6.6.6. For $\alpha \in \,]0, +\infty[$, the function

$$\ell_{\alpha(\mathbf{x})} = \sum_{\emptyset \neq I \subset \{1, \ldots, d\}} (-1)^{\mathrm{card}(I)+1} \left(\sum_{j \in I} x_j^{-\alpha} \right)$$

satisfies the conditions of Theorem 6.6.5, and hence it generates a copula. The resulting copula

$$C_\alpha(\mathbf{u}) = \exp\left(-\ell_\alpha \left(-\ln(u_1), \ldots, -\ln(u_d) \right) \right)$$

is an extreme-value copula, called *Galambos copula*. Its origin dates back to [Galambos, 1975]. ∎

Because of its homogeneity the stable tail dependence function ℓ can be also expressed via the *Pickands dependence function* $A : \Gamma_{d-1} \to [1/d, 1]$ defined by

$$A(\mathbf{w}) := \frac{1}{x_1 + \cdots + x_d} \, \ell(x_1, \ldots, x_d) \quad \text{where} \quad w_j := \frac{x_j}{x_1 + \cdots + x_d},$$

if $x_1 + \cdots + x_d \neq 0$. Replacing this latter function into (6.6.4) leads to the representation of the extreme-value copula C by

$$C(\mathbf{u}) = \exp\left\{ \left(\sum_{j=1}^{d} \ln u_j \right) A\left(\frac{u_1}{\sum_{j=1}^{d} \ln u_j}, \ldots, \frac{u_d}{\sum_{j=1}^{d} \ln u_j} \right) \right\}.$$

Pickands dependence function A is also convex and satisfies

$$\max\{w_1, \ldots, w_d\} \leq A(\mathbf{w}) \leq 1.$$

These two properties characterise the class of Pickands dependence functions only in the case $d = 2$ (see the counterexample in [Beirlant et al., 2004, page 257]).

In the case $d = 2$ the simplex $\Gamma_2 = \{(t, 1-t) : t \in \mathbb{I}\}$ may be identified with the unit interval $\mathbb{I}$. Thus one has the following

Theorem 6.6.7. *For a copula* $C \in \mathscr{C}_2$ *the following statements are equivalent:*

(a) C *is an extreme-value copula;*

(b) C *has the representation*

$$C_A(u, v) = \exp\left(A\left(\frac{\ln v}{\ln u + \ln v} \right) \ln(uv) \right) \qquad (u, v) \in \,]0, 1[^2 , \qquad (6.6.7)$$

where $A : \mathbb{I} \to [1/2, 1]$ is convex and satisfies the inequality

$$\max\{1 - t, t\} \leq A(t) \leq 1.$$

The upper and lower bounds in condition (b) of the previous theorem have a probabilistic meaning. If $A(t) = 1$ in $\mathbb{I}$, then $C_A = \Pi_2$, while $C_A = M_2$ if $A(t) = \max\{t, 1 - t\}$. In general, the inequality $A(t) \leq 1$ implies that every extreme-value 2-copula is PQD.

The Kendall's tau and Spearman's rho associated with an extreme-value copula $C_A \in \mathscr{C}_2$ are given, respectively, by

$$\tau(C_A) = \int_0^1 \frac{t(1 - t)}{A(t)} \, dA'(t)$$

and

$$\rho(C_A) = 12 \int_0^1 \frac{1}{(1 + A(t))^2} \, dt - 3.$$

Notice that the Stieltjes integrator $dA'(t)$ is well defined since A is convex on $\mathbb{I}$; moreover, if A is twice differentiable, it can be replaced by $A''(t) \, dt$. For more details, see [Hürlimann, 2003].

The Kendall d.f. associated with an extreme-value copula $C \in \mathscr{C}_2$ is given, for every $t \in \mathbb{I}$, by

$$\kappa_C(t) = t - (1 - \tau_C) t \ln(t),$$

where τ_C denotes the Kendall's tau of C (see, e.g., [Ghoudi et al., 1998]).

The tail dependence coefficients of an extreme-value copula $C_A \in \mathscr{C}_2$ take the form

$$\lambda_L(C_A) = 0, \qquad \lambda_U(C_A) = 2(1 - A(1/2)).$$

Example 6.6.8. Here we collect some examples of extreme-value 2-copulas obtained by specific forms of the Pickands dependence function.

(a) Let α and β belong to $\mathbb{I}$ and consider the Pickands dependence function defined by

$$A(t) := 1 - \min\{\beta t, \alpha(1 - t)\};$$

then (6.6.7) yields the Marshall-Olkin copula.

(b) For $\alpha \geq 1$ consider

$$A_\alpha(t) := (t^\alpha + (1 - t)^\alpha)^{1/\alpha};$$

then one recovers the Gumbel-Hougaard copulas. These are the only copulas in $\mathscr{C}_2$ that are both Archimedean and extreme-value [Genest and Rivest, 1989].

(c) The family of Pickands functions

$$A_\alpha(t) := \alpha t^2 - \alpha t + 1 \qquad (\alpha \in \mathbb{I})$$

occurs in Tawn's model [Tawn, 1988]. This family includes extreme-value 2-copulas that may be non-exchangeable.

Other parametric families of extreme-value copulas include the Hüsler-Reiss copulas [Hüsler and Reiss, 1989] and the t-extreme-value copulas [Demarta and McNeil, 2005].   ■

Remark 6.6.9. Capéraà et al. [2000] introduced a family of bivariate copulas that includes both the extreme-value and the Archimedean copulas, and which they called the *Archimax copulas*. This family has been extended to d-dimension by Charpentier et al. [2014]. The Archimax d-copulas are defined by

$$C_{\varphi,\ell}(\mathbf{u}) := (\varphi \circ \ell) \left(\varphi^{(-1)}(u_1), \ldots, \varphi^{(-1)}(u_d) \right), \qquad (6.6.8)$$

where φ is a generator of a d-dimensional Archimedean copula and $\ell : [0, +\infty[^d \to [0, +\infty[$ is a d-dimensional tail dependence function of an extreme-value d-copula D.   ■

6.7 Elliptical copulas

Elliptical copulas are obtained from the multivariate elliptical distributions by applying the inverse transformation (2.2.2) related to Sklar's theorem.

We recall that a random vector $\mathbf{X} = (X_1, \ldots, X_d)$ is said to have an *elliptical distribution* if it can be expressed in the form

$$\mathbf{X} \stackrel{d}{=} \mu + R\mathbf{A}\mathbf{U}, \qquad (6.7.1)$$

where $\mu \in \mathbb{R}^d$, $\mathbf{A} \in \mathbb{R}^{d \times k}$ with $\Sigma := \mathbf{A}\mathbf{A}^{\mathbf{T}} \in \mathbb{R}^{d \times d}$ and $rank(\Sigma) = k \leq d$, $\mathbf{U}$ is a d-dimensional random vector uniformly distributed on the sphere $\mathbb{S}^{d-1} = \{ \mathbf{u} \in \mathbb{R}^d : u_1^2 + \cdots + u_d^2 = 1 \}$ and R is a positive random variable independent of $\mathbf{U}$. We use the notation $\mathbf{X} \sim E_d(\mu, \Sigma, g)$.

Because of the transformation matrix $\mathbf{A}$ the uniform r.v. $\mathbf{U}$ produces elliptically contoured density level surfaces (provided the existence of the density), whereas the generating r.v. R gives the distribution shape; in particular it determines the tails of the distribution.

Definition 6.7.1. An *elliptical copula* is any copula that can be obtained from an elliptical distribution using the inversion method of eq. (2.2.2).   ◇

Since a copula is invariant under strictly increasing transformations of the marginals and since the characterisation of an elliptical distribution in terms of (μ, Σ, φ) is not unique, all the classes of elliptical copulas can be obtained from the elliptical distributions of type $E_d(0, P, \varphi)$, where $P = (p_{ij}) \in \mathbb{R}^{d \times d}$ is such that $p_{ij} = \Sigma_{ij} / \left(\sqrt{\Sigma_{ii}} \sqrt{\Sigma_{jj}} \right)$ (see [Mai and Scherer, 2012b, Remark 4.3]).

Typically it is not possible to write the expression of an elliptical copula in closed form; the values assumed by these copulas are usually obtained via numerical approximation.

Elliptical copulas are radially symmetric and, hence, their lower and upper bivariate tail dependence coefficients coincide (the general expression of such coefficients is provided in [Schmidt, 2002]).

Example 6.7.2. The *Gaussian copula* is the copula of an elliptical r.v. $\mathbf{X}$ that follows a Gaussian distribution, i.e.,

$$\mathbf{X} \stackrel{d}{=} \mathbf{A}\mathbf{Z},$$

where $\mathbf{A} \in \mathbb{R}^{d \times k}$, $\Sigma := \mathbf{A}\mathbf{A}^{\mathbf{T}} \in \mathbb{R}^{d \times d}$ is the covariance matrix, $rank(\Sigma) = k \leq d$ and $\mathbf{Z}$ is a d-dimensional random vector whose independent components have univariate standard Gaussian law. We write $\mathbf{X} \sim N_d(\mu, \Sigma)$.

The bivariate Gaussian copula is given by

$$C_\rho^{\mathbf{Ga}}(u, v) = \int_{-\infty}^{\Phi^{-1}(u)} ds \int_{-\infty}^{\Phi^{-1}(v)} \frac{1}{2\pi\sqrt{1 - \rho^2}} \exp\left(-\frac{s^2 - 2\rho st + t^2}{2(1 - \rho^2)} \right) dt,$$

where ρ is in $]-1, 1[$, and Φ^{-1} denotes the inverse of the standard Gaussian distribution $N(0, 1)$.

The measures of association related to $C_\rho^{\mathbf{Ga}}$ are given by

$$\tau\left(C_\rho^{\mathbf{Ga}}\right) = \frac{2}{\pi} \arcsin(\rho), \qquad \rho\left(C_\rho^{\mathbf{Ga}}\right) = \frac{6}{\pi} \arcsin(\rho/2). \tag{6.7.2}$$

Bivariate Gaussian copulas have both upper and lower tail dependence coefficients equal to 0. For more details, see [Meyer, 2013]. ∎

Example 6.7.3. The *Student's t-copula* is the copula of an elliptical r.v. $\mathbf{X}$ that follows a multivariate Student's t-distribution, i.e.,

$$\mathbf{X} \stackrel{d}{=} \mu + \Sigma^{1/2}\sqrt{W}\mathbf{Z},$$

where $\mathbf{Z} \sim N_d(0, \mathbf{I}_d)$ is a Gaussian distribution, $\Sigma := \Sigma^{1/2}\Sigma^{1/2}$ is positive definite. Moreover, W and $\mathbf{Z}$ are independent, and W follows an inverse Gamma distribution with parameters $(\nu/2, \nu/2)$.

The bivariate Student's t-copula is given by

$$C_{\rho,\nu}^{\mathbf{t}}(u, v) = \mathbf{t}_{\rho,\nu}\left(t_\nu^{-1}(u), t_\nu^{-1}(v)\right),$$

where ρ is in $]-1, 1[$, and $\nu > 1$, while $\mathbf{t}_{\rho,\nu}$ is the bivariate Student t-distribution with zero mean, the correlation matrix having off-diagonal element ρ, and ν degrees of freedom, while t_ν^{-1} denotes the inverse of the standard t-distribution. The Student t-copula becomes a Gaussian copula in the limit $\nu \to \infty$.

The measures of association related to $C_{\rho,\nu}^{t}$ are given by (6.7.2), i.e. they do not depend on the degrees of freedom ν. As a consequence of the radial symmetry, the lower and upper tail dependence coefficients are identical and are given by

$$\lambda_L\left(C_{\rho,\nu}^{t}\right) = \lambda_U\left(C_{\rho,\nu}^{t}\right) = 2 \cdot t_{\nu+1}\left(-\sqrt{\frac{(\nu+1)(1-\rho)}{1+\rho}}\right).$$

∎

Further readings 6.7.4. For more details about elliptical distributions and elliptical copulas, we refer to [Fang et al., 1990, 2002; Frahm et al., 2003; Genest et al., 2007; Mai and Scherer, 2012b]. ∎

6.8 Invariant copulas under truncation

In this section, we present a family of copulas that is generated by an invariance property with respect to a special kind of conditioning.

Let C be a copula in $\mathscr{C}_{d+1}$ and let $(U, V_1, \ldots, V_d)$ be a r.v. distributed according to C. For every $\alpha \in \ ]0,1[$, we consider the copula $C_{[\alpha]}$ of the conditional d.f. of $(U, V_1, \ldots, V_d)$ given $U \leq \alpha$. Under the previous notations, the following definition is given.

Definition 6.8.1. A copula $C \in \mathscr{C}_{d+1}$ is *invariant under univariate truncation* (or, simply, truncation-invariant) with respect to the first coordinate if, for every $\alpha \in \ ]0,1[$, $C = C_{[\alpha]}$. The class of these copulas will be denoted by $\mathscr{C}_{d+1}^{\mathbf{LT}}$. ◇

Example 6.8.2. If C is an Archimedean $(d+1)$-copula generated by φ, then, for every $\alpha \in \ ]0,1[$, $C_{[\alpha]}$ is also Archimedean with generator

$$\varphi_\alpha = \frac{1}{\alpha}\varphi\left(t + \varphi^{(-1)}(\alpha)\right).$$

See, for instance, [Mesiar et al., 2008]. ∎

Remark 6.8.3. Let C be a copula. Then, it can be proved that, if the limit of $C_{[\alpha]}$, as $\alpha \to 0$ exists, then it is a copula in $\mathscr{C}_2^{\mathbf{LT}}$ (see, e.g., [Jágr et al., 2010]). In other words, truncation-invariant copulas can be used in order to approximate the dependence structure of a random vector $(X, Y_1, \ldots, Y_d)$ when the first component takes very small values (compare with [Jaworski, 2013b]). This fact is potentially very useful, for instance, in risk management, when the considered r.v.'s are profit-loss distributions and one wants to estimate the behaviour of the vector assuming a stress scenario corresponding to large losses of one of its components. ∎

The following result characterises all the copulas that are truncation-invariant (see [Jaworski, 2013a]).

Theorem 6.8.4. *The copula C belongs to $\mathscr{C}_{d+1}^{\mathbf{LT}}$ if, and only if,*

$$C(u, v_1, \ldots, v_d) = \int_0^u C_0 \left(\partial_1 C_1(\xi, v_1), \ldots, \partial_1 C_d(\xi, v_d) \right) \, \mathrm{d}\xi \tag{6.8.1}$$

for a d-copula C_0 and 2-copulas $C_1, \ldots, C_d \in \mathscr{C}_2^{\mathbf{LT}}$.

Thus, the elements of $\mathscr{C}_{d+1}^{\mathbf{LT}}$ can be fully described once all the elements of $\mathscr{C}_2^{\mathbf{LT}}$ are known. This latter characterisation has been obtained by Durante and Jaworski [2012] and is reproduced here.

Theorem 6.8.5. *A copula $C \in \mathscr{C}_2^{\mathbf{LT}}$ if, and only if, there exist*

(a) *a collection $(]a_i, b_i[)_{i \in \mathscr{I}}$ of nonempty disjoint subintervals of $\mathbb{I}$ ($\mathscr{I}$ finite or countable);*

(b) *a collection of functions $(f_i)_{i \in \mathscr{I}}$, where, for every $i \in \mathscr{I}$, $f_i : [0, +\infty] \to \mathbb{I}$, and either f_i or $1 - f_i$ is onto, increasing and concave;*

so that C can be expressed in the form

$$C(u, v) = \begin{cases} a_i u + (b_i - a_i) u f_i \left(\dfrac{1}{u} f_i^{(-1)} \left(\dfrac{v - a_i}{b_i - a_i} \right) \right), & v \in]a_i, b_i[, \\ uv, & \text{otherwise,} \end{cases}$$

where $f^{(-1)} : \mathbb{I} \to [0, \infty]$ denotes the pseudo-inverse of f.

Distinguished elements in $\mathscr{C}_2^{\mathbf{LT}}$ are those copulas C such that $C \neq \Pi_2$ on $]0, 1[^2$; these copulas are also called *irreducible*; their class will be denoted by $\mathscr{C}_2^{\mathbf{LTI}}$. Explicitly, the elements of $\mathscr{C}_2^{\mathbf{LTI}}$ are given below (see [Durante and Jaworski, 2012]).

Theorem 6.8.6. *The function $C_f : \mathbb{I}^2 \to \mathbb{I}$ defined by*

$$C_f(u, v) = \begin{cases} u f \left(\dfrac{f^{(-1)}(v)}{u} \right), & u \neq 0, \\ 0, & \text{otherwise,} \end{cases} \tag{6.8.2}$$

where $f : [0, +\infty] \to \mathbb{I}$ is onto and monotone, is a copula in $\mathscr{C}_2^{\mathbf{LTI}}$ if, and only if, f is concave and increasing (or, convex and decreasing).

Elements of $\mathscr{C}_2^{\mathbf{LTI}}$ are, for instance, the Hoeffding-Fréchet upper and lower bounds generated, respectively, by $f(t) = \min(t, 1)$ and $f(t) = \max(0, 1 - t)$ via eq. (6.8.2).

Analogously to Archimedean copulas, irreducible truncation-invariant copulas are generated by a one-dimensional function f, which also characterises their dependence properties. In fact, if f is concave and increasing, then C_f is positively quadrant dependent; while, if f is convex and decreasing, then C_f is negatively quadrant dependent (see [Durante et al., 2011]).

As a consequence, if f generates an Archimedean 2-copula, $f \in \mathscr{G}_2$, then f generates a negatively quadrant dependent copula in $\mathscr{C}_2^{\mathrm{LTI}}$. Moreover, it may be easily proved that if $\bar{f} = 1 - f$ for some $f \in \mathscr{G}_2$, then $C_{\bar{f}} \in \mathscr{C}_2^{\mathrm{LTI}}$ is positively quadrant dependent. Notice that it holds that $C_{\bar{f}}(u, v) = u - C_f(u, 1 - v)$, i.e. $C_{\bar{f}}$ and C_f are connected via the symmetry of Corollary 2.4.4.

While Archimedean 2-copulas are exchangeable, the only exchangeable copulas in $\mathscr{C}_2^{\mathrm{LTI}}$ are the comonotonicity copula M_2 and Clayton copulas.

Example 6.8.7. Consider $f_\theta(t) = \min(t^\theta, 1)$ for $\theta \in \,]0, 1[$. Then $C_f \in \mathscr{C}_2^{\mathrm{LTI}}$ is given by

$$
C_f(u, v) = \begin{cases} u, & v > u^\theta, \\ u^{1-\theta} v, & \text{otherwise,} \end{cases}
$$

which is a (non-exchangeable) member of the Marshall-Olkin family. ∎

Example 6.8.8. Consider the piecewise linear function $f : [0, +\infty] \to \mathbb{I}$, given by $f(x) = \max\{0, 1 - 2x, (1 - x)/2\}$. Then $f \in \mathscr{G}_2$ and it generates via eq. (6.8.2) the singular copula C_f given by:

$$
C_f(u, v) = \begin{cases} 0, & v \le f(u), \\ W_2(u, v), & v \ge 1 - 2u/3, \\ v + \frac{u-1}{2}, & \frac{3-u}{6} \le v \le \frac{1}{3}, \\ \frac{u}{2} + \frac{v-1}{4}, & \text{otherwise.} \end{cases}
$$

The support of C_f consists of the 3 segments connecting, respectively, $(0, 1)$ with $(1/3, 1/3)$, $(1/3, 1/3)$ with $(1, 0)$, and $(1, 1)$ with $(1/3, 1/3)$. ∎

It is possible to establish a connexion between $\mathscr{C}_2^{\mathrm{LTI}}$ and the class of Archimedean 2-copulas as the following result shows.

Theorem 6.8.9. *Let* $(\Omega, \mathscr{F}, \mathbb{P})$ *be a probability space.*

(a) *Let* (U, V) *be a pair of continuous r.v.'s distributed according to an Archimedean copula* A_f. *Then* $C_f \in \mathscr{C}_2^{\mathrm{LTI}}$ *is the distribution function of the random pair* (X, Y), *where almost surely*

$$
X = \frac{f^{(-1)}(V)}{f^{(-1)}(U) + f^{(-1)}(V)}, \qquad Y = V.
$$

(b) *Let* (X, Y) *be a pair of continuous r.v.'s distributed according to* $C_f \in \mathscr{C}_2^{\mathrm{LTI}}$, *where* $f \in \mathscr{G}_2$. *Then the Archimedean copula* A_f *is the distribution function of the random pair* (U, V), *where almost surely*

$$
U = f\left(\frac{f^{(-1)}(Y)}{X} - f^{(-1)}(Y) \right), \qquad V = Y.
$$

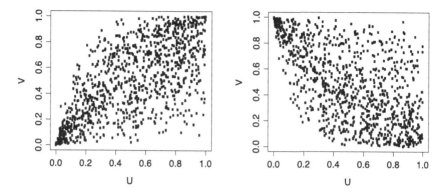

Figure 6.1 *Bivariate sample clouds of* 1000 *points from the copula* $C_f \in \mathscr{C}_2^{\mathbf{LTI}}$, *where (left)* $f \in \mathscr{G}_2$ *is the generator of the Frank copula with parameter* 0.5; *(right)* $f = 1 - f_2$, *where* f_2 *is the generator of the Gumbel copula with parameter* 1.2.

The previous result provides a powerful method for generating copulas in $\mathscr{C}_2^{\mathbf{LTI}}$ starting with available algorithms for sampling Archimedean copulas. For possible examples, see Figure 6.1.

Remark 6.8.10. If $C_f \in \mathscr{C}_2^{\mathbf{LT}}$, then the transpose $C_f^T(u,v) = C_f(v,u)$ is a copula that is invariant under truncation with respect to the second coordinate (compare with Durante and Jaworski [2012]). Moreover, if $C \in \mathscr{C}_2$ is invariant under truncation with respect to the first and the second coordinate, then, in view of [Durante and Jaworski, 2012, Theorem 4.1], either $C = M$ or C is a Clayton copula.

Notice that the Clayton family of copulas (with the limiting elements Π_2 and M_2) includes all the copulas that are invariant with respect to a double truncation, i.e. if $(U, V) \sim C$, then C is also the copula of the conditional distribution of (U, V) given $U \le \alpha$ and $V \le \beta$ for all $\alpha, \beta \in \,]0, 1]$. This latter result is known from [Oakes, 2005; Charpentier and Juri, 2006] and has been proved in full generality in [Durante and Jaworski, 2012]. ∎

Chapter 7

Generalisations of copulas:
quasi-copulas

In recent years, copulas have been generalised in various ways in answering different requirements. Far from being just an academic exercise these generalisations have proved to be necessary in a variety of fields going, for instance, from survival analysis to fuzzy set theory; moreover, they have highlighted properties of copulas that are of a different nature (probabilistic, algebraic, analytical, geometrical, etc.).

Roughly speaking, generalisations of copulas fall into two types. One type consists of all the functions that share some (but not all) properties of a copula (e.g., boundary conditions, continuity, Lipschtizianity, etc.). Examples of such concepts are quasi-copulas and semi-copulas. Another type deals with functions that are defined on a slightly different domain (see, for instance, subcopulas and discrete copulas). Mixtures of the two previous approaches are, of course, present in the literature: see, e.g., the concept of discrete quasi-copulas considered by Quesada Molina and Sempi [2005].

The present chapter and the next one are devoted to the presentation of these generalisations. After an overview of these concepts, special emphasis is devoted to the historical motivations that have been suggested to modify the concept of copulas in order to obtain a larger class of functions.

Most of these generalisations are also related to the theory of aggregation functions, which are monotonic functions $A : \mathbb{I}^d \to \mathbb{I}$ that represent the way a multi-source information represented by the components of a vector in $\mathbb{I}^d$ may be aggregated into one single output (belonging to the range $\mathbb{I}$). For the reader interested in these general aspects, we recommend the monograph by Grabisch et al. [2009].

7.1 Definition and first properties

The first ingredient in the definition of a quasi-copula is the notion of track in $\mathbb{I}^d$.

Definition 7.1.1. A *track* B in $\mathbb{I}^d$ is a subset of unit square $\mathbb{I}^d$ that can be written in the form

$$B := \{(F_1(t), \ldots, F_d(t)) : t \in \mathbb{I}\}, \tag{7.1.1}$$

where F_1, ..., F_d are continuous distribution functions such that $F_j(0) = 0$ and $F_j(1) = 1$ for every j in $\{1, \ldots, d\}$. $\diamond$

Definition 7.1.2. A *d-quasi-copula* is a function $Q : \mathbb{I}^d \to \mathbb{I}$ such that for every track B in $\mathbb{I}^d$ there exists a d-copula C_B that coincides with Q on B, namely such that, for every point $\mathbf{u} \in B$, $Q(\mathbf{u}) = C_B(\mathbf{u})$. $\diamond$

It follows immediately that every copula $C \in \mathscr{C}_d$ is a quasi-copula, since it suffices to take $C_B = C$ for every track B. However, it will be seen later (see Example 7.2.5) that there are quasi-copulas that are not copulas; therefore, if $\mathscr{Q}_d$ denotes the set of d-quasi-copulas one has the strict inclusion $\mathscr{C}_d \subset \mathscr{Q}_d$. A quasi-copula that is not a copula will be called *proper*.

The need to describe quasi-copulas in formal terms started with the study of operations on distribution functions deriving from corresponding operations on the same probability space. This study was first conducted by Alsina et al. [1993] for the bivariate case and by Nelsen et al. [1996] when $d > 2$. The main motivation is presented below. For simplicity's sake, we restrict ourselves to the bivariate case; the general case is treated by Nelsen et al. [1996].

We start with two definitions that underline different ways of combining (and interpreting) operations on the space $\mathscr{D}$ of univariate d.f.'s.

Definition 7.1.3. A binary operation φ on $\mathscr{D}$ is said to be *induced pointwise* by a function $\psi : \mathbb{I}^2 \to \mathbb{I}$ if, for all F and G in $\mathscr{D}$, the value of $\varphi(F, G)$ at t is a function of $F(t)$ and $G(t)$, namely $\varphi(F, G)(t) = \psi(F(t), G(t))$ for all $t \in \mathbb{R}$. $\diamond$

A typical example of operation on $\mathscr{D}$ that is induced pointwise is the convex combination of two d.f.'s, namely a mixture in the language of Statistics, which can be derived by the function $\psi(u, v) = \alpha u + (1 - \alpha)v$ for every $\alpha \in \mathbb{I}$.

However, not all operations on $\mathscr{D}$ are induced pointwise. Consider, for instance, the *convolution* of $F, G \in \mathscr{D}$ defined, for every $x \in \mathbb{R}$, by

$$F \otimes G(x) = \int_{\mathbb{R}} F(x - t) \, \mathrm{d}G(t).$$

Now, since the value of the convolution of two d.f.'s F and G at the point t generally depends on more than just the values of F and G at t, convolution is not induced pointwise by any two-place function. However, as is well known, the convolution is interpretable in terms of r.v.'s, since $F \otimes G$ can be seen as the d.f. of the sum of independent r.v.'s X and Y such that $X \sim F$ and $Y \sim G$. The general class of operations on $\mathscr{D}$ that allow a probabilistic interpretation is defined below.

Definition 7.1.4. A binary operation φ on $\mathscr{D}$ is said to be *derivable* from a function on random variables if there exists a Borel-measurable function $V : \mathbb{R}^2 \to \mathbb{R}$ satisfying the following condition: for every pair of d.f.'s F and G in $\mathscr{D}$ there exist r.v.'s X and Y, defined on a common probability space $(\Omega, \mathscr{F}, \mathbb{P})$, such that F is the d.f. of X, G is the d.f. of Y and $\varphi(F, G)$ is the d.f. of $V(X, Y)$. $\diamond$

Now, as a matter of fact, the convex combination of two d.f.'s is not derivable from a binary operation on random variables defined on the same probability space as shown below (see Alsina and Schweizer [1988]). Here, ε_a denotes the unit step function at a, i.e.,

$$\varepsilon_a(t) := \begin{cases} 0, & t < a, \\ 1, & t \geq a. \end{cases}$$

Example 7.1.5. Assume that a convex combination φ of d.f.'s with parameter $\alpha \in \,]0, 1[$ is derivable from a suitable function V on corresponding r.v.'s.

For all $a, b \in \mathbb{R}$, $a \neq b$, let F and G be the unit step functions at a and b, respectively. Then F and G are, respectively, the d.f.'s of the r.v.'s X and Y, which are defined on a common probability space and are equal, respectively, to a and b almost surely. Hence $V(X, Y)$ is a r.v. defined on the same probability space and equal to $V(a, b)$ almost surely. Thus the distribution function of $V(X, Y)$ is the unit step function at $V(a, b)$. But because φ is derivable from V, the distribution function of $V(X, Y)$ must be equal to $\alpha F + (1 - \alpha)G$ that is not a step function with a single jump. ∎

As a consequence, it was, in general, of interest to characterise the class of those binary operations on $\mathscr{D}$ that are both induced pointwise and derivable from operations on random variables. This class turns out to be quite small, as the following theorem shows.

Theorem 7.1.6. *Let φ be a binary operation on $\mathscr{D}$ that is both induced pointwise by a two-place function $\psi : \mathbb{I}^2 \to \mathbb{I}$ and derivable from a function V on random variables defined on a common probability space. Then precisely one of the following holds:*

(a) *$V = \max$ and ψ is a quasi-copula;*

(b) *$V = \min$ and $\psi(x, y) = x + y - Q(x, y)$, where Q is a quasi-copula;*

(c) *V and φ are trivial in the sense that, for all x and y in $\mathbb{R}$ and for all a and b in $\mathbb{I}$, either $V(x, y) = x$ and $\psi(a, b) = a$ or $V(x, y) = y$ and $\psi(a, b) = b$.*

Proof. Since $\varphi(F, G)$, F and G are d.f.'s, the function ψ is right-continuous and increasing in each place; moreover, $\psi(0, 0) = 0$ and $\psi(1, 1) = 1$. Consider $F = \varepsilon_x$ and $G = \varepsilon_y$ the unit step functions at x and y, respectively, with $x, y \in \mathbb{R}$. Then F and G are the d.f.'s of two random variables X and Y that may be defined on a common probability space $(\Omega, \mathscr{F}, \mathbb{P})$ and which are respectively equal to x and y almost surely. It follows that $V(X, Y)$ is also a random variable on $(\Omega, \mathscr{F}, \mathbb{P})$ and that it takes the value $V(x, y)$ almost surely; hence its d.f. is the step function $\varepsilon_{V(x,y)}$, which, since φ is derivable from V, must be $\varphi(\varepsilon_x, \varepsilon_y)$. Thus, for every $t \in \mathbb{R}$, one has

$$\varepsilon_{V(x,y)}(t) = \varphi(\varepsilon_x, \varepsilon_y)(t) = \psi\left(\varepsilon_x(t), \varepsilon_y(t)\right)$$
$$= \begin{cases} \psi(1, 0)\,\varepsilon_x(t) + (1 - \psi(1, 0))\,\varepsilon_y(t), & x \leq y, \\ (1 - \psi(0, 1))\,\varepsilon_x(t) + \psi(0, 1)\,\varepsilon_y(t), & x \geq y. \end{cases}$$

But $\varepsilon_{V(x,y)}$ takes only the values 0 and 1, so that one of the following four cases holds:

(a) $\psi(1,0) = 0$, $\psi(0,1) = 0$ and $V(x,y) = \max\{x,y\}$;

(b) $\psi(1,0) = 1$, $\psi(0,1) = 1$ and $V(x,y) = \min\{x,y\}$;

(c_1) $\psi(1,0) = 1$, $\psi(0,1) = 0$ and $V(x,y) = x$;

(c_2) $\psi(1,0) = 0$, $\psi(0,1) = 1$ and $V(x,y) = y$.

Now let $B = \{(\alpha(t), \beta(t)) : t \in \mathbb{I}\}$ be a track and let F and G be given by

$$F(t) = \begin{cases} 0, & t < 0, \\ \alpha(t), & t \in \mathbb{I}, \\ 1, & t > 1, \end{cases} \qquad G(t) = \begin{cases} 0, & t < 0, \\ \beta(t), & t \in \mathbb{I}, \\ 1, & t > 1. \end{cases}$$

Let X, Y and $V(X,Y)$ be r.v.'s having d.f.'s F, G and $\varphi(F,G)$, respectively, and let H and C be the joint d.f. and the (uniquely defined) copula of X and Y. In case (a), one has $V(X,Y) = \max\{X,Y\}$, so that $F_{V(X,Y)}$, the d.f. of $V(X,Y)$, is given by $F_{V(X,Y)}(t) = H(t,t)$; then

$$\psi(F(t), G(t)) = \varphi(F,G)(t) = F_{V(X,Y)}(t) = H(t,t)$$
$$= C(F(t), G(t)).$$

Thus ψ agrees with a copula on the track B, which is arbitrary; therefore ψ is a quasi-copula.

In case (b), one has $V(X,Y) = \min\{X,Y\}$, so that $F_{V(X,Y)}(t) = F(t) + G(t) - H(t,t)$. Therefore

$$\psi(F(t), G(t)) = \varphi(F,G)(t) = F_{V(X,Y)}(t)$$
$$= F(t) + G(t) - C(F(t), G(t)),$$

which is the assertion in this case.

In case (c_1) one has $V(X,Y) = X$, whence $\psi(F(t), G(t)) = F_{V(X,Y)}(t) = F(t)$; thus $\psi(u,v) = u$; the remaining case (c_2) is completely analogous. $\square$

Basically, the only possibility of combining in a non-trivial sense two d.f.'s by preserving the derivability from an operation on suitable corresponding r.v.'s relies on the use of quasi-copulas or related functions.

7.2   Characterisations of quasi-copulas

It is immediately apparent that it may be very hard to check whether a given function $Q : \mathbb{I}^d \to \mathbb{I}$ is a d-quasi-copula according to the original definition 7.1.2. Therefore, two characterisations have been provided by Genest et al. [1999] in the case $d = 2$ and by Cuculescu and Theodorescu [2001] for $d > 2$ in order to give an operationally simpler way of dealing with d-quasi-copulas. This characterisation will be introduced in the wake of the following theorem.

Theorem 7.2.1. *A d-quasi-copula Q satisfies the following properties:*

(a) *for every* $j \in \{1, \ldots, d\}$, $Q(1, \ldots, 1, u_j, 1, \ldots, 1) = u_j$;

(b) Q *is increasing in each of its arguments, viz. for every* $j \in \{1, \ldots, d\}$ *and for every point* $(u_1, \ldots, u_{j-1}, u_{j+1}, \ldots, u_d)$ *in* $\mathbb{I}^{d-1}$, *the function* $t \mapsto Q(\mathbf{u}_j(t))$ *is increasing;*

(c) Q *satisfies the following Lipschitz condition: if* $\mathbf{u}$ *and* $\mathbf{v}$ *are in* $\mathbb{I}^d$, *then*

$$|Q(\mathbf{v}) - Q(\mathbf{u})| \leq \sum_{j=1}^{d} |v_j - u_j|. \tag{7.2.1}$$

Proof. If B_1 is a track that includes, as one of its parts, the segment having $(1, \ldots, 1, u_j, 1, \ldots, 1)$ and $\mathbf{1} = (1, \ldots, 1)$ as endpoints, then there exists a copula C_{B_1} such that

$$Q(1, \ldots, 1, u_j, 1, \ldots, 1) = C_{B_1}(1, \ldots, 1, u_j, 1, \ldots, 1) = u_j.$$

This proves (a).

(b) Fix $(u_1, \ldots, u_{j-1}, u_{j+1}, \ldots, u_d) \in \mathbb{I}^{d-1}$ and let $0 < t < t' < 1$. If B_2 is a track passing through the points $\mathbf{u}_j(t)$ and $\mathbf{u}_j(t')$,

$$Q(u_j(t)) = C_{B_2}(u_j(t)) \leq C_{B_2}(u_j(t')) = Q(u_j(t')),$$

which establishes (b).

(c) For $j = 1, \ldots, d$, let B'_j be a track including the segment S_j of endpoints

$$(v_1, \ldots, v_{j-1}, u_j, u_{j+1}, \ldots, u_d) \quad \text{and} \quad (v_1, \ldots, v_{j-1}, v_j, u_{j+1}, \ldots, u_d).$$

Then there exists d copulas $C_{B'_j}$ $(j = 1, \ldots, d)$ in $\mathscr{C}_d$ such that

$$Q(v_1, \ldots, v_{j-1}, u_j, u_{j+1}, \ldots, u_d) = C_{B'_j}(v_1, \ldots, v_{j-1}, u_j, u_{j+1}, \ldots, u_d),$$

$$Q(v_1, \ldots, v_{j-1}, v_j, u_{j+1}, \ldots, u_d) = C_{B'_j}(v_1, \ldots, v_{j-1}, v_j, u_{j+1}, \ldots, u_d).$$

Notice that the second endpoint of S_j coincides with the first endpoint of S_{j+1}. Therefore

$$
\begin{aligned}
|Q(\mathbf{v}) - Q(\mathbf{u})| = |Q(\mathbf{u}) - Q(\mathbf{v})| &\leq |Q(\mathbf{u}) - Q(v_1, u_2, \ldots, u_d)| \\
&\quad + |Q(v_1, u_2, \ldots, u_d) - Q(v_1, v_2, u_3, \ldots, u_d)| \\
&\quad + |Q(v_1, v_2, u_3, u_4 \ldots, u_d) - Q(v_1, v_2, v_3, u_4, \ldots, v_d)| \\
&\quad + \cdots \quad | \; |Q(v_1, \ldots, v_{d-1}, u_d) - Q(\mathbf{v})| \\
&= |C_{B'_1}(\mathbf{u}) - C_{B'_1}(v_1, u_2, \ldots, u_d)| \\
&\quad + |C_{B'_2}(v_1, u_2, \ldots, u_d) - C_{B'_2}(v_1, v_2, u_3, \ldots, u_d)| \\
&\quad + \cdots + \left| C_{B'_d}(v_1, \ldots, v_{d-1}, u_d) - C_{B'_d}(\mathbf{v}) \right| \\
&\leq \sum_{j=1}^{d} |v_j - u_j|,
\end{aligned}
$$

where the last inequality follows from the Lipschitz condition of eq. (1.5.1). $\qquad \square$

Properties (a) and (b) of Theorem 7.2.1 together imply $Q(\mathbf{u}) = 0$ when at least one of the components of $\mathbf{u}$ vanishes, viz. when $\min\{u_1, \ldots, u_d\} = 0$; moreover, a quasi-copula satisfies the same general bounds (1.7.2) as a copula.

Theorem 7.2.2. *For every d-quasi-copula Q and for every point $\mathbf{u} \in \mathbb{I}^d$*

$$W_d(\mathbf{u}) \leq Q(\mathbf{u}) \leq M_d(\mathbf{u}). \tag{7.2.2}$$

Proof. Given a point $\mathbf{u} \in \mathbb{I}^d$, consider any track B going through $\mathbf{u}$ and a copula C_B that coincides with Q on B; then

$$W_d(\mathbf{u}) \leq C_B(\mathbf{u}) \leq M_d(\mathbf{u}).$$

Since $Q(\mathbf{u}) = C_B(\mathbf{u})$, the inequalities (7.2.2) have been proved. □

Example 7.2.3. It easily follows from (7.2.2) that every quasi-copula $Q \in \mathscr{Q}_2$ satisfies

$$\max\{u, v\} \leq u + v - Q(u, v) \leq \min\{u + v, 1\} \tag{7.2.3}$$

for all $(u, v) \in \mathbb{I}^2$. Since neither the arithmetic mean $\psi(u, v) = \alpha u + (1 - \alpha)v$, $\alpha \in \,]0, 1[$, nor the geometric mean $\psi(u, v) = \sqrt{uv}$ satisfies (7.2.2) and (7.2.3), we have at once that mixtures are not derivable and that the geometric mean is not derivable as a consequence of Theorem 7.1.6. ■

The properties of Theorem 7.2.1 will be seen to characterise quasi-copulas, as stated below. The proof of this result is given in detail in Subsection 7.2.1

Theorem 7.2.4. *A function $Q : \mathbb{I}^d \to \mathbb{I}$ is a d-quasi-copula if, and only if, it satisfies the following conditions:*

(a) *for every $j \in \{1, \ldots, d\}$, $Q(1, \ldots, 1, u_j, 1, \ldots, 1) = u_j$;*

(b) *Q is increasing in each of its arguments;*

(c) *Q satisfies Lipschitz condition (7.2.1).*

Example 7.2.5. Using the characterisation of Theorem 7.2.4 one immediately proves that W_d is a d-quasi-copula, and, since it is already known (see Section 1.7) that it is not a d-copula if $d \geq 3$, it is a proper d-quasi-copula for $d \geq 3$. The lower and upper bounds provided by (7.2.2) are, obviously, the best possible in $\mathscr{Q}_d$. ■

Remark 7.2.6. Thanks to Theorem 7.2.4, the notion of quasi-copula has been recently used in the field of aggregation functions. We recall that a d-dimensional aggregation function is a monotonic mapping from $\mathbb{I}^d$ to $\mathbb{I}$ that is used in order to aggregate a set of d inputs $x_1, \ldots, x_d$ from $\mathbb{I}$ into a single output $\mathbb{I}$, possibly preserving some features of interest for decision theory. In particular, property (c) of Theorem 7.2.4 is interpreted in this context in terms of some kind of robustness of the aggregation process. In fact, it implies that small differences in the input values correspond to a small difference in the output value. For more details about aggregation functions, see [Grabisch et al., 2009]. ■

A different characterisation of quasi-copulas is possible and sometimes useful.

Theorem 7.2.7. *For a function $Q : \mathbb{I}^d \to \mathbb{I}$ the following statements are equivalent:*
(a) *Q is a d-quasi-copula;*
(b) *Q satisfies property* (a) *of Theorem 7.2.1, and the functions*

$$t \mapsto Q(\mathbf{u}_j(t)) \qquad (j = 1, \ldots, d) \tag{7.2.4}$$

are absolutely continuous for every choice of $\mathbf{u} = (u_1, \ldots, u_d)$ in $\mathbb{I}^{d-1}$ and

$$0 \leq \partial_j Q(\mathbf{u}_j(t)) \leq 1 \qquad (j = 1, \ldots, d) \tag{7.2.5}$$

for almost every $t \in \mathbb{I}$ and for all $(u_1, \ldots, u_d) \in \mathbb{I}^{d-1}$.

Proof. (a) $\Longrightarrow$ (b) A d-quasi-copula Q obviously satisfies the boundary conditions
(a) of Theorem 7.2.1. Moreover, since it satisfies the Lipschitz condition (7.2.1), the
functions (7.2.4) are absolutely continuous for every $\mathbf{u} \in \mathbb{I}^d$ and the partial derivatives of Q, where they exist, satisfy (7.2.5).

(b) $\Longrightarrow$ (a) Since the partial derivatives of Q are a.e. bounded above by 1, it follows that Q satisfies the Lipschitz condition (7.2.1). Thus it remains to prove that
all the functions (7.2.4) are increasing. Since $t \mapsto Q(\mathbf{u}_j(t))$ $(j = 1, \ldots, d)$ is absolutely continuous, one has, for t and t' in $\mathbb{I}$ with $t < t'$, and for every choice of
$(u_1, \ldots, u_{j-1}, u_{j+1}, \ldots, u_d)$ in $\mathbb{I}^{d-1}$,

$$Q(\mathbf{u}_j(t')) - Q(\mathbf{u}_j(t)) = \int_t^{t'} \partial_j Q(\mathbf{u}_j(s)) \, du_1 \ldots du_{j-1} \, du_{j+1} \ldots du_d \,,$$

which is positive since so is the integrand. $\square$

Example 7.2.8. Consider the function $Q : \mathbb{I}^2 \to \mathbb{I}$ defined by

$$Q(u, v) := uv + f(v) \sin(2\pi u) \,,$$

where $f : \mathbb{I} \to [0, 1/24]$ is defined by

$$f(t) := \begin{cases} 0, & t \in [0, 1/4] \,, \\ \frac{4t-1}{24}, & t \subset [1/4, 1/2] \,, \\ \frac{1-t}{12}, & t \in [1/2, 1] \,. \end{cases}$$

It is clear that Q satisfies the properties of Theorem 7.2.7 and, hence, it is a quasi-
copula. However, this function cannot be a copula, because the mixed second partial
derivative of $\partial_{12}Q(x, y)$ is strictly negative when $1/4 < y < 1/2$ and $\cos(2\pi x)$ is
sufficiently close to -1. ∎

Just as for copulas one may introduce the ordinal sum construction for quasi-
copulas.

Example 7.2.9. Let N be a finite or countable subset of the natural numbers $\mathbb{N}$, let $(Q_k)_{k \in N}$ be a family of quasi-copulas indexed by N and let $(]a_k, b_k[)_{k \in N}$ be a family of subintervals of $\mathbb{I}$, also indexed by N, such that any two of them have at most an endpoint in common. Then the *ordinal sum* Q of $(Q_k)_{k \in N}$ with respect to family of intervals $(]a_k, b_k[)_{k \in N}$ is defined, for every $\mathbf{u} \in \mathbb{I}^d$, by

$$
Q(\mathbf{u}) := \begin{cases} a_k + (b_k - a_k) \, Q_k \left(\frac{\min\{u_1, b_k\} - a_k}{b_k - a_k}, \ldots, \frac{\min\{u_d, b_k\} - a_k}{b_k - a_k} \right), \\ \qquad \text{if } \min\{u_1, u_2, \ldots, u_d\} \in \cup_{k \in N} \,]a_k, b_k[, \\ \min\{u_1, u_2, \ldots, u_d\}, \qquad \text{elsewhere.} \end{cases} \tag{7.2.6}
$$

It can be proved that such a Q is a quasi-copula, usually denoted by $Q = ((]a_k, b_k[, Q_k))_{k \in N}$. $\diamond$

Historical remark 7.2.10. It should be noticed that quasi-copulas already appear in Definition 7.1.5 by Schweizer and Sklar [1983]; there they are not given a name, but their set is denoted by $\mathscr{I}_C$. As was said in the text, the term "quasi-copula" was introduced by Alsina et al. [1993]. Their definition applied to the case $d = 2$ but it was soon extended, in a natural way, to the case $d > 2$ by Nelsen et al. [1996]. Later, their definition was proved to be equivalent to that of Theorem 7.2.4 by Genest et al. [1999] for the case $d = 2$ and by Cuculescu and Theodorescu [2001] for the case $d > 2$. ∎

7.2.1 Proof of Theorem 7.2.4

The remaining part of this section will be devoted to proving Theorem 7.2.4 in a series of lemmata in which only polygonal tracks will be considered.

For every point $\mathbf{a} \in \mathbb{I}^d$, set $\|\mathbf{a}\|_1 = \sum_{i=1}^d |a_i|$. In the next definition the endpoints of the segments of the polygonal tracks considered will be introduced.

Definition 7.2.11. A *system* will be a finite sequence of points in $\mathbb{I}^d$

$$
\mathbf{0} = \mathbf{a}^{(0)} \leq \mathbf{a}^{(1)} \leq \cdots \leq \mathbf{a}^{(n)} = \mathbf{1},
$$

together with a finite sequence, of the same length n, of numbers in $\mathbb{I}$, $0 = q_0 \leq q_1 \leq \cdots \leq q_n = 1$, such that

$$
\max\{0, a_1^{(j)} + a_2^{(j)} + \cdots + a_d^{(j)} - d + 1\} \leq q_j \leq \min\{a_1^{(j)}, a_2^{(j)}, \ldots, a_d^{(j)}\}
$$

and

$$
q_{j+1} - q_j \leq \left\| \mathbf{a}^{(j+1)} - \mathbf{a}^{(j)} \right\|_1 = \sum_{i=1}^d \left(a_i^{(j+1)} - a_i^{(j)} \right).
$$

Such a system will be denoted by

$$
\Sigma = \left(\mathbf{a}^{(0)}, \mathbf{a}^{(1)}, \ldots, \mathbf{a}^{(n)}; q_0, q_1, \ldots, q_n \right),
$$

and it will be said to be *increasing* if $\mathbf{a}^{(j)} < \mathbf{a}^{(j+1)}$ for every $j = 0, 1, \ldots, n - 1$. Finally a system will be said to be *simple* if

- it is increasing;
- it is of even length;
- $q_{j+1} = q_j$ for every even index j, while $q_{j+1} - q_j = \|\mathbf{a}^{(j+1)} - \mathbf{a}^{(j)}\|_1$ for every odd index.

The *track of a system* is the polygonal line that consists of all the segments S_j with endpoints $\mathbf{a}^{(j)}$ and $\mathbf{a}^{(j+1)}$ ($j = 1, \ldots, n - 1$). The *function of a system* is the piecewise linear function f defined by $f(\mathbf{a}^{(j)}) := q_j$ and linear on S_j. One speaks of a d-copula C as the *solution* for a system when $C(\mathbf{a}^{(j)}) = q_j$ for $j = 0, 1, \ldots, n$. ◇

The idea of the proof is to show that every system has a solution. Thus, given a d-quasi-copula Q and a track, one can consider n points along this track and the polygonal having those points as the endpoints of the segments constituting it. For n sufficiently large, the polygonal approximates the given track; moreover, one can find a d-copula agreeing with Q at those points. The proof will be achieved in a series of steps.

Lemma 7.2.12. *Let Σ be an increasing system. If k is such that $2 \leq k \leq n - 2$, assume the following:*

(a) *the system Σ_k obtained from Σ by deleting the elements with indices $1, \ldots, k-1$, has a solution C_k;*

(b) *there exists a measure μ on the Borel sets of $[\mathbf{0}, \mathbf{a}^{(k)}]$ such that $\mu([\mathbf{0}, \mathbf{a}^{(j)}]) = q_j$ for $j \leq k$ and such that its i-th margin μ_i satisfies*

$$\mu_i \leq \mathbf{1}_{[0, a_i^{(k)}]} \cdot \lambda = \lambda\Big|_{[0, a_i^{(k)}]}.$$

Then Σ has a solution.

Proof. Let $\mathscr{A}$ be the partition of $\mathbb{I}^d$ induced by the points $\mathbf{a}^{(j)}$, $j = 0, k, k+1, \ldots, n$,

$$\mathscr{A} := \left\{ \prod_{i=1}^{d} \left[a_i^{j(i)}, a_i^{s(j(i))} \right] : j(i) = 0, k, k+1, \ldots, n - 1 \right\},$$

where $s : \{0, k, k+1, \ldots, n-1\} \to \{k, k+1, \ldots, n\}$ with $s(0) = k$ and $s(i) = i+1$ for $i \neq 0$. Notice that the interiors of any two different rectangles in $\mathscr{A}$ are disjoint.

Since $q_k \leq a_i^{(k)}$, for every $i = 1, \ldots, d$, it is possible to define

$$\eta_i := \begin{cases} 0, & \text{if } q_k = a_i^{(k)}, \\ \left(\mathbf{1}_{[0, a_i^{(k)}]} \cdot \lambda - \nu_i \right) / \left(a_i^{(k)} - q_k \right), & \text{if } q_k < a_i^{(k)}. \end{cases} \quad (7.2.7)$$

If $q_k < a_i^{(k)}$, then η_i is a probability measure on the Borel sets of the interval $\left[0, a_i^{(k)} \right]$.

Given a rectangle R in $\mathscr{A}$, define a measure μ_R on $(R, \mathscr{B}(R))$, as follows:

- if $R = [\mathbf{0}, \mathbf{a}^{(k)}]$, then set $\mu_R = \mu$, by assumption (b);

- otherwise, if μ_{C_k} denotes the measure generated by the copula C_k, whose existence is asserted by assumption (a), set

$$\mu_R = \mu_{C_k}(R)\,(\pi_1 \otimes \pi_2 \otimes \cdots \otimes \pi_d)\,,$$

where $\pi_i = \eta_i$ if $a_i^{(j(i))} = 0$, while π_i is the uniform distribution on $[a_i^{(j(i))}, a_i^{(j(i)+1)}]$ if $a_i^{(j(i))} > 0$.

For every Borel set B in $\mathbb{I}^d$ define

$$\nu(B) := \sum_{R \in \mathscr{A}} \mu_R\,(B \cap R)\,.$$

The set function ν thus defined is a measure on $(\mathbb{I}^d, \mathscr{B}(\mathbb{I}^d))$.

Notice that, by definition, $\mu_R(R) = \mu_{C_k}(R)$, so that

$$\nu(\mathbb{I}^d) = \sum_{R \in \mathscr{A}} \mu_{C_k}(R)\,;$$

therefore $\nu(\mathbb{I}^d) = 1$, and, hence, ν is a probability measure on $(\mathbb{I}^d, \mathscr{B}(\mathbb{I}^d))$.

Next we prove that ν is a d-fold stochastic measure; as a consequence, by Theorem 3.1.2, there will exist a copula C such that $\nu = \mu_C$. Let ν_i be the i-th marginal of ν, and consider its restriction $1_{[a_i^{j(i)}, a_i^{s(j(i))}]} \cdot \nu_i$ to the interval $[a_i^{j(i)}, a_i^{s(j(i))}]$. Two cases will have to be considered.

If $a_i^{j(i)} > 0$, then $1_{[a_i^{j(i)}, a_i^{s(j(i))}]} \cdot \nu_i$ equals the uniform distribution multiplied by the sum of the measures $\mu_{C_k}(R)$ of those rectangles included in

$$\mathbb{I}^{i-1} \times \left[a_i^{j(i)}, a_i^{s(j(i))}\right] \times \mathbb{I}^{d-i}\,.$$

But this sum is $a_i^{s(j(i))} - a_i^{j(i)}$, since μ_{C_k} is a d-fold stochastic measure.

If $a_i^{j(i)} = 0$, then, necessarily, $a_i^{s(j(i))} = a_i^{(k)}$; in this case the same restriction $1_{[a_i^{j(i)}, a_i^{s(j(i))}]} \cdot \nu_i$ equals ν_i plus η_i multiplied by a sum analogous to that of the previous case but without the term corresponding to $[0, \mathbf{a}^{(k)}]$. This sum is now equal to $a_i^{(k)} - q_k$. Therefore, one has

$$1_{[0, a_i^{(k)}]} \cdot \nu_i = \nu_i + \left(a_i^{(k)} - q_k\right) \cdot \eta_i = 1_{[0, a_i^{(k)}]} \cdot \lambda\,,$$

which proves that ν is indeed a d-fold stochastic measure.

The unique copula C that corresponds to the measure ν is a solution for the system Σ. In fact, for $j \leq k$, one has

$$\left[0, a_i^{(j)}\right] \subseteq \left[0, a_i^{(k)}\right]\,,$$

so that

$$\mu_C\left([0, a_i^{(j)}]\right) = \mu\left([0, a_i^{(j)}]\right) = q_j\,.$$

If $j > k$, then, by the same argument used above, one has that $\mu_C([0, a_i^{(j)}])$ equals the sum of all the measures $\mu_C(R)$ of the rectangles R included in $[0, a_i^{(j)}]$, namely $\mu_C([0, a_i^{(j)}]) = q_j$, which concludes the proof.   □

Corollary 7.2.13. *Let Σ be as in Lemma 7.2.12 and assume that a solution exists for the system Σ_k' obtained from Σ by deleting the elements with index $j \geq k + 1$. Then condition (b) of Lemma 7.2.12 holds.*

Proof. Let C_k' be the solution for Σ_k'; then, for every $j \in \{1, \ldots, k\}$, one has

$$C_k'(\mathbf{a}^{(j)}) = q_j,$$

so that, if $\mu_{C_k'}$ denotes the measure associated with C_k',

$$\mu_{C_k'}([\mathbf{0}, \mathbf{a}^{(j)}]) = q_j.$$

Since $\mu_{C_k'}$ is a d-fold stochastic measure, its marginals are $\nu_i = \mathbf{1}_{[0, a_i^{(j)}]} \cdot \lambda$, for $i = 1, \ldots, d$.   □

Condition (b) of Lemma 7.2.12 is satisfied for $k = 2$ for every simple system.

Lemma 7.2.14. *Let Σ be a simple system*

$$\Sigma = \left(\mathbf{a}^{(0)}, \mathbf{a}^{(1)}, \ldots, \mathbf{a}^{(n)}; q_0, q_1, \ldots, q_n \right).$$

Then there exists a measure μ on the Borel sets of $[\mathbf{0}, \mathbf{a}^{(2)}]$ such that

$$\mu([\mathbf{0}, \mathbf{a}^{(1)}]) = q_1 \quad and \quad \mu([\mathbf{0}, \mathbf{a}^{(2)}]) = q_2, \tag{7.2.8}$$

and such that for every one of its margins μ_i, one has $\mu_i \leq \lambda$.

Proof. Since Σ is simple one has $q_1 = 0$ and $q_2 = \sum_{i=1}^d \left(a_i^{(2)} - a_i^{(1)} \right)$. Consider the partition of $[\mathbf{0}, \mathbf{a}^{(2)}]$ into the rectangles

$$R_i := [0, a_1^{(1)}] \times \cdots \times [0, a_{i-1}^{(1)}] \times [a_i^{(1)}, a_i^{(2)}] \times [0, a_{i+1}^{(1)}] \times \ldots [0, a_d^{(1)}],$$

and define an absolutely continuous measure μ on the Borel sets of $[\mathbf{0}, \mathbf{a}^{(2)}]$ by means of its density

$$f_\mu := \sum_{i=1}^d \left(a_i^{(2)} - a_i^{(1)} \right) \mathbf{1}_{R_i}.$$

Then, by construction, condition (7.2.8) is satisfied with $q_1 = 0$. The density of the i-th margin of μ is given by

$$f_i = \left(\sum_{k \neq i} \left(a_k^{(2)} - a_k^{(1)} \right) \right) \mathbf{1}_{[0, a_k^{(1)}]} + \left(a_i^{(2)} - a_i^{(1)} \right) \mathbf{1}_{[a_i^{(1)}, a_i^{(2)}]}.$$

Notice that

$$\sum_{k \neq i} \left(a_k^{(2)} - a_k^{(1)} \right) = q_2 - \left(a_i^{(2)} - a_i^{(1)} \right) = \left(q_2 - a_i^{(2)} \right) + a_i^{(1)} \leq a_i^{(1)},$$

since

$$q_2 \leq \min \left\{ a_1^{(2)}, a_2^{(2)}, \dots, a_d^{(2)} \right\} \leq a_i^{(2)}.$$

This concludes the proof.   □

Two operations will be introduced on a system Σ; each of them yields a new system as a result.

Straightening: this operation consists in deleting some elements of a system Σ; specifically, one deletes the elements of index j if either one of the following conditions holds:

- $q_{j-1} = q_j = q_{j+1}$;
- $\left\| \mathbf{a}^{(j)} - \mathbf{a}^{(j-1)} \right\|_1 = q_j - q_{j-1}$ and $\left\| \mathbf{a}^{(j+1)} - \mathbf{a}^{(j)} \right\|_1 = q_{j+1} - q_j$.

Smashing: Consider all the segments of endpoints $\mathbf{a}^{(j)}$ and $\mathbf{a}^{(j+1)}$ such that

$$q_j < q_{j+1} < q_j + \left\| \mathbf{a}^{(j+1)} - \mathbf{a}^{(j)} \right\|_1,$$

and on each such segment insert a new point $\mathbf{b}^{(j)}$ determined by the condition

$$\left\| \mathbf{a}^{(j+1)} - \mathbf{b}^{(j)} \right\|_1 = q_{j+1} - q_j;$$

the function of Σ then takes the value q_j at $\mathbf{b}^{(j)}$ and is linear on the segment of endpoints $\mathbf{b}^{(j)}$ and $\mathbf{a}^{(j+1)}$.

Lemma 7.2.15. *If C is a solution for a system Σ, it is also a solution for the system obtained from Σ by either smashing or straightening it, or both.*

Proof. The assertion is obvious when the system is straightened. Consider then its smashing. For $k \leq j$ and for $k \geq j+1$, the system keeps the same endpoints and the same q_k's, so that

$$C(\mathbf{a}^{(k)}) = q_k \qquad (k \leq j; k \geq j+1).$$

At $\mathbf{b}^{(j)}$, one has, by definition,

$$C(\mathbf{b}^{(j)}) = q_j.$$

Therefore C is a solution for the straightened system.   □

Lemma 7.2.16. *Let*

$$\Sigma = \left(\mathbf{a}^{(0)}, \mathbf{a}^{(1)}, \mathbf{a}^{(2)}, \mathbf{a}^{(3)}, \mathbf{1}; 0, 0, q, q, 1 \right)$$

be a simple system of length 4; then the system

$$\Sigma' = \left(\mathbf{a}^{(0)}, \mathbf{a}^{(2)}, \mathbf{a}^{(3)}, \mathbf{1}; 0, q, q, 1 \right)$$

obtained from Σ by deleting the elements with index 1 has a solution.

Proof. Let Σ'' be the system obtained by adding to Σ' the point $\widetilde{\mathbf{a}}^{(1)} := q\mathbf{1}$ with $\widetilde{q}_1 := q$. Next, let Σ_0 be the system obtained by smashing and then straightening Σ''.

$$\Sigma_0 := \left(\mathbf{a}^{(0)}, \mathbf{b}^{(1)}, \widetilde{\mathbf{a}}^{(1)}, \mathbf{a}^{(3)}, \mathbf{1}; 0, 0, q, q, 1 \right),$$

where $\mathbf{b}^{(1)}$ is the point on the segment of endpoints $\mathbf{0}$ and $\mathbf{a}^{(1)}$ determined by the condition

$$\left\| \widetilde{\mathbf{a}}^{(1)} - \mathbf{b}^{(1)} \right\|_1 = q.$$

It is immediately checked that Σ_0 is a system and that it is simple.

Now apply Lemma 7.2.14 to Σ_0, in order to obtain a measure $\widetilde{\mu}$ on the Borel sets of $[\mathbf{0}, \widetilde{\mathbf{a}}^{(1)}]$ such that

$$\widetilde{\mu}([\mathbf{0}, \widetilde{\mathbf{a}}^{(1)}]) = q \quad \text{and} \quad \widetilde{\mu}([\mathbf{0}, \mathbf{b}^{(1)}]) = 0.$$

By construction, the point $\widetilde{\mathbf{a}}^{(1)}$ lies on the main diagonal of $\mathbb{I}^d$ so that all the marginals of $\widetilde{\mu}$ are equal to $\mathbf{1}_{[0,q]} \cdot \lambda$.

On the other hand, it is easily checked that

$$\left(\mathbf{0}, \mathbf{a}^{(3)} - \widetilde{\mathbf{a}}^{(1)}, \mathbf{1} - \widetilde{\mathbf{a}}^{(1)}, \mathbf{1}; 0, 0, 1 - q, 1 \right)$$

is a system, which is increasing.

Smashing this latter system one obtains a further system given by

$$\Sigma_1 = \left(\mathbf{0}, \mathbf{a}^{(3)} - \widetilde{\mathbf{a}}^{(1)}, \mathbf{1} - \widetilde{\mathbf{a}}^{(1)}, \mathbf{b}^{(2)}, \mathbf{1}; 0, 0, 1 - q, 1 - q, 1 \right),$$

where $\mathbf{b}^{(2)}$ is the point on the segment of endpoints $\mathbf{1} - \widetilde{\mathbf{a}}^{(1)}$ and $\mathbf{1}$ determined by the condition

$$\left\| \mathbf{1} - \mathbf{b}^{(2)} \right\|_1 = 1 - (1 - q) = q.$$

The system Σ_1 is simple.

Applying to it Lemma 7.2.14 one has a measure $\widetilde{\nu}$ on the Borel sets of the rectangle $[\mathbf{0}, \mathbf{1} - \widetilde{\mathbf{a}}^{(1)}]$ such that

$$\widetilde{\nu}([\mathbf{0}, \mathbf{a}^{(3)} - \widetilde{\mathbf{a}}^{(1)}]) = 0 \quad \text{and} \quad \widetilde{\nu}([\mathbf{0}, \mathbf{1} - \widetilde{\mathbf{a}}^{(1)}]) = 1 - q.$$

By construction, the point $\widetilde{\mathbf{a}}^{(1)}$ lies on the main diagonal of $\mathbb{I}^d$ so that all the marginals of $\widetilde{\nu}$ are equal to $\mathbf{1}_{[0,1-q]} \cdot \lambda$.

Let ζ be defined for every Borel set A of $\mathbb{I}^d$ by

$$\zeta(A) := \tilde{\nu}\left((A - \tilde{\mathbf{a}}^{(1)}) \cap \mathbb{I}^d\right).$$

Now $\mu := \tilde{\mu} + \zeta$ is a d-fold stochastic measure. In fact

$$\mu\left(\mathbb{I}^d\right) = \tilde{\mu}\left(\mathbb{I}^d\right) + \zeta\left(\mathbb{I}^d\right) = \tilde{\mu}\left(\left[0, \tilde{\mathbf{a}}^{(1)}\right]\right) + \zeta\left(\left[\tilde{\mathbf{a}}^{(1)}, 1\right]\right)$$
$$= \tilde{\mu}([0, \tilde{\mathbf{a}}^{(1)}]) + \tilde{\nu}([0, 1 - \tilde{\mathbf{a}}^{(1)}]) = q + (1 - q) = 1;$$

while the i-th margin of μ is given by

$$\mu_i = \mathbf{1}_{[0, \tilde{\mathbf{a}}^{(1)}]} \cdot \lambda + \mathbf{1}_{[\tilde{\mathbf{a}}^{(1)}, 1]} \cdot \lambda = \lambda.$$

Let C be the unique copula associated with the measure μ. Then

$$C(\mathbf{b}^{(1)}) = \mu([0, \mathbf{b}^{(1)}]) = \tilde{\mu}([0, \mathbf{b}^{(1)}]) + \zeta([0, \mathbf{b}^{(1)}]) = 0,$$
$$C(\tilde{\mathbf{a}}^{(1)}) = \mu([0, \tilde{\mathbf{a}}^{(1)}]) + \zeta([0, \tilde{\mathbf{a}}^{(1)}]) = q,$$
$$C(\mathbf{a}^{(3)}) = \mu([0, \mathbf{a}^{(3)}]) + \zeta([0, \mathbf{a}^{(3)}]) = q.$$

Thus C is a solution for Σ_0 and, hence, also for Σ'. This concludes the proof. $\square$

Lemmata 7.2.14 and 7.2.16 together yield

Corollary 7.2.17. *Every simple system of length* 4 *has a solution.*

The final auxiliary result is provided by the following

Lemma 7.2.18. *Every system* Σ *has a solution.*

Proof. Let Σ be a system. Assume, at first, that it is increasing. In fact, by smashing and then straightening Σ, it is possible to reduce the proof to the case in which Σ is simple. Thus, let Σ be a simple system; the proof of the existence of a solution of Σ will be given by induction on $n = 2p$.

The result is known to hold for $p = 1$ and $p = 2$ by Lemma 7.2.14 and Corollary 7.2.17, respectively. Assume now that there exists $p' \geq 3$ such that every simple system of length $2p$ with $p < p'$ has a solution. Let Σ_{p+1} be a simple system of length $2(p+1)$. Set $k := p'$ if p' is odd and $k := p' - 1$ if p is even.

Smash Σ_k by adding a point strictly preceding $\mathbf{a}^{(k)}$; then straighten the resulting system by removing $\mathbf{a}^{(k)}$. The system thus obtained is simple and has length $2p' - (k-1) < 2p'$, since $k \geq p' - 1 \geq 2$. This system has therefore a solution according to the induction assumption.

In general, let Σ be any system, and, for α in $]0, 1[$, let Σ_α be the simple system obtained from Σ by replacing $\mathbf{a}^{(j)}$ by $(1 - \alpha)\mathbf{a}^{(j)} + \alpha^2 j \cdot \mathbf{1}$ and q_j by $(1 - \alpha)q_j + \alpha^2 j$. By what has been proved above, the system Σ_α has a solution C_α. Since $\mathscr{C}_d$ is compact, there exists an infinitesimal sequence (α_n) such that (C_{α_n}) converges to a copula C. Therefore, for every $n \in \mathbb{N}$ and for every point $(1 - \alpha_n) \mathbf{a}^{(j)} + \alpha_n^2 j \cdot \mathbf{1}$, one has

$$C_{\alpha_n} \left((1 - \alpha_n) \mathbf{a}^{(j)} + \alpha_n^2 j \cdot \mathbf{1} \right) = (1 - \alpha_n) q_j + \alpha_n^2 j . \qquad (7.2.9)$$

Letting n tend to $+\infty$ yields

$$C(\mathbf{a}^{(j)}) = q_j$$

for every index j. Thus C is a solution of the given system Σ. $\qquad \square$

One can then prove the announced characterisation of quasi-copulas.

Theorem 7.2.4. Because of Theorem 7.2.1, it is already known that a quasi-copula Q satisfies the properties of the statement.

Conversely, let a function $Q : \mathbb{I}^d \to \mathbb{I}$ be given that satisfies properties (a), (b) and (c), and let

$$B = \{ (F_1(t), \ldots, F_d(t)) : t \in \mathbb{I} \}$$

be a track in $\mathbb{I}^d$. For $n = 2^m$ consider the points in $\mathbb{I}^d$

$$\mathbf{u}^{(j)} = \left(F_1 \left(\frac{j}{2^m} \right), \ldots, F_d \left(\frac{j}{2^m} \right) \right),$$

and set $q_j := Q \left(\mathbf{u}^{(j)} \right)$ $(j = 1, 2, \ldots, 2^m)$. As a consequence of Lemma 7.2.18, for every $m \in \mathbb{N}$, there exists a copula C_m that coincides with Q at each of these points, viz., $C_m(\mathbf{u}^{(j)}) = Q(\mathbf{u}^{(j)})$.

Since $\mathscr{C}_d$ is compact, the sequence (C_m) of copulas obtained in this fashion contains a subsequence $\left(C_{m(k)} \right)_{k \in \mathbb{N}}$ that converges to a copula C. This limiting copula coincides with Q on B. To see this, fix $t \in \mathbb{I}$ and for each $k \in \mathbb{N}$, let $j = j(m(k))$ be the largest integer smaller than, or, at most equal to, $2^{m(k)} - 1$ for which

$$\left| t - \frac{j}{2^{m(k)}} \right| \leq \frac{1}{2^{m(k)}} .$$

Because of the continuity of the functions F_i $(i = 1, \ldots, d)$ it is then possible to take $m(k)$ large enough to ensure that

$$\left| F_i(t) - F_i \left(\frac{j}{2^m} \right) \right| < \frac{\varepsilon}{3 d}$$

simultaneously for every index $(i = 1, \ldots, d)$ and for a given $\varepsilon > 0$. Since $(C_{m(k)})$ converges to C on $\mathbb{I}^d$, for k sufficiently large, one has

$$\left| C_{m(k)}(F_1(t), \ldots, F_d(t)) - C(F_1(t), \ldots, F_d(t)) \right| < \frac{\varepsilon}{3} .$$

Keeping in mind that $C_{m(k)}(\mathbf{u}^{(j)}) = Q(\mathbf{u}^{(j)})$ $(j = 1, \ldots, 2^{m(k)})$ and that both Q

and $C_{m(k)}$ satisfy the Lipschitz condition, then one has

$$|Q(F_1(t), \ldots, F_d(t)) - C(F_1(t), \ldots, F_d(t))|$$
$$\leq \left| Q(F_1(t), \ldots, F_d(t)) - Q(\mathbf{u}^{(j)}) \right|$$
$$+ \left| C_{m(k)}(\mathbf{u}^{(j)}) - C_{m(k)}(F_1(t), \ldots, F_d(t)) \right|$$
$$+ \left| C_{m(k)}(F_1(t), \ldots, F_d(t)) - C(F_1(t), \ldots, F_d(t)) \right|$$
$$< d \frac{\varepsilon}{3d} + d \frac{\varepsilon}{3d} + \frac{\varepsilon}{3} = \varepsilon.$$

Because of the arbitrariness of ε, Q coincides with the copula C on the track B and is, therefore, a d-quasi-copula. $\qquad\square$

7.3 The space of quasi-copulas and its lattice structure

It is easily checked that $\mathscr{Q}_d$, the space of d-quasi-copulas, is a closed subset of the space $(\Xi(\mathbb{I}^d), d_\infty)$ of continuous functions on $\mathbb{I}^d$ endowed with the topology of uniform convergence. Moreover, the following analogue of Theorem 1.7.7 can be proved.

Theorem 7.3.1. *The set $\mathscr{Q}_d$ of d-quasi-copulas is a compact and convex subset of $(\Xi(\mathbb{I}^d), d_\infty)$.*

We recall a few definitions from lattice theory (see, e.g., [Davey and Priestley, 2002]). Given a partial ordered set (=poset) $(P, \leq)$, and two elements x and y in P, $x \vee y$ denotes their *join*, namely their least upper bound when it exists, while $x \wedge y$ denotes their *meet*, namely their greatest lower bound, when it exists. If S is a subset of P, $\vee S$ and $\wedge S$ are defined in a similar way.

Now, as in the case of copulas (section 1.7.1), one can consider the classical pointwise order among functions in $\mathscr{Q}_d$, namely, for $Q_1, Q_2 \in \mathscr{Q}_d$, $Q_1 \leq Q_2$ if, and only if, $Q_1(\mathbf{u}) \leq Q_2(\mathbf{u})$ for all $\mathbf{u} \in \mathbb{I}^d$. In particular, we may define

$$Q_1 \vee Q_2 := \inf\{Q \in \mathscr{Q}_d : Q_1 \leq Q, Q_2 \leq Q\},$$
$$Q_1 \wedge Q_2 := \sup\{Q \in \mathscr{Q}_d : Q \leq Q_1, Q \leq Q_2\}.$$

When the join, or meet, is found in a particular poset P, we write $\vee_P S$ and $\wedge_P S$. Given two posets A and B, A is said to be join-dense (respectively, meet-dense) in B if for every $D \in B$, there exists a set $S \subseteq A$, such that $D = \vee_B S$ (respectively, $D = \wedge_B S$). A poset $P \neq \emptyset$ is said to be a *lattice* if for all x and y in P, both $x \vee y$ and $x \wedge y$ are in P; and P is a *complete lattice* if both $\vee S$ and $\wedge S$ are in P for every subset S of P. If $\varphi : P \to L$ is an order-preserving injection of a poset $(P, \prec)$ into a complete lattice $(L, \prec_1)$, then L is said to be a *completion* of P; in particular, if φ maps P onto L, then it is an *order-isomorphism*.

Definition 7.3.2. A complete lattice $(L, \prec_1)$ is said to be the *Dedekind-MacNeille completion* of a poset $(P, \prec)$ (also referred to as the *normal completion* or the *completion by cuts*) if $(P, \prec)$ is both join-dense and meet-dense in $(L, \prec_1)$.   ◇

In view of Theorem 7.2.2, upper and lower bounds in $\mathscr{Q}_d$ coincide with the Hoeffding–Fréchet bounds; however, more can be said. The relevant result is the following

Theorem 7.3.3. *The set $\mathscr{Q}_d$ of d-quasi-copulas is a complete lattice.*

Proof. It is enough to prove that both the join and the meet, respectively $\bigvee \mathbf{Q}$ and $\bigwedge \mathbf{Q}$, of every set $\mathbf{Q} \subseteq \mathscr{Q}_d$ are d-quasi-copulas.

Since it is easily proved that $\bigvee \mathbf{Q}$ is increasing in each place and satisfies the boundary conditions of a quasi-copula, it remains only to prove the Lipschitz condition. For every $\varepsilon > 0$, there exists a d-quasi-copula Q_ε such that, for every $\mathbf{u} \in \mathbb{I}^d$,

$$Q_\varepsilon(\mathbf{u}_j(t)) > \bigvee \mathbf{Q}(\mathbf{u}_j(t)) - \varepsilon \,.$$

Thus, since $Q_\varepsilon(\mathbf{u}) \leq \bigvee \mathbf{Q}(\mathbf{u})$, one has

$$\bigvee \mathbf{Q}(\mathbf{u}_j(t)) - \bigvee \mathbf{Q}(\mathbf{u}) < Q_\varepsilon(\mathbf{u}_j(t)) + \varepsilon - Q_\varepsilon(\mathbf{u}_j(s)) \leq t - s + \varepsilon \,,$$

which, on account of the arbitrariness of ε, yields, for every $j \in \{1, \dots, d\}$,

$$\bigvee \mathbf{Q}(\mathbf{u}_j(t)) - \bigvee \mathbf{Q}(\mathbf{u}_j)(s)) \leq t - s \,.$$

Now, if $\mathbf{u}$ and $\mathbf{v}$ are in $\mathbb{I}^d$ with $\mathbf{u} \leq \mathbf{v}$ (componentwise), then

$$\bigvee \mathbf{Q}(\mathbf{v}) - \bigvee \mathbf{Q}(\mathbf{u}) = \bigvee \mathbf{Q}(\mathbf{v}) - \bigvee \mathbf{Q}(u_1, v_2, \dots, v_d)$$

$$+ \bigvee \mathbf{Q}(u_1, v_2, \dots, v_d) - \bigvee \mathbf{Q}(u_1, u_2, v_3 \dots, v_d)$$

$$+ \cdots + \bigvee \mathbf{Q}(u_1, u_2, \dots, u_{d-1}, v_d) - \bigvee \mathbf{Q}(\mathbf{u})$$

$$\leq (v_1 - u_1) + (v_2 - u_2) + \cdots + (v_d - u_d) = \sum_{j=1}^{n} (v_j - u_j) \,,$$

which proves the Lipschitz condition. Therefore $\bigvee \mathbf{Q}$ is a d-quasi-copula.

The proof that $\bigwedge \mathbf{Q}$ is a d-quasi-copula is very similar and will not be repeated here.   □

Since the set $\mathscr{C}_d$ of d-copulas is included in $\mathscr{Q}_d$, one immediately has the following corollary.

Corollary 7.3.4. *Both the join and the meet of every set of d-copulas are d-quasi-copulas.*

However, proper subsets of $\mathcal{Q}_d$ may not be closed under supremum and infimum operations; two important cases in point are given in the next result.

Theorem 7.3.5. *For every $d \geq 2$, neither the set $\mathscr{C}_d$ of copulas nor the set $\mathcal{Q}_d \setminus \mathscr{C}_d$ of proper quasi-copulas is a lattice.*

Proof. We start by considering the assertion in the case $d = 2$. In order to prove that $\mathscr{C}_2$ is not a lattice, it is enough to exhibit two 2-copulas C_1 and C_2 such that their join, or their meet, is a proper quasi-copula.

For every $\theta \in \mathbb{I}$ consider the 2-copula

$$C_\theta(u,v) = \begin{cases} \min\{u, v - \theta\}, & (u,v) \in [0, 1-\theta] \times [\theta, 1], \\ \min\{u + \theta - 1, v\}, & (u,v) \in [1-\theta, 1] \times [0, \theta], \\ W_2(u,v), & \text{elsewhere.} \end{cases}$$

If U and V are uniform random variables on $(0,1)$ with $V = U + \theta$ (mod 1); then C_θ is their distribution function. Set $\mathbf{C} = \{C_{1/3}, C_{2/3}\}$; then $\bigvee \mathbf{C}$ is given by

$$\bigvee \mathbf{C}(s,t) = \begin{cases} \max\{0, s - 1/3, t - 1/3, s + t - 1\}, & -1/3 \leq t - s \leq 2/3, \\ W_2(s,t), & \text{elsewhere.} \end{cases}$$

It is known from Corollary 7.3.4 that $\bigvee \mathbf{C}$ is a quasi-copula. Since

$$V_{\bigvee \mathbf{C}}\left([1/3, 2/3]^2\right) = -1/3 < 0,$$

$\bigvee \mathbf{C}$ is not a copula.

In the case $d > 2$, consider the following copulas

$$C_1(\mathbf{u}) = C_{1/3}(u_1, u_2)\, u_3 \cdots u_d,$$
$$C_2(\mathbf{u}) = C_{2/3}(u_1, u_2)\, u_3 \cdots u_d.$$

Then, $C_1 \vee C_2$ is not a copula, since its $(1,2)$-bivariate margin is not a copula. So, $\mathscr{C}_d$ is not join dense.

Analogously, in order to prove that $\mathcal{Q}_2 \setminus \mathscr{C}_2$ is not a lattice, it is enough to exhibit two proper 2-quasi-copulas Q_1 and Q_2 such that their join, or their meet, is not a proper quasi-copula. Let Q be the proper quasi-copula $C_{1/3} \vee C_{2/3}$, C_θ having been defined above, and define the proper quasi-copulas

$$Q_1(u,v) := \begin{cases} \frac{1}{2} Q(2u, 2v), & (u,v) \in [0, 1/2]^2, \\ M_2(u,v), & \text{elsewhere,} \end{cases}$$

and

$$Q_2(u,v) := \begin{cases} \frac{1}{2} \left(1 + Q(2u - 1, 2v - 1)\right), & (u,v) \in [1/2, 1]^2, \\ M_2(u,v), & \text{elsewhere.} \end{cases}$$

It is left as an (easy) exercise for the reader to check that Q_1 and Q_2 are indeed proper quasi-copulas; for instance,

$$V_{Q_1}\left([1/8, 3/8]^2\right) = -\frac{5}{24} < 0 \quad \text{and} \quad V_{Q_2}\left([5/8, 7/8]^2\right) = -\frac{1}{6} < 0.$$

On the other hand, $Q_1 \vee Q_2 = M_2$, which is a copula.

In the case $d > 2$, it suffices to exhibit two proper d-quasi-copulas Q_1 and Q_2 whose join (or meet) is a d-copula. Let Q be a proper 2-quasi-copula, and define

$$Q_1(\mathbf{u}) = \begin{cases} Q(2u_1, 2u_2)\, u_3 \cdots \frac{u_d}{2}, & \mathbf{u} \in \left[0, \frac{1}{2}\right]^2 \times \mathbb{I}^{d-2}, \\ \min\{u_1, \ldots, u_d\}, & \text{elsewhere}, \end{cases}$$

and

$$Q_2(\mathbf{u}) = \begin{cases} (1 + Q(2u_1 - 1, 2u_2 - 1)\, u_3 \cdots u_d)/2, & \mathbf{u} \in \left[\frac{1}{2}, 1\right]^2 \times \mathbb{I}^{d-2}, \\ \min\{u_1, \ldots, u_d\}, & \text{elsewhere}. \end{cases}$$

Then Q_1 and Q_2 are proper quasi-copulas, since its $(1, 2)$-bivariate margin is not a copula. Finally, $Q_1 \vee Q_2 = M_d$ and, hence, it is not a proper quasi-copula. $\square$

Next we examine the question of the Dedekind-MacNeille completion of $\mathscr{C}_d$ distinguishing the two cases $d = 2$ and $d > 2$. As a preliminary, a few lemmata will be needed.

In the next result, which refers to bivariate copulas, sharper bounds, both upper and lower, are given related to the class of all 2-copulas taking a specified value at a point in the interior of the unit square.

Lemma 7.3.6. *Let the 2-copula C take the value θ at the point $(a, b) \in \,]0, 1[^2$, i.e. $C(a, b) = \theta$, where θ belongs to the interval $[\max\{a + b - 1, 0\}, \min\{a, b\}]$. Then, for every (u, v) in $\mathbb{I}^2$,*

$$C_L(u, v) \le C(u, v) \le C_U(u, v), \tag{7.3.1}$$

where C_L and C_U are defined by

$$C_L(u, v) := \begin{cases} \max\{0, u - a + v - b + \theta\}, & (u, v) \in [0, a] \times [0, b], \\ \max\{0, u + v - 1, u - a + \theta\}, & (u, v) \in [0, a] \times [b, 1], \\ \max\{0, u + v - 1, v - b + \theta\}, & (u, v) \in [a, 1] \times [0, b], \\ \max\{\theta, u + v - 1\}, & (u, v) \in [a, 1] \times [b, 1], \end{cases}$$

and

$$C_U(u, v) = \begin{cases} \min\{u, v, \theta\}, & (u, v) \in [0, a] \times [0, b], \\ \min\{u, v - b + \theta\}, & (u, v) \in [0, a] \times [b, 1], \\ \min\{u - a + \theta, v\}, & (u, v) \in [a, 1] \times [0, b], \\ \min\{u, v, u - a + v - b + \theta\}, & (u, v) \in [a, 1] \times [b, 1], \end{cases}$$

respectively. The bounds in (7.3.1) are the best possible.

Proof. First of all notice that, by definition, $C_L(a, b) = C_U(a, b) = \theta$.

In order to show that C_U is the best upper bound for C, consider four cases.

Case 1: (u, v) is in $[0, a] \times [0, b]$. Since C is increasing in each place, $C(u, v) \le C(a, b) = \theta$ so that $C(u, v) \le \min\{u, v, \theta\} = C_U(u, v)$.

Case 2: (u, v) is in $[a, 1] \times [b, 1]$. Since $[u, 1] \times [v, 1]$ is included in $[a, 1] \times [b, 1]$, it follows that

$$V_C\left([u, 1] \times [v, 1]\right) \le V_C\left([a, 1] \times [b, 1]\right),$$

which is equivalent to $C(u, v) \le u - a + v - b + \theta$; hence

$$C(u, v) \le \min\{u, v, u - a + v - b + \theta\} = C_U(u, v).$$

Case 3: (u, v) is in $[0, a] \times [b, 1]$. Then, considering the C-volume of $[a, 1] \times [b, v]$, one has $C(u, v) \le C(a, v) \le v - b + C(a, b) = v - b + \theta \le t$ so that

$$C(u, v) \le \min\{u, v - b + \theta\} = C_U(u, v).$$

Case 4: (u, v) is in $[a, 1] \times [0, b]$. Then, considering the C-volume of $[a, u] \times [b, 1]$, one has $C(u, v) \le C(u, b) \le u - a + C(a, b) = u - a + \theta \le s$ so that

$$C(u, v) \le \min\{u - a + \theta, v\} = C_U(u, v).$$

This ends the proof of the second inequality in (7.3.1).

Similarly we show that C_L is the best lower bound for C.

Case 1: (u, v) is in $[0, a] \times [0, b]$. Since $[u, 1] \times [v, 1]$ includes $[a, 1] \times [b, 1]$, it follows that

$$V_C\left([u, 1] \times [v, 1]\right) \ge V_C\left([a, 1] \times [b, 1]\right),$$

which is equivalent to $C(u, v) \ge u - a + v - b + \theta$, whence

$$C(u, v) \ge \max\{0, u - a + v - b + \theta\} = C_L(u, v).$$

Case 2: (u, v) is in $[a, 1] \times [b, 1]$. Since C is increasing in each place, one has $C(u, v) \ge C(a, b) = \theta$, whence $C(u, v) \ge \max\{\theta, u + v - 1\} = C_L(u, v)$.

Case 3: (u, v) is in $[0, a] \times [b, 1]$. By considering the C-volume of $[a, 1] \times [v, b]$, one has $C(u, v) \ge C(a, v) \ge v - b + \theta$, so that

$$C(u, v) \ge \max\{0, u + v - 1, t - b + \theta\} = C_L(u, v).$$

Case 4: (u, v) is in $[a, 1] \times [0, b]$. Then, considering the C-volume of $[a, 1] \times [v, b]$, one has $C(u, v) \ge C(a, v) \ge v - b + \theta$, whence

$$C(u, v) \ge \max\{0, u + v - 1, v - b + \theta\} = C_L(u, v).$$

This proves that C_L is the best lower bound for C and concludes the proof. □

Lemma 7.3.7. *For $(a, b) \in]0, 1[^2$ and for $\theta \in [W_2(a, b), M_2(a, b)]$ let $S_{a,b,\theta}$ denote the set of quasi-copulas that take the value θ at (a, b),*

$$S_{a,b,\theta} := \{Q \in \mathscr{Q}_2 : Q(a, b) = \theta\} .$$

Then both $\vee S_{a,b,\theta}$ and $\wedge S_{a,b,\theta}$ are copulas and

$$\bigvee S_{a,b,\theta}(u, v) = \min\{M_2(u, v), \theta + (u - a)^+ + (v - b)^+\} ,$$

$$\bigwedge S_{a,b,\theta}(u, v) = \max\{W_2(u, v), \theta - (a - u)^+ - (b - v)^+\} .$$

Proof. The properties of a quasi-copula yield the inequalities

$$-(a - u)^+ \le Q(u, v) - Q(a, v) \le (u - a)^+ ,$$
$$-(b - v)^+ \le Q(a, v) - Q(a, b) \le (v - b)^+ ,$$

whence, by summing,

$$\theta - (a - u)^+ - (b - v)^+ \le Q(u, v) \le \theta + (u - a)^+ + (v - b)^+ .$$

Therefore $\vee S_{a,b,\theta}(u, v) \le Q(u, v) \le \wedge S_{a,b,\theta}(u, v)$ for every point (u, v) in $\mathbb{I}^2$. That $\vee S_{a,b,\theta}$ and $\wedge S_{a,b,\theta}$ are copulas follows from Lemma 7.3.6. □

Lemma 7.3.8. *For every quasi-copula $Q \in \mathscr{Q}_2$, one has $Q = \vee S_1(Q)$, where $S_1(Q) := \{C \in \mathscr{C}_2 : C \le Q\}$, and $Q = \wedge S_2(Q)$, where $S_2(Q) := \{C \in \mathscr{C}_2 : C \ge Q\}$.*

Proof. It suffices to prove only one of the two assertions since the other is proved in an analogous manner. Let (a, b) be a point in $]0, 1[^2$ and set $\theta := Q(a, b)$. Then $\sup\{C(a, b) : C \in S_1(Q)\} = Q(a, b)$ since, by the previous Lemma, $\wedge S_{a,b,\theta}$ is a copula that belongs to $S_1(Q)$. □

In particular, it follows that 2-quasi-copulas can be characterised in terms of copulas. The following result is now obvious.

Lemma 7.3.9. *The set of bivariate copulas $\mathscr{C}_2$ is both join-dense and meet-dense in $\mathscr{Q}_2$.*

Theorem 7.3.10. *The complete lattice $\mathscr{Q}_2$ of bivariate quasi-copulas is order-isomorphic to the Dedekind-MacNeille completion of the poset $\mathscr{C}_2$ of bivariate copulas.*

Proof. This is an immediate consequence of Definition 7.3.2 and of Lemma 7.3.9. Infact, let $\mathrm{DM}(\mathscr{C}_2)$ be the Dedekin-McNeille completion of $\mathscr{C}_2$. Then the order-isomorphism $\varphi : \mathscr{Q}_2 \to \mathrm{DM}(\mathscr{C}_2)$ is realised by the function defined by $\varphi(Q) := \{C \in \mathscr{C}_2 : C \le Q\}$. □

The sets of copulas $\mathscr{C}_2$ and $\mathscr{C}_d$ with $d > 2$ differ with respect to their Dedekind-MacNeille completion: in fact, while $\mathscr{Q}_2$ is the Dedekind-MacNeille completion of $\mathscr{C}_2$, $\mathscr{Q}_d$ is not the Dedekind-MacNeille completion of $\mathscr{C}_d$ for $d > 2$.

Theorem 7.3.11. *For $d > 2$, the complete lattice $\mathscr{Q}_d$ of d-quasi-copulas is not order-isomorphic to the Dedekind-MacNeille completion of the poset $\mathscr{C}_d$ of d-copulas.*

Proof. It is enough to consider that W_d is a proper quasi-copula and that, because of Theorem 7.2.2, W_d cannot be the upper bound of any set of d-copulas.   □

A non-trivial example that shows that, for $d > 2$, the set $\mathscr{C}_d$ of d-copulas is not join-dense in $\mathscr{Q}_d$ is provided below.

Example 7.3.12. Consider the following function

$$Q(\mathbf{u}) := \frac{1}{2} + \frac{1}{2} W_d \left(2 \left(u_1 - \tfrac{1}{2} \right), \ldots, 2 \left(u_d - \tfrac{1}{2} \right) \right),$$

if $\min\{u_1, \ldots, u_d\} \in \,]1/2, 1[$, while $Q(\mathbf{u}) = \min\{u_1, \ldots, u_d\}$ elsewhere. The function Q is actually a quasi-copula: it is the ordinal sum (see Example 7.2.9) of the copula M_d and of the proper quasi-copula W_d with respect to the intervals $]0, 1/2[$ and $]1/2, 1[$. If $\mathscr{C}_d$ were join-dense in $\mathscr{Q}_d$, then there would exist a subset $\mathscr{S} \subseteq \mathscr{C}_d$ such that $\vee_{S \in \mathscr{S}} S = \mathscr{Q}_d$. Now, either (a) there exists a d-copula C such that $C(1/2, \ldots, 1/2) = 1/2$, or (b) by the compactness of $\mathscr{C}_d$ there is a sequence (C_n) of d-copulas with $C_n \leq Q$ for every $n \in \mathbb{N}$ such that

$$C_n(1/2, \ldots, 1/2) \xrightarrow[n \to +\infty]{} 1/2.$$

By the compactness of $\mathscr{C}_d$ a subsequence $(C_{n(k)})$ of (C_n) converges to a d-copula C; then necessarily $C(1/2, \ldots, 1/2) = 1/2$. Now $C \leq Q$, which means that, for every $\mathbf{u} \in [1/2, 1]^d$, one has, as a consequence of the very definition of Q,

$$C(\mathbf{u}) \leq \frac{1}{2} + \frac{1}{2} \sum_{j=1}^{d} 2 \left(u_j - \frac{1}{2} \right) - (d-1) = \sum_{j=1}^{d} u_j - \frac{3}{2} (d-1) < W_d(\mathbf{u}).$$

Since this contradicts the lower Hoeffding–Fréchet bound, the proper quasi-copula Q is not the join of any subfamily of $\mathscr{C}_d$.   ■

Further readings 7.3.13. Theorem 7.3.3 shows the relevance of quasi-copulas in studying copulas. It was proved by Nelsen et al. [2004] in the bivariate case $d = 2$ and by Rodríguez-Lallena and Úbeda Flores [2004a] in the general case. In this latter paper, one also finds the generalisation of Lemma 7.3.6 to the case $d > 2$.

The different behaviour of the set of quasi-copulas with respect to the Dedekind-MacNeille completion was highlighted by Nelsen and Úbeda Flores [2005] and Fernández-Sánchez et al. [2011a].   ■

7.3.1 Bounds for copulas with given diagonal

Upper and lower bounds for quasi-copulas may be improved in the presence of additional information on the values assumed in specific regions of the domain. Following analogous investigations in the space of copulas, here we focus our attention on the class of quasi-copulas $Q \in \mathcal{Q}_d$ with a given diagonal section defined by $\delta_Q(t) := Q(t, \ldots, t)$.

First of all, notice that the statements of Theorem 2.6.1 hold also for the diagonals of a quasi-copula $Q \in \mathcal{Q}_d$, namely the diagonal section of a quasi-copula can be characterised by its analytical properties.

A specific construction yields a quasi-copula for every given 2-diagonal δ.

Theorem 7.3.14. *For every* 2-*diagonal* δ, *i.e. any function* $\delta : \mathbb{I} \to \mathbb{I}$ *satisfying* (a)-(d) *of Theorem 2.6.1 for* $d = 2$, *the function* $A_\delta : \mathbb{I}^2 \to \mathbb{I}$ *defined by*

$$A_\delta(u, v) := \min \left\{ u, v, \max\{u, v\} - \max\{\hat{\delta}(t) : t \in [u \wedge v, u \vee v]\} \right\} \qquad (7.3.2)$$

$$= \begin{cases} \min \left\{ u, v - \max_{t \in [u,v]}\{t - \delta(t)\} \right\}, & u < v, \\ \min \left\{ v, u - \max_{t \in [v,u]}\{t - \delta(t)\} \right\}, & v \leq u, \end{cases}$$

is a symmetric 2-*quasi-copula having diagonal equal to* δ.

Proof. The equality $A_\delta(t, t) = \delta(t)$ and the symmetry of A_δ are immediate. We prove first that A_δ satisfies the boundary conditions; for every $t \in \mathbb{I}$, one has

$$A_\delta(0, t) = \min \left\{ 0, \max\{t - \max\{\hat{\delta}(s) : s \in [0, t]\} \right\},$$

where $\hat{\delta}(t) := t - \delta(t)$. Let t_1 be the point at which $\hat{\delta}$ takes its maximum in $[0, t]$; then $\hat{\delta}(t_1) \leq t_1 \leq t$, so that $A_\delta(0, t) = 0$ for every $t \in \mathbb{I}$; by symmetry also $A_\delta(t, 0) = 0$.

For the other boundary condition, one has

$$A_\delta(1, t) = \min \left\{ t, \max\{1 - \hat{\delta}(s) : s \in [t, 1]\} \right\}.$$

Let t_2 be the point at which $\hat{\delta}$ takes its maximum in $[t, 1]$; then

$$1 - \hat{\delta}(t_2) > 1 - (1 - t_2) = t_2 \geq t,$$

so that $A_\delta(1, t) = t$ for every $t \in \mathbb{I}$. Again, by symmetry, $A_\delta(t, 1) = t$.

We prove that A_δ is increasing in each argument and that it satisfies the Lipschitz condition (7.2.1).

Let $u_1 < u_2$, and assume $u_1 \leq v \leq u_2$. Let $t_0 \in [v, u_2]$ be such that $\hat{\delta}(t_0) = t_0 - \delta(t_0) = \max\{\hat{\delta}(t) : t \in [v, u_2]\}$. Then

$$\max\{\hat{\delta}(t) : t \in [v, u_2]\} - \max\{\hat{\delta}(t) : t \in [u_1, v]\}$$
$$\leq t_0 - \delta(t_0) - v + \delta(v) \leq u_2 - \delta(v) - v + \delta(v) = u_2 - v,$$

so that

$$v - \max\{\hat{\delta}(t) : t \in [u_1, v]\} \leq u_2 - \max\{\hat{\delta}(t) : t \in [v, u_2]\};$$

hence $A_\delta(u_1, v) \leq A_\delta(u_2, v)$.

If $A_\delta(u_1, v) = u_1$, then $A_\delta(u_2, v) - A_\delta(u_1, v) = A_\delta(u_2, v) - u_1 \leq u_2 - u_1$. If, on the other hand, $A_\delta(u_1, v) = v - \hat{\delta}(t_1)$, then, since $\hat{\delta}$ satisfies the Lipschitz condition (2.6.4),

$$A_\delta(u_2, v) - A_\delta(u_1, v) \leq u_2 - \hat{\delta}(t_2) - v + \hat{\delta}(t_1) \leq u_2 - \hat{\delta}(v) - v + \hat{\delta}(t_1)$$

$$= u_2 - v - \left(\hat{\delta}(v) - \hat{\delta}(t_1)\right) \leq u_2 - v + v - t_1$$

$$= u_2 - t_1 \leq u_2 - u_1,$$

which concludes the proof in this case.

Assume now $u_1 < u_2 \leq v$ and set $\hat{\delta}(t_j) := \max\{\hat{\delta}(t) : t \in [u_j, v]\}$ $(j = 1, 2)$; then $\hat{\delta}(t_1) \geq \hat{\delta}(t_2)$, so that $v - \hat{\delta}(t_1) \leq v - \hat{\delta}(t_2)$, which immediately yields $A_\delta(u_1, v) \leq A_\delta(u_2, v)$.

As above, if $A_\delta(u_1, v) = u_1$, then $A_\delta(u_2, v) - A_\delta(u_1, v) \leq u_2 - u_1$. If $A_\delta(u_1, v) = v - \hat{\delta}(t_1)$, then, because of (2.6.4),

$$A_\delta(u_2, v) - A_\delta(u_1, v) \leq \hat{\delta}(t_1) - \hat{\delta}(t_2) \leq \hat{\delta}(t_1) - \hat{\delta}(u_2) \leq u_2 - t_1 \leq u_2 - u_1,$$

which concludes the proof in this case.

The last case to be considered is $v \leq u_1 < u_2$. Set $\hat{\delta}(t_j) := \max\{\hat{\delta}(t) : t \in [v, u_j]\}$ $(j = 1, 2)$. Then, on account of the Lipschitz property of $\hat{\delta}$,

$$u_1 - \hat{\delta}(t_1) \leq u_1 - \hat{\delta}(u_1) \leq t_2 - \hat{\delta}(t_2) \leq u_2 - \hat{\delta}(t_2),$$

which proves that, also in this case, one has $A_\delta(u_1, v) \leq A_\delta(u_2, v)$.

As in the two previous cases, if $A_\delta(u_1, v) = u_1$, then $A_\delta(u_2, v) - A_\delta(u_1, v) \leq u_2 - u_1$. Assume, now, $A_\delta(u_1, v) = u_1 - \hat{\delta}(t_1)$. Then, since $\hat{\delta}(t_1) \leq \hat{\delta}(t_2)$,

$$A_\delta(u_2, v) - A_\delta(u_1, v) \leq u_2 - \hat{\delta}(t_2) - u_1 + \hat{\delta}(t_1) \leq u_2 - u_1.$$

The proof is now complete. $\square$

The Bertino copula $C_\delta^{\mathbf{Ber}}$ of equation (2.6.5) and the quasi-copula A_δ of equation (7.3.2) are the lower and upper bounds, respectively, for the sets of copulas and quasi-copulas with diagonal δ, as is shown below.

Theorem 7.3.15. *For every 2-diagonal δ and for every quasi-copula $Q \in \mathcal{Q}_\delta$ one has*

$$C_\delta^{\mathbf{Ber}} \leq Q \leq A_\delta. \tag{7.3.3}$$

Proof. The first inequality in (7.3.3) is proved in exactly the same manner as in Theorem 7.3.15.

Consider now u and v in $\mathbb{I}$ with $u \leq v$. For every $t \in [u, v]$, one has $Q(t, v) - Q(t, t) \leq v - t$, so that $Q(u, v) \leq Q(t, v) \leq v - t + \delta(t)$, from which, taking the minimum for t in $[u, v]$, one has

$$Q(u, v) \leq v - \max\{\hat{\delta}(t) : t \in [u, v]\}\,;$$

since one also has $Q(u, v) \leq \min\{u, v\} = u$ it follows that

$$Q(u, v) \leq \min\{u, v - \max\{\hat{\delta}(t) : t \in [u, v]\}\}\,,$$

namely $Q(u, v) \leq A_\delta(u, v)$. A similar argument holds for $u > v$, so that the assertion is completely proved. $\qquad\square$

Further readings 7.3.16. Fernández-Sánchez and Trutschnig [2015] gave necessary and sufficient conditions on the graph of δ that ensure that the quasi-copula A_δ is a copula (see also [Úbeda Flores, 2008] and [Nelsen et al., 2004; Klement and Kolesárová, 2007]). $\qquad\blacksquare$

7.4 Mass distribution associated with a quasi-copula

Given a d-quasi-copula Q and a box $B \subseteq \mathbb{I}^d$, one may consider the Q-volume of B in perfect analogy with the procedure used in the case of copulas (i.e., by using formula (1.2.4)). Roughly speaking, we refer to V_Q as the mass distribution associated with Q (on boxes), and to $V_Q(B)$ as the mass accumulated by Q on the box B.

Moreover, one can define a finitely additive set function μ_Q on the ring of $\mathscr{R}$ of finite disjoint unions of boxes, i.e., if $S = \cup_{i=1}^n B_i$, where the interiors of the boxes B_i do not overlap, then

$$\mu_Q(S) := \sum_{i=1}^{n} \mu_Q(B_i)\,.$$

If Q is a copula, then μ_Q can be extended to a measure on the Borel sets of $\mathbb{I}^d$, as seen, for instance, in the proof of Theorem 3.1.2.

Analogously, if Q is a proper quasi-copula, one may wonder whether μ_Q can be extended to a real measure on the Borel sets of $\mathbb{I}^d$; we recall that a *real measure*, often called a *signed measure*, $\mu : \mathscr{F} \to \mathbb{R}$ is a σ-additive set function defined on the measurable space $(\Omega, \mathscr{F})$ with the condition that $\mu(\emptyset) = 0$ and μ assumes at most one of the values ∞ and $-\infty$ (see, e.g., Halmos [1974]). Equivalently, μ is the difference between two (positive) measures μ_1 and μ_2 (defined on the same measure space), such that at least one of them is finite.

As a consequence of the definition of quasi-copulas, if Q induces a signed measure μ_Q, this measure should be d-fold stochastic, i.e. the image measure of μ_Q under any projection is equal to the Lebesgue measure λ. In particular, any d-fold stochastic signed measure μ on $\mathscr{B}(\mathbb{I}^d)$ should satisfy $\mu(\mathbb{I}^d) = 1$ and, in view of [Halmos, 1974, page 119], $|\mu(E)| < \infty$ for every $E \in \mathscr{B}(\mathbb{I}^d)$.

As we shall see below, the connexion between quasi-copulas and signed measures may not be true for every $d \geq 2$. To develop these investigations, it is convenient to start with preliminary results about the volume associated with a 2-quasi-copula.

As in the case of copulas, any quasi-copula Q may be characterised in terms of properties of the Q-volume of (suitable) rectangles of $\mathbb{I}^d$. This characterisation has been given by Rodríguez-Lallena and Úbeda Flores [2009] for the general case. The analogous characterisation of 2-quasi-copulas was first proved by Genest et al. [1999].

Theorem 7.4.1. *For a function* $Q : \mathbb{I}^2 \to \mathbb{I}$ *the following statements are equivalent:*

(a) Q *is a 2-quasi-copula;*

(b) Q *satisfies the boundary conditions, namely properties* (a) *and* (b) *of Theorem 1.4.1, and it satisfies*

$$V_Q\left([u, u'] \times [v, v']\right) = Q(u', v') - Q(u', v) - Q(u, v') + Q(u, v) \geq 0, \ (7.4.1)$$

whenever $u \leq u'$ *and* $v \leq v'$, *and at least one of* u, u', v *and* v' *is equal to either* 0 *or* 1.

Proof. (a) $\Longrightarrow$ (b) Since a quasi-copula satisfies the same boundary conditions of a copula, it is enough to consider what happens when one of the following holds: $u = 0$, $u' = 1$, $v = 0$ or $v' = 1$.

If $u = 0$, then (7.4.1) reduces to $Q(u', v') \geq Q(u', v)$, namely property (b) of the characterisation Theorem 7.2.4. Next, suppose $u' = 1$; it must then be shown that $Q(u, v) + v' \geq Q(u, v') + v$, which is an immediate consequence of the Lipschitz condition. Similar arguments may be invoked when $v = 0$ or $v' = 1$.

(b) $\Longrightarrow$ (a) One has to show that properties (b) and (c) of Theorem 7.2.4 may be deduced from the boundary conditions of Theorem 1.4.1 and (7.4.1). In order to prove that $s \mapsto Q(u, v)$ is increasing for every $t \in \mathbb{I}$, notice that, for all s and s' in $\mathbb{I}$ with $s \leq s'$, the mass associated to the rectangle $[s, s'] \times [0, t]$ is positive, whence, by Theorem 1.4.1, one has $Q(s', t) \geq Q(s, t)$. The proof that $t \mapsto Q(s, t)$ is increasing for every $s \in \mathbb{I}$ is analogous.

Finally, fix $t \in \mathbb{I}$ and look at the mass assigned by Q to the rectangle $[s, s'] \times [t, 1]$, with $0 \leq s \leq s' \leq 1$. By assumption, one has

$$Q(s', t) - Q(s, t) \leq s' - s,$$

which shows that $(s, t) \mapsto Q(s, t)$ satisfies the Lipschitz condition in s for every given $t \in \mathbb{I}$. Arguing similarly for $t \mapsto Q(s, t)$, one may than deduce that

$$|Q(s', t') - Q(s, t)| \leq |Q(s', t') - Q(s, t')| + |Q(s, t') - Q(s, t)|$$
$$\leq |s' - s| + |t' - t|,$$

which is precisely condition (c) of Theorem 7.2.4. $\qquad\square$

Thus, a quasi-copula $Q \in \mathcal{Q}_2$ assigns positive mass to every rectangle $[s, s'] \times$

$[t, t']$ with the property that at least one of its sides lies on a side of the unit square. However, the Q-volume of a rectangle entirely contained in the interior of the unit square $\mathbb{I}^2$ may be negative, as shown by the next example.

Example 7.4.2. For every $(u, v) \in \mathbb{I}^2$, let $Q(u, v) = \int_{[0,u] \times [0,v]} q \, d\lambda_2$, where q is the function so defined: $q = 3$ on the rectangles $R_1, \ldots, R_4$ of Figure 7.1, $q = -3$ on R_5, while $q = 0$, elsewhere. Then it can be proved that Q is a quasi-copula. Moreover, $V_Q([1/3, 2/3]^2) = -1/3$. ∎

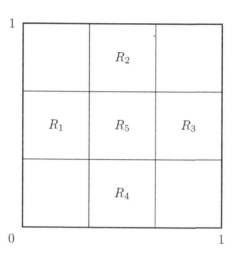

Figure 7.1: *The function q defined in Example 7.4.2.*

The following result provides information about the bounds of $V_Q(R)$.

Theorem 7.4.3. *For every 2-quasi-copula Q and for every rectangle R contained in* $\mathbb{I}^2$, *one has*

$$-\frac{1}{3} \leq V_Q(R) \leq 1.$$

Proof. Let $R = [u_1, u_2] \times [v_1, v_2]$; then the Q-volume of R is given by

$$V_Q(R) = Q(u_2, v_2) - Q(u_2, v_1) - Q(u_1, v_2) + Q(u_1, v_1).$$

The inequalities

$$Q(u_2, v_2) - Q(u_2, v_1) \leq v_2 - v_1 \leq 1 \quad \text{and} \quad -Q(u_1, v_2) + Q(u_1, v_1) \leq 0$$

yield $V_Q(R) \leq 1$.

Because of Theorem 7.2.7, if at least one of u_1, u_2, v_1 and v_2 equals either 0 or 1, then $V_Q(R) \geq 0$. Therefore, assume $0 = u_0 < u_1 < u_2 < u_3 = 1$ and $0 = v_0 < v_1 < v_2 < v_3 = 1$ and divide the unit square $\mathbb{I}^2$ into the nine rectangles

$$R_{ij} := [u_{i-1}, u_i] \times [v_{j-1}, v_j] \qquad i, j = 1, 2, 3.$$

Take now $R = R_{22}$ as the given rectangle R; Theorem 7.4.1 ensures that $\nu_{ij} \geq 0$, where $\nu_{ij} := V_Q(R_{ij})$, if $(i, j) \neq (2, 2)$ and that $\nu_{12} + \nu_{22} \geq 0$ and $\nu_{22} + \nu_{32} \geq 0$.

Assume, if possible, $\nu_{22} < -1/3$; then, $\nu_{12} \geq -\nu_{22} > 1/3$ and $\nu_{32} \geq -\nu_{22} > 1/3$. Hence, $u_1 > 1/3$ and $1 - u_2 > 1/3$, which implies $u_2 - u_1 < 1/3$. On the other hand

$$\nu_{22} = u_2 - u_1 - \nu_{23} - \nu_{21}.$$

Since $\nu_{23} \leq \min\{1 - v_2, u_2 - u_1\}$ and $\nu_{21} \leq \min\{v_1, u_2 - u_1\}$, one has

$$\nu_{22} \geq u_2 - u_1 - \min\{1 - v_2, u_2 - u_1\} - \min\{v_1, u_2 - u_1\} \geq -(u_2 - u_1) > 1/3,$$

a contradiction; thus $\nu_{22} \geq -1/3$, which concludes the proof. $\qquad\square$

The following theorem complements the previous one.

Theorem 7.4.4. *For a 2-quasi-copula Q and a rectangle $R = [u_1, u_2] \times [v_1, v_2]$ contained in $\mathbb{I}^2$, one has*

(a) $V_Q(R) = 1$ *if, and only if,* $R = \mathbb{I}^2$;

(b) *if* $V_Q(R) = -1/3$, *then* $R = [1/3, 2/3]^2$.

Proof. (a) It is immediate that $V_Q(\mathbb{I}^2) = 1$. Conversely, since

$$V_Q(R) \leq \min\{u_2 - u_1, v_2 - v_1\},$$

one has $V_Q(R) < 1$ if $R \neq \mathbb{I}^2$.

(b) Assume $V_Q(R) = -1/3$; then, as in the previous theorem, and with the same notation, $u_1 \geq 1/3$ and $u_2 \leq 2/3$, which implies $u_2 - u_1 \leq 1/3$. As above

$$v_{22} = -1/3 \geq u_2 - u_1 - \min\{1 - v_2, u_2 - u_1\} - \min\{v_1, u_2 - u_1\} \geq -(u_2 - u_1),$$

which yields $u_2 - u_1 \geq 1/3$; thus $u_2 - u_1 = 1/3$ and, as consequence, $u_1 = 1/3$ and $u_2 = 2/3$. Similarly, one proves that $v_1 = 1/3$ and $v_2 = 2/3$. $\qquad\square$

Remark 7.4.5. Summarising one may say that, for $d = 2$, there exists a unique rectangle on which the minimal mass (which turns out to be $-1/3$) can be spread, as well as a unique rectangle (the unit square itself) on which the maximal mass 1 can be accumulated. The situation is slightly different in higher dimensions. For the case $d = 3$, De Baets et al. [2007] showed that there still exists a unique 3-box on which the minimal mass (which now turns out to be $-4/5$) can be spread, while there exist multiple 3-boxes on which the maximal mass 1 can be spread. In principle, the methodology exposed can be applied to the case $d > 3$ as well (apart from an increasing complexity of the formulation). $\qquad\blacksquare$

Thus, in general, the area of rectangles contained in $\mathbb{I}^2$ with given Q-volume is subject to specific bounds.

Theorem 7.4.6. *Let $R = [u_1, u_2] \times [v_1, v_2]$ be a rectangle contained in $\mathbb{I}^2$ and let θ be in $[-1/3, 1]$. If there is a 2-quasi-copula Q for which $V_Q(R) = \theta$, then the area $A(R)$ of R satisfies*

$$\theta^2 \le A(R) \le \left(\frac{1+\theta}{2}\right)^2. \tag{7.4.2}$$

Moreover, when $A(R)$ attains either bound, then R is necessarily a square.

Proof. Use the same notation as in the proof of Theorem 7.4.3 and assume $\nu_{22} = \theta$. We look for the extrema of the product $A(R) = (u_2 - u_1)(v_2 - v_1)$. Since $\nu_{ij} \ge 0$ if (i, j) is different from $(2, 2)$, one has $(u_2 - u_1) + (v_2 - v_1) \le 1 + \theta$, so that the arithmetic-geometric mean inequality implies

$$(u_2 - u_1)(v_2 - v_1) \le \left(\frac{1+\theta}{2}\right)^2,$$

which establishes the upper bound in (7.4.2).

Assume $\theta \ge 0$; then $u_2 - u_1 \ge \theta$ and $v_2 - v_1 \ge \theta$, whence $(u_2 - u_1)(v_2 - v_1) \ge \theta^2$. For $\theta \in [-1/3, 0[$, one has $\nu_{21} \ge -\theta$ and $\nu_{23} \ge -\theta$, and hence $u_2 - u_1 \ge -\theta$. Similarly, $v_2 - v_1 \ge -\theta$, so that again $(u_2 - u_1)(v_2 - v_1) \ge \theta^2$.

Equality holds only if $u_2 - u_1 = v_2 - v_1$. $\square$

The following examples show that the upper and lower bounds in (7.4.2) may be attained for every $\theta \in [-1/3, 1]$.

Example 7.4.7. Let θ belong to $]0, 1[$ and consider the 2-copula C obtained by spreading the mass θ uniformly on the segment with endpoints $(0, \theta)$ and $(\theta, 0)$ and the mass $1 - \theta$ uniformly on the segment with endpoints $(\theta, 1)$ and $(1, \theta)$. Then one has

$$V_C([0, \alpha]^2) = \theta,$$

for every $\alpha \in \left[\theta, \frac{1+\theta}{2}\right]$. But one also has that the area of $[0, \alpha]^2$ is

$$A([0, \alpha]^2) = \alpha^2 \in \left[\theta^2, \left(\frac{1+\theta}{2}\right)^2\right].$$

■

Example 7.4.8. Let θ be in $[-1/3, 0[$ and consider the 2-quasi-copula Q obtained by spreading uniformly the mass

- θ on the segment with endpoints $\left(\frac{1+\theta}{2}, \frac{1-\theta}{2}\right)$ and $\left(\frac{1-\theta}{2}, \frac{1+\theta}{2}\right)$;
- $\frac{1-\theta}{2}$ on the segment with endpoints $\left(0, \frac{1-\theta}{2}\right)$ and $\left(\frac{1-\theta}{2}, 0\right)$;
- $\frac{1-\theta}{2}$ on the segment with endpoints $\left(\frac{1+\theta}{2}, 1\right)$ and $\left(1, \frac{1+\theta}{2}\right)$.

Then one has

$$V_Q\left(\left[\frac{1-\alpha}{2}, \frac{1+\alpha}{2}\right]^2\right) = \theta,$$

for every $\alpha^2 \in \left[\theta^2, \left(\frac{1+\theta}{2}\right)^2\right]$. But one also has that the area of $\left[\frac{1-\alpha}{2}, \frac{1+\alpha}{2}\right]^2$ is

$$A\left(\left[\frac{1-\alpha}{2}, \frac{1+\alpha}{2}\right]^2\right) = \alpha^2 \in \left[\theta^2, \left(\frac{1+\theta}{2}\right)^2\right].$$

∎

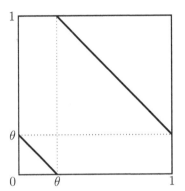 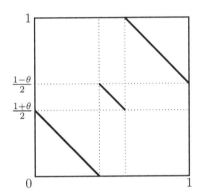

Figure 7.2 *Mass distribution of the copula C defined in Example 7.4.7 (left) and mass distribution of the quasi-copula Q defined in Example 7.4.8 (right).*

In Chapter 4, we showed that any copula can be approximated by special classes of copulas. It is possible to extend this result to the class of quasi-copulas; but it is interesting to observe that the approximating quasi-copulas may have a negative mass that is as large as one wishes. The first result deals with a special case, the approximation of the copula Π_2.

Theorem 7.4.9. *For all given $\varepsilon > 0$ and $H > 0$, there exist a 2-quasi-copula Q and a Borel subset S of $\mathbb{I}^2$ such that*

(a) $\mu_Q(S) < -H$;

(b) *for all u and v in $\mathbb{I}$, $|Q(u,v) - \Pi_2(u,v)| < \varepsilon$.*

Proof. Let the odd number $m \in \mathbb{N}$ be such that

$$m \geq \frac{4}{\varepsilon} \quad \text{and} \quad \frac{(m-1)^2}{4m} > H.$$

Divide the unit square $\mathbb{I}^2$ into the m^2 squares

$$R_{ij} := \left[\frac{i-1}{m}, \frac{i}{m}\right] \times \left[\frac{j-1}{m}, \frac{j}{m}\right] \qquad (i, j = 1, 2 \ldots, m).$$

In its turn each such square is divided into the m^2 squares

$$R_{ijkl} := \left[\frac{(i-1)m+k-1}{m^2}, \frac{(i-1)m+k}{m^2}\right] \times \left[\frac{(j-1)m+l-1}{m^2}, \frac{(j-1)m+l}{m^2}\right],$$

where $k, l = 1, 2, \ldots, m$. The unit mass is then distributed on $\mathbb{I}^2$ in the following manner.

Let $r = (m+1)/2$; for every (i, j) $(i, j = 1, 2 \ldots, m)$, spread

- a mass equal to $1/m^3$ uniformly on each R_{ijkl} with $k = 1, 2, \ldots, r$ and $l = r - k + 1, r - k + 2, \ldots, r + k - 1$;
- a mass equal to $-1/m^3$ uniformly on each R_{ijkl} with $k = 2, 3, \ldots, r$ and $l = r - k + 2, r - k + 3, \ldots, r + k - 2$;
- no mass in the remaining squares R_{ijkl} with $k = 1, 2, \ldots, r$;
- mass symmetrically on the squares R_{ijkl} with $k > r$, so that the mass on R_{ijkl} with $k > r$ equals that on $R_{ij(m+1-k)l}$.

The positive mass spread on the square R_{ij} is therefore

$$[2(1 + 2 + \cdots + r - 1) + r]\frac{1}{m^3} = \frac{r^2}{m^3} = \frac{(m+1)^2}{4m^3},$$

while the negative mass on the square R_{ij} is

$$[2(1 + 2 + \cdots + r - 2) + r - 1]\left(-\frac{1}{m^3}\right) = -\frac{(m-1)^2}{4m^3}.$$

As a consequence the total mass spread on the set

$$S := \bigcup \{R_{ijkl} : \mu(R_{ijkl}) = -1/m^3\}$$

is

$$-\frac{(m-1)^2}{4m} < -H.$$

For $(u, v) \in \mathbb{I}^2$, define $Q(u, v) := \mu([0, u] \times [0, v])$. Then, in view of Theorem 7.4.1, Q is a quasi copula. Moreover $\mu_Q(R_{ij}) = 1/m^2 = \mu_{\Pi_2}(R_{ij})$ for all i and j. Thus, for $i, j = 1, 2, \ldots, m$ one has

$$Q\left(\frac{i}{m}, \frac{j}{m}\right) = \Pi_2\left(\frac{i}{m}, \frac{j}{m}\right).$$

For every (u, v) in $\mathbb{I}^2$ there are indices i and j with $i, j = 1, 2, \ldots, m$ such that

$$\left|u - \frac{i}{m}\right| < \frac{1}{m} \quad \text{and} \quad \left|v - \frac{j}{m}\right| < \frac{1}{m}.$$

Finally

$$
|Q(u,v) - \Pi_2(u,v)| \leq \left| Q(u,v) - Q\left(\frac{i}{m},\frac{j}{m}\right) \right|
$$
$$
+ \left| Q\left(\frac{i}{m},\frac{j}{m}\right) - \Pi_2\left(\frac{i}{m},\frac{j}{m}\right) \right| + \left| \Pi_2\left(\frac{i}{m},\frac{j}{m}\right) - \Pi_2(u,v) \right|
$$
$$
\leq 2\left| u - \frac{i}{m} \right| + 2\left| v - \frac{j}{m} \right| < \frac{4}{m} < \varepsilon,
$$

which concludes the proof.   □

We are now in the position to give the result announced above.

Theorem 7.4.10. *For all given $\varepsilon > 0$ and $H > 0$, and for every 2-quasi-copula $\widetilde{Q}$, there exist a 2-quasi-copula Q and a subset S of $\mathbb{I}^2$ such that*

(a) $\mu_Q(S) < -H$;

(b) *for all u and v in $\mathbb{I}$, $|Q(u,v) - \widetilde{Q}(u,v)| < \varepsilon$.*

Proof. Let m and the squares R_{ij} and R_{ijkl} be as in the previous theorem. For all i and j, set $q_{ij} := V_{\widetilde{Q}}(R_{ij})$. Because of Theorem 7.2.7, one has $q_{ij} \geq 0$ whenever either i or j is equal to either 1 or m. Moreover, $\sum_{i,j=1}^{m} q_{ij} = 1$.

In the square R_{ij}, spread the mass on the subsquares R_{ijkl} as in the previous theorem, but replace the values $1/m^3$ and $-1/m^3$ by q_{ij}/m^3 and $-q_{ij}/m^3$, respectively.

If $q_{ij} > 0$, the positive and negative masses spread on R_{ij} are

$$
\frac{q_{ij}(m+1)^2}{4m} \quad \text{and} \quad -\frac{q_{ij}(m-1)^2}{4m},
$$

respectively. If $q_{ij} < 0$, the positive and negative masses spread on R_{ij} are

$$
-\frac{q_{ij}(m-1)^2}{4m} \quad \text{and} \quad \frac{q_{ij}(m+1)^2}{4m},
$$

respectively. If $q_{ij} = 0$, then no mass is spread on R_{ij}.

For $(u,v) \in \mathbb{I}^2$, define $Q(u,v) := \mu([0,u] \times [0,v])$. Then, in view of Theorem 7.4.1, Q is a quasi-copula. Set

$$
S := \bigcup \{ R_{ijkl} : \mu(R_{ijkl}) < 0 \} .
$$

It is possible to evaluate

$$
\mu_Q(S) = \sum_{q_{ij}<0} q_{ij}\frac{(m+1)^2}{4m} + \sum_{q_{ij}>0}\left(-q_{ij}\frac{(m-1)^2}{4m}\right)
$$
$$
\leq -\frac{(m-1)^2}{4m}\sum_{i,j=1}^{m} q_{ij} = -\frac{(m-1)^2}{4m} < -H .
$$

Part (b) is proved in a manner very similar to that of the analogous part in Theorem 7.4.9.   □

Now, we are able to investigate a possible connexion between quasi-copulas and signed measures. For $d = 2$, the following result holds.

Theorem 7.4.11. *A 2-quasi-copula Q exists that does not induce a doubly stochastic signed measure on $\mathbb{I}^2$.*

Proof. For every $n \in \mathbb{N}$ let Q_n be a quasi-copula that admits a set whose Q_n-volume is less than -2^{n+1} (the existence of such a Q_n is guaranteed by Theorem 7.4.9). Consider the ordinal sum

$$Q = \left(\left\langle \left] \tfrac{2^n-1}{2^n}, \tfrac{2^{n+1}-1}{2^{n+1}} \right[, Q_n \right\rangle \right)_{n \in \mathbb{N}}.$$

It follows that, on the square $\left[\tfrac{2^n-1}{2^n}, \tfrac{2^{n+1}-1}{2^{n+1}} \right]^2$, there exists a set with Q-volume less than -1. Thus, on the square $\left[0, \tfrac{2^{n+1}-1}{2^{n+1}} \right]^2$, there exists a set with Q-volume less than $-n$. Thus, the negative mass is not bounded, which implies that Q cannot generate a doubly stochastic signed measure. $\qquad \square$

In the general case, the connexion between quasi-copulas and signed measures was investigated by Nelsen [2005]; Nelsen et al. [2010]. Here, we report the short proof presented by Fernández-Sánchez and Úbeda Flores [2014].

Theorem 7.4.12. *For every $d \geq 3$, W_d does not induce a d-fold stochastic signed measure on $\mathbb{I}^d$.*

Proof. First, suppose, if possible, that W_3 induces a signed measure μ_{W_3} on $\mathbb{I}^3$. Now, the mass distribution of W_3 is spread on the triangle T_1 whose vertices are $A = (1,1,0)$, $B = (1,0,1)$ and $C = (0,1,1)$. Let T_2 be the triangle whose vertices are $A' = (1,1,1)$, B and C, and consider the projection $\pi : T_1 \to T_2$ given by $\pi(u_1, u_2, u_3) = (u_1, u_2, 1)$. Since $W_3(u_1, u_2, 1) = W_2(u_1, u_2)$, it follows that π is an isomorphism between measurable spaces. Thus, $\mu_{W_3}(S) = \mu_{W_2}(\pi(S))$ for every Borel set S with $S \subseteq T_1$. Since W_2 is a copula whose mass is spread along the segment $\overline{CB}$ which joins the points C and B, and $\pi(\overline{CB}) = \overline{CB}$, then the mass of W_3 is distributed along the segment $\overline{CB}$. Thus $W_3(u_1, u_2, u_3) = 0$ when $u_3 < 1$, which is a contradiction.

Now, suppose, if possible, that W_d induces a signed measure on $\mathbb{I}^d$ for every $d \geq 4$. Then, each of its 3-margins, which are W_3, induces a signed measure on $\mathbb{I}^3$, which contradicts the previous part of the proof. $\qquad \square$

Further readings 7.4.13. For more details about the results reported in this section we refer to Nelsen et al. [2002]. The proof of Theorem 7.4.11 was communicated to the authors by J. Fernández-Sánchez. Another proof can be found in [Fernández-Sánchez et al., 2011b]. $\qquad \blacksquare$

Chapter 8

Generalisations of copulas: semi-copulas

The definition of a semi-copula extends that of a copula; semi-copulas have appeared in contexts like reliability theory [Bassan and Spizzichino, 2005b], fuzzy set theory and multivalued logic [Durante et al., 2007a; Hájek and Mesiar, 2008], non-additive measures and integrals [Klement et al., 2010], and probabilistic metric spaces [Saminger-Platz and Sempi, 2008, 2010].

Specifically, Section 8.1 presents the main properties of semi-copulas. Bivariate semi-copulas are discussed in detail in Section 8.2. The link between semi-copulas and capacities (i.e. generalised probabilities) is underlined in Section 8.3, where a Sklar-type theorem for capacities is also presented. The remaining part of the chapter (Sections 8.5 and 8.6) is devoted to the use of semi-copulas for describing notions of multivariate ageing for vectors of exchangeable lifetimes.

8.1 Definition and basic properties

We start with the definition of a semi-copula.

Definition 8.1.1. A function $S : \mathbb{I}^d \to \mathbb{I}$ is called (d-dimensional) *semi-copula* if it satisfies the following properties:

(a) for every $j \in \{1, \ldots, d\}$, $S(1, \ldots, 1, u_j, 1, \ldots, 1) = u_j$;

(b) S is increasing in each argument, i.e., for every $j \in \{1, \ldots, d\}$ and for all u_1 , $\ldots, u_{j-1}, u_{j+1}, \ldots u_d \in \mathbb{I}$, $t \mapsto S(\mathbf{u}_j(t))$ is increasing on $\mathbb{I}$. $\diamondsuit$

Notice that properties (a) and (b) of Definition 8.1.1 together imply $S(\mathbf{u}) = 0$ when at least one of the components of $\mathbf{u}$ vanishes.

The term "semi-copula" has been used in order to stress the fact that semi-copulas share some (but not all!) of the properties that characterise a copula, as can be immediately seen. Notice that the definition of a semi-copula also generalises that of a quasi-copula and that a semi-copula need not be continuous. For instance, the so-called *drastic semi-copula*

$$Z_d(\mathbf{u}) = \begin{cases} 0, & \mathbf{u} \in]0, 1[^d , \\ \min\{u_1, \ldots, u_d\}, & \text{otherwise,} \end{cases}$$

is not continuous on the boundary of $\mathbb{I}^d$.

Example 8.1.2. The function $S : \mathbb{I}^2 \to \mathbb{I}$ defined by

$$S(x, y) := \begin{cases} xy^2, & x \leq y, \\ x^2 y, & \text{elsewhere,} \end{cases}$$

is a continuous semi-copula, but not a quasi-copula. In fact, one has

$$S_2(8/10, 9/10) - S_2(8/10, 8/10) = 136/1000 > 1/10;$$

thus S is not 1-Lipschitz.   ∎

If S is a d-semi-copula, then any m-marginal, $2 \leq m \leq (d-1)$ of S is an m-semi-copula.

Let $\mathscr{S}_d$ denote the space of d-semi-copulas. As is easily seen, $\mathscr{S}_d$ is a subset of the space $X(\mathbb{I}^d)$ of all the functions on $\mathbb{I}^d$ with values in $[0, 1]$. This is in contrast with what happens for both copulas and quasi-copulas that are continuous. Moreover, the following result holds.

Theorem 8.1.3. *The class $\mathscr{S}_d$ of semi-copulas is a convex and compact (under the topology of pointwise convergence) subset of $X(\mathbb{I}^d)$.*

Proof. Convexity is easily established. Since $X(\mathbb{I}^d)$ is a product of compact spaces, it is well known from the Tychonoff's theorem (see, e.g., [Kelley, 1955]) that $X(\mathbb{I}^d)$ is compact. The proof will be completed by showing that $\mathscr{S}_d$ is a closed subset of $X(\mathbb{I}^d)$, namely, that, given a sequence $(S_n)_{n \in \mathbb{N}}$ in $\mathscr{S}_d$, if S_n converges pointwise to S, then S belongs to $\mathscr{S}_d$. Now, it can be easily shown that the limit S satisfies properties (a) and (b) of Definition 8.1.1, which yields the desired conclusion.   □

Now, as in the case of copulas, one can introduce a pointwise order, denoted by $\leq$, in $\mathscr{S}_d$. Moreover, for all semi-copulas S_1 and S_2 in $\mathscr{S}_d$, let $S_1 \vee S_2$ and $S_1 \wedge S_2$ denote, respectively, the pointwise join and the meet of S_1 and S_2. The relevant result is the following

Theorem 8.1.4. *The set $\mathscr{S}_d$ is a complete lattice, i.e. both the upper and the lower bounds, $\bigvee \mathbf{S}$ and $\bigwedge \mathbf{S}$, of every set $\mathbf{S}$ of d-semi-copulas are d-semi-copulas, viz., $\bigvee \mathbf{S} \in \mathscr{S}_d$ and $\bigwedge \mathbf{S} \in \mathscr{S}_d$. Moreover, for every $S \in \mathscr{S}_d$ and for every $\mathbf{u} \in \mathbb{I}^d$, one has*

$$Z_d(\mathbf{u}) \leq S(\mathbf{u}) \leq \min\{u_1, \dots, u_d\}. \tag{8.1.1}$$

Proof. It can be easily proved that $\mathscr{S}_d$ is a complete lattice, since properties (S1) and (S2) are preserved under the operations of taking supremum and infimum. In order to prove eq. (8.1.1), consider $S \in \mathscr{S}_d$. For every $\mathbf{u} \in \mathbb{I}^d$ and for every $i \in \{1, \dots, d\}$, one has $S(\mathbf{u}) \leq S(1, \dots, 1, u_i, 1, \dots, 1) = u_i$, from which it follows that $S(\mathbf{u}) \leq \min\{u_1, \dots, u_d\}$. Since it is obvious that $S(\mathbf{u}) \geq Z_d(\mathbf{u})$, the desired assertion follows.   □

8.2 Bivariate semi-copulas, triangular norms and fuzzy logic

In this section, we present aspects of semi-copulas that are related to several concepts that have played an important role in fuzzy set theory and fuzzy logic. Loosely speaking, a fuzzy logic is usually considered as a multivalued propositional logic in which the class of truth values is modelled by the unit interval $\mathbb{I}$, and which forms an extension of the classical Boolean logic (see, e.g., Hájek [1998]). In these logics, a key role is played by the so-called *conjunction*, i.e., a binary operation on $\mathbb{I}$ that is used to extend the Boolean conjunction from $\{0, 1\}$ to $[0, 1]$. In other words, a conjunction is any operation S on $\mathbb{I}$ such that its restriction to $\{0, 1\} \times \{0, 1\}$ coincides with the truth table of the Boolean conjunction.

Now, it is not difficult to check that any bivariate semi-copula S is a conjunction that is, in addition, increasing in each variable and admits neutral element 1, i.e. $S(u, 1) = S(1, u) = u$ for all $u \in \mathbb{I}$. For this reason, (special) semi-copulas have played the role of conjunction in multivalued logics. For instance, the three basic semi-copulas (which are also copulas), namely W_2, M_2 and Π_2, have been used to generate the so-called Łukasiewicz, Gödel and the product logic (see [Hájek, 1998] and the references therein).

It should be stressed that, in order to provide logical structures that exhibit interesting features, further requirements on the conjunction have been imposed in the literature. To this end, a subclass of semi-copulas that is frequently encountered in the literature, both in theory and in applications, is that of triangular norms defined below.

Definition 8.2.1. A binary operation T on $\mathbb{I}$, i.e., a function $T : \mathbb{I}^2 \to \mathbb{I}$, is said to be *associative* if, for all s, t and u in $\mathbb{I}$,

$$T\left(T(s, t), u\right) = T\left(s, T(t, u)\right) . \tag{8.2.1}$$

It is said to be *commutative* if $T(s, t) = T(t, s)$ for all $(s, t) \in \mathbb{I}^2$.

A *triangular norm*, or, briefly, a *t-norm* T, is a semi-copula $T : \mathbb{I}^2 \to \mathbb{I}$ that is both associative and commutative. $\diamond$

Historically, triangular norms have been an indispensable tool for the interpretation of the conjunction in fuzzy logics and, subsequently, for the intersection of fuzzy sets (see [Zadeh, 1965]). They were first introduced in the context of probabilistic metric spaces (see, e.g., [Schweizer and Sklar, 1983]), whose presentation was based on Menger's ideas [Menger, 1942].

Further readings 8.2.2. It is worth mentioning that in many practical applications of fuzzy logic one needs more flexibility in the choice of the conjunction: in particular, the associativity and/or the commutativity of a conjunction may be relaxed [Suárez-García and Gil-Álvarez, 1986; Fodor and Keresztfalvi, 1995]. For instance, associative (but not necessarily commutative) semi-copulas have been used by Flondor et al. [2001]. These operations are usually called *pseudo triangular norms*. For other considerations about non-associative logics, see also the works by Durante et al. [2007a] and Hájek and Mesiar [2008].   ■

The copulas W_2, Π_2 and M_2 are also t-norms, as well as the drastic t-norm T_D of Example 8.2.13. However, not all semi-copulas are triangular norms.

Example 8.2.3. The function $S : \mathbb{I}^2 \to \mathbb{I}$ given by $S(u,v) = uv \max\{u,v\}$ is a semi-copula, but not a t-norm, because it is not associative.   ∎

Example 8.2.4. The 2-copula C defined by

$$C(u,v) = uv + uv\,(1-u)\,(1-v)$$

is commutative but is not a t-norm since it is not associative.   ∎

Triangular norms that are copulas were characterised by Moynihan [1978].

Theorem 8.2.5. *For a t-norm T the following statements are equivalent:*

(a) *T is a 2-copula;*

(b) *T satisfies the Lipschitz condition:*

$$T(x',y) - T(x,y) \leq x' - x \tag{8.2.2}$$

for all x, x' and y in $\mathbb{I}$ with $x \leq x'$.

Proof. Since any copula satisfies the Lipschitz condition with constant 1, only the reverse implication (b) $\Longrightarrow$ (a) has to be proved. Thus assume (8.2.2) for T. Since T is symmetric, (8.2.2) implies that T is continuous. Choose x, x', y and y' in $\mathbb{I}$ with $x \leq x'$ and $y \leq y'$. Thus, as $T(0,y') = 0$ and $T(1,y') = y'$, there exists $s \in \mathbb{I}$ such that $T(s,y') = y$. On account of the symmetry and the associativity of T, one has

$$\begin{aligned}
T(x',y) - T(x,y) &= T\left(x', T(s,y')\right) - T\left(x, T(s,y')\right) \\
&= T\left(T(x',y'), s\right) - T\left(T(x,y'), s\right) \leq T(x',y') - T(x,y'),
\end{aligned}$$

which shows that T is 2-increasing and, hence, a copula.   □

In the language of algebra, T is a t-norm if, and only if, $(\mathbb{I}, T, \leq)$ is a fully ordered commutative semigroup with neutral element 1 and annihilator (zero element) 0. Therefore, it is natural to consider additional algebraic properties a t-norm may have. The Archimedean property is particularly important and has therefore received special attention in the literature [Schweizer and Sklar, 1983; Klement et al., 2005].

Definition 8.2.6. Let T be an associative operation on $\mathbb{I}$. The T-*powers* of an element $x \in \mathbb{I}$ under T are defined recursively by

$$x_T^{(1)} := x \quad \text{and} \quad x_T^{(n+1)} := T\left(x_T^{(n)}, x\right), \tag{8.2.3}$$

for every $n \in \mathbb{N}$.   ◇

Definition 8.2.7. Let T be an associative 2-semi-copula on $\mathbb{I}$. Then T is said to be *Archimedean* if, for all x and y in $]0, 1[$, there is $n \in \mathbb{N}$ such that $x_T^{(n)} < y$. ◇

Intuitively, the Archimedean property is the translation in a semigroup setting of the Archimedean property of reals, stating that for any two positive reals x, y there exists $n \in \mathbb{N}$ such that $nx > y$.

The following theorem originally proved by Ling [1965] is of paramount importance in the theory of t-norms, since it characterises the class of continuous and Archimedean t-norms in terms of univariate functions, called *Archimedean generators* that are defined below.

Definition 8.2.8. A decreasing and continuous function $\psi : [0, +\infty] \to \mathbb{I}$ which satisfies the conditions $\psi(0) = 1$, $\psi(+\infty) = 0$ and is strictly decreasing on $[0, \inf\{t : \psi(t) = 0\}]$ is called an *Archimedean generator*. Its quasi-inverse $\psi^{(-1)} : \mathbb{I} \to \overline{\mathbb{R}}_+$ is given by $\psi^{(-1)}(t) = \psi^{-1}(t)$ for $t \in]0, 1]$, while $\psi^{-1}(0) = \inf\{t : \psi(t) = 0\}$. ◇

Theorem 8.2.9 (Representation of continuous and Archimedean t-norms). *The following statements are equivalent for a t-norm T:*

(a) *T is continuous and Archimedean;*

(b) *T admits the representation*

$$T(x, y) = \psi\left(\psi^{(-1)}(x) + \psi^{(-1)}(y)\right), \tag{8.2.4}$$

where $\psi : \overline{\mathbb{R}}_+ \to \mathbb{I}$ is an Archimedean generator.

An Archimedean generator is not unique since, if ψ is a generator of T, and $\psi^{(-1)}$ its quasi-inverse in the same representation, so are the functions ψ_1 and $\psi_1^{(-1)}$ defined, for $x \in \mathbb{I}$, by

$$\psi_1(x) := \psi(\alpha x), \qquad \text{and} \qquad \psi_1^{(-1)}(x) := \psi^{(-1)}(x)/\alpha,$$

with $\alpha > 0$, as is immediately checked.

Remark 8.2.10. The representation (8.2.4) is also presented in the literature in the equivalent form

$$T(x, y) = \varphi^{(-1)}(\varphi(x) + \varphi(y)), \tag{8.2.5}$$

where $\varphi : \mathbb{I} \to [0, +\infty]$ is strictly decreasing with $\varphi(1) = 0$, and its quasi-inverse $\varphi^{(-1)} : [0, +\infty] \to \mathbb{I}$ is given by $\varphi^{(-1)}(t) = \varphi^{-1}(t)$ for $t \in [0, \varphi(0)]$, while $\varphi(t) = 0$, elsewhere. In the following, we will adopt either representations depending on the context.

Remark 8.2.11. By recourse to the transformations

$$h(t) := \psi(-\ln t) \qquad \text{and} \qquad k(t) := \exp\left(-\psi^{(-1)}(t)\right),$$

one can write eq. (8.2.4) as

$$T(u, v) = h(k(u)\, k(v))\,. \tag{8.2.6}$$

This latter representation is usually referred to as the *multiplicative representation* of a continuous and Archimedean t-norm T, while the representation (8.2.4) is called the *additive representation* of T. In the same vein, the function h of (8.2.6) is called the *multiplicative Archimedean generator*, while the function ψ of (8.2.4) is also called the *additive Archimedean generator*.

The class of continuous Archimedean t-norms can be divided in two subclasses by means of a simple algebraic property that is recalled below.

Definition 8.2.12. An element $a \in\,]0, 1[$ is said to be a *nilpotent element* of the t-norm T if there exists $n \in \mathbb{N}$ such that $a_T^{(n)} = 0$. $\diamond$

A continuous and Archimedean t-norm T that admits a nilpotent element is said to be *nilpotent*. Otherwise, it is said to be *strict*, since it can be proved that, for every $x \in\,]0, 1[, t \mapsto T(x, t)$ and $t \mapsto T(t, x)$ are strictly increasing on $]0, 1[$.

The t-norms Π_2 and W_2 provide examples of a strict and of a nilpotent t-norm, respectively. While it is immediately seen that Π_2 is strict, in order to see that W_2 is nilpotent, one easily calculates, for $a \in\,]0, 1[$,

$$a_{W_2}^{(n)} = \max\{na - (n - 1), 0\}\,,$$

so that $a_{W_2}^{(n)} = 0$ for $n \geq 1/(1 - a)$.

The distinction between nilpotent and strict t-norms can be made also with the help of their additive generators. Indeed, one has:

- additive generators of strict t-norms satisfy $\psi(x) > 0$ for every $x \in \mathbb{R}$ and $\lim_{x \to +\infty} \psi(x) = 0$;
- while additive generators of nilpotent t-norms satisfy $\psi(x) = 0$ for some $x \in \mathbb{R}$.

Example 8.2.13. The copula Π_2 is continuous, Archimedean and strict; its additive generator is $\psi_{\Pi_2}(t) = \exp(-t)$ $(t \in \mathbb{I})$. W_2 is continuous, Archimedean but not strict; it is additively generated by

$$\psi_{W_2}(t) := \max\{1 - t, 0\} \quad (t \in \mathbb{R}_+)\,.$$

The t-norm M_2 is continuous but not Archimedean.   ∎

Not every continuous and Archimedean t-norm T is a copula, since T may fail to be 2-increasing.

Example 8.2.14. Consider the continuous and Archimedean t-norm T given by

$$T(x, y) = \max\left\{\frac{x + y - 1}{x + y - xy}, 0\right\}\,.$$

Then T is a t-norm, but not a 2-copula, since

$$T(0.6, 0.9) + T(0.4, 0.6) - T(0.6, 0.6) - T(0.4, 0.9) < 0.$$

For other examples, see [Alsina et al., 2006, page 210].   ∎

Continuous Archimedean t-norms that are copulas are characterised below; the proof will be a special case of Theorem 6.5.7.

Theorem 8.2.15. *For a continuous and Archimedean t-norm T of type (8.2.4), the following statements are equivalent:*

(a) *T is a 2-copula;*

(b) *ψ is convex.*

Contrary to the class of copulas and quasi-copulas, continuous Archimedean t-norms do not form a convex set, since the convex combination of two t-norms need not be associative. For example, consider non-trivial convex combinations of Π_2 and W_2.

Moreover, it can be easily proved that pointwise maximum and minimum of two t-norms need not be a t-norm. However, it is important to stress that, somehow, semi-copulas provide a natural lattice-theoretic extension of t-norms, as illustrated below.

Theorem 8.2.16. *The class of triangular norms is join-dense and meet-dense in the class of commutative semi-copulas. Moreover, every continuous commutative semi-copula is the supremum (respectively, infimum) of a countable set of t-norms.*

Proof. See Durante et al. [2008c].   □

Historical remark 8.2.17. The history of the representation of associative functions begins with N.H. Abel; his first published paper [Abel, 1826] in Crelle's journal was devoted to this subject, under considerably stronger assumptions. The hypotheses were then weakened by Brouwer [1909], É. Cartan [1930] and Aczél [1949]. The representation (8.2.4) is due to Ling [1965]; the same representation holds under weaker assumptions (see, for instance, [Krause, 1981; Sander, 2005]). A large literature is nowadays available on triangular norms; the reader should bear in mind at least three fundamental references: Chapter 5 in the book by Schweizer and Sklar [1983], and the monographs by Klement et al. [2000], Klement and Mesiar [2005] and by Alsina et al. [2006]. The reader interested in the history of t-norms ought to consult these works.   ∎

8.3 Relationships between capacities and semi-copulas

Many areas of applications require the use of set functions that, like measures, are monotone with respect to set inclusion, but, unlike measures, need not be additive, not even finitely additive. Choquet's theory of capacities lies at the mathematical core of these investigations, even if it should be noticed that the original definition of a capacity presented in [Choquet, 1954] does not coincide with the later use of this term. Here we follow the current trend of the literature and consider the so-called *capacities* (or *fuzzy measures*), defined as follows (see, for instance, [Denneberg, 1994; Pap, 2002]).

Definition 8.3.1. Let Ω be a non-empty set and let $\mathscr{A}$ be a σ-field in the power set 2^{Ω}. A mapping $\nu : \mathscr{A} \to \mathbb{I}$ is called a *capacity* (sometimes also called *fuzzy measure* or *monotone set function*) if it is *monotone*, viz. $\nu(A) \leq \nu(B)$ for all A and B in $\mathscr{A}$ with $A \subseteq B$, and *normalised*, namely $\nu(\emptyset) = 0$ and $\nu(\Omega) = 1$. $\diamond$

Here, following Scarsini [1996], we present some properties of capacities on a finite dimensional space by means of the distribution function of a capacity. Then, in perfect analogy with the idea of Sklar's theorem, we show how semi-copulas may be used in the description (from a distributional viewpoint) of a capacity. Before continuing it is important to stress that, due to the lack of additivity, the distribution function of a capacity does not characterise a capacity. Nevertheless it can provide useful information when all that is needed is the probability of lower (or upper) intervals.

Since we are interested in the use of semi-copulas in survival analysis [Durante and Spizzichino, 2010], we focus our attention on (probability) measures concentrated on $\mathbb{R}_+^d$, where $\mathbb{R}_+ = [0, +\infty[$. Specifically, in the sequel, we consider capacities ν defined on the Borel σ-field $\mathscr{B}(\mathbb{R}_+^d)$. It is clear that all the statements can be formulated analogously for capacities on $\mathscr{B}(\mathbb{R}^d)$.

In analogy with the case of probability measures, to each capacity ν we may associate its *survival function* $\overline{F}_\nu : \mathbb{R}_+^d \to \mathbb{I}$ defined by

$$\overline{F}_\nu(x_1, \ldots, x_d) = \nu\left( ]x_1, +\infty[\times \cdots \times]x_d, +\infty[\right) . \tag{8.3.1}$$

If ν is a probability measure, then $\overline{F}_\nu$ is the usual probability survival function.

For any survival function $\overline{F} = \overline{F}_\nu : \mathbb{R}_+^d \to \mathbb{I}$ associated with a capacity ν and for every $i \in \{1, \ldots, d\}$, we denote by $\overline{F}_i : \mathbb{R}_+ \to \mathbb{I}$ the *i-th marginal survival function* of $\overline{F}$ given by $\overline{F}_i(x) = \overline{F}(\widetilde{x})$, where $\widetilde{x}$ is a vector of $\mathbb{R}_+^d$ having all the components equal to 0 except for the i-th one that is equal to x. Note that, given a capacity ν with survival function $\overline{F}$, $\overline{F}_i$ is the survival function of ν_i, the i-th projection of ν, usually called *i-th marginal capacity* of ν.

Any survival function $\overline{F} = \overline{F}_\nu$ associated with a capacity ν satisfies:

(a) $\overline{F} : \mathbb{R}_+^d \to \mathbb{I}$ is decreasing in each argument;

(b) $\overline{F}(0, \ldots, 0) = 1$;

(c) $\overline{F}(x_1, \ldots, x_d) \to 0$ when $\max\{x_1, \ldots, x_d\} \to +\infty$.

However, in general, $\overline{F}$ does not satisfy the d-increasing property (and, so, it may not be a probability survival function).

Conversely, each function $\overline{F}$ satisfying (a), (b) and (c) listed above may be used to construct a capacity ν whose survival function is exactly $\overline{F}$. But, contrary to the case of probability measures, such a capacity is not uniquely determined by $\overline{F}$.

Example 8.3.2. Given a probability survival function $\overline{F}$, consider the set function defined, for all Borel sets $A \subseteq \mathbb{R}^d_+$, by

$$\nu^*_{\overline{F}}(A) = \sup_{\mathbf{x} \in A} \overline{F}(\mathbf{x}). \tag{8.3.2}$$

This set function is easily proved to be a capacity with survival function equal to $\overline{F}$. On the other hand, any probability survival function $\overline{F}$ can be used in order to construct a probability measure $\mathbb{P}_{\overline{F}}$ on $\mathscr{B}(\mathbb{R}^d_+)$ defined on any d-dimensional box $B =]\mathbf{x}, \mathbf{y}]$ in $\mathbb{R}^d_+$ via $\mathbb{P}_{\overline{F}}(B) = V_{\overline{F}}(B)$, where $V_{\overline{F}}$ is the $\overline{F}$-*volume* of B, and extended by using standard techniques to the whole σ-field $\mathscr{B}(\mathbb{R}^d_+)$. It is clear that the survival function of $\mathbb{P}_{\overline{F}}$ is $\overline{F}$. However, there exists also another capacity, namely $\nu^*_{\overline{F}}$, having survival function $\overline{F}$ but generally not coinciding with $\mathbb{P}_{\overline{F}}$. For instance, consider the survival function $\overline{F}$ given on $\mathbb{I}^2$ by $\overline{F}(x,y) = (1-x)(1-y)$. Under the previous notations, it is easy to show that

$$\nu^*_{\overline{F}}(A) = 9/16 \quad \text{and} \quad \mathbb{P}_{\overline{F}}(A) = 1/4,$$

when $A = [1/4, 3/4] \times [1/4, 3/4]$. Thus $\nu^*_{\overline{F}} \neq \mathbb{P}_{\overline{F}}$. ∎

If suitable assumptions are satisfied, every semi-copula links the survival function of a capacity to its marginal survival functions.

Theorem 8.3.3. *Let ν be a capacity on the space $(\mathbb{R}^d_+, \mathscr{B}(\mathbb{R}^d_+))$, $\overline{F}$ its associated survival function and $\overline{F}_1, \ldots, \overline{F}_d$ the marginal survival functions. Suppose that, for every $i \in \{1, \ldots, d\}$, $\overline{F}_i$ is strictly decreasing on $[0, +\infty[$ and continuous. Then there exists a unique semi-copula $S = S_\nu : \mathbb{I}^d \to \mathbb{I}$ such that, for every $(x_1, \ldots, x_d) \in \mathbb{R}^d_+$,*

$$\overline{F}(x_1, \ldots, x_d) = S\left(\overline{F}_1(x_1), \ldots, \overline{F}_d(x_d)\right). \tag{8.3.3}$$

Proof. Given a survival function $\overline{F}$ associated with a capacity ν, consider the function $S : \mathbb{I}^d \to \mathbb{I}$ defined by

$$S(u_1, \ldots, u_d) := \overline{F}\left(\overline{F}_1^{-1}(u_1), \ldots, \overline{F}_d^{-1}(u_d)\right).$$

Now, since ν is a capacity, it follows that, for every $\mathbf{u} \in \mathbb{I}^d$ such that $u_j = 1$ for every $j \in \{1, \ldots, d\}, j \neq i$,

$$S(u_1, \ldots, u_d) = \overline{F}\left(0, \ldots, \overline{F}_i^{-1}(u_i), \ldots, 0\right) = \overline{F}_i(\overline{F}_i^{-1}(u_i)) = u_i.$$

Moreover, obviously, S is increasing in each variable. Thus, S is a semi-copula. The uniqueness of S follows by construction. □

Remark 8.3.4. Although the term semi-copula was not mentioned, Scarsini [1996] proved an analogous version of Theorem 8.3.3 under less restrictive assumptions on the marginals, but with an additional requirement on the capacity. ∎

Remark 8.3.5. Under additional assumptions, more can be said about the semi-copula associated with some capacities.

If $\overline{F}$ is *supermodular*, i.e., for all $\mathbf{x}, \mathbf{y} \in \mathbb{R}_+^d$

$$\overline{F}(\mathbf{x} \vee \mathbf{y}) + \overline{F}(\mathbf{x} \wedge \mathbf{y}) \geq \overline{F}(\mathbf{x}) + \overline{F}(\mathbf{y}) \,,$$

then the associated semi-copula S is a quasi-copula, as can be proved by using Lemma 5.2 in [Burchard and Hajaiej, 2006].

Obviously, if $\overline{F}$ is a probability survival function, then the associated semi-copula S is a copula; see Theorem 2.2.13. ∎

Given a semi-copula S (respectively, a copula) and the marginal survival functions $\overline{F}_1, \ldots, \overline{F}_d$, it is easily proved that $\overline{F} : \mathbb{R}_+^d \to \mathbb{I}$ defined as in (8.3.3) is a survival function (respectively, a probability survival function) that may be used to construct a capacity ν on $(\mathbb{R}_+^d, \mathscr{B}(\mathbb{R}_+^d))$ whose survival function is $\overline{F}$.

The semi-copula associated with a capacity ν on $(\mathbb{R}_+^d, \mathscr{B}(\mathbb{R}_+^d))$ takes a special form when ν is the product of its projections ν_i, i.e., for every $A_i \in \mathscr{B}(\mathbb{R}_+)$, $i \in \{1, \ldots, d\}$,

$$\nu(A_1 \times \cdots \times A_d) = \prod_{i=1}^d \nu_i(A_i) \,.$$

In fact, the following result is easily proved.

Corollary 8.3.6. *Under the assumptions of Theorem* 8.3.3, *suppose that ν is the product of its projections ν_i. Then the semi-copula S associated with ν is the independence copula Π_d.*

Remark 8.3.7. Note that, for given capacities $\nu_1, \ldots, \nu_d$ on $\mathscr{B}(\mathbb{R}_+)$, there exist several different capacities ν on $\mathscr{B}(\mathbb{R}_+^d)$ whose marginal capacities are $\nu_1, \ldots, \nu_d$, while their associated semi-copula is always Π_d. For instance, consider Example 8.3.2 and observe that the semi-copula S associated with the probability survival function $\overline{F}$ is equal to Π_2. ∎

As will be seen in the sequel, among various capacities, the so-called *distorted probabilities* are of special interest (see [Denneberg, 1994] and the references therein).

Definition 8.3.8. Let Ω be a non-empty set and let $\mathscr{A}$ be a σ-field on Ω. A capacity $\delta : \mathscr{A} \to \mathbb{I}$ is called a *distorted probability* if there exists a probability measure $\mathbb{P}$ on $\mathscr{A}$ and an increasing bijection $h : \mathbb{I} \to \mathbb{I}$ such that $\delta = h \circ \mathbb{P}$. ◇

The following interesting relation holds for the semi-copula that can be associated with a distorted probability.

Theorem 8.3.9. *Let $\mathbb{P}$ be a probability measure on $(\mathbb{R}_+^d, \mathscr{B}(\mathbb{R}_+^d))$, $\overline{F}$ its associated survival function and $\overline{F}_1, \ldots, \overline{F}_d$ its marginal survival functions. Let $h : \mathbb{I} \to \mathbb{I}$ be* *an increasing bijection. Then* $D = h \circ \overline{F}$ *is the survival function associated with the distorted probability* $h \circ \mathbb{P}$.

Moreover, if, for every $i \in \{1, \ldots, d\}$, $\overline{F}_i$ *is strictly decreasing on* $[0, +\infty[$ *and continuous, and* C *is the copula of* $\overline{F}$, *then*

$$S(u_1, \ldots, u_d) = h(C(h^{-1}(u_1), \ldots, h^{-1}(u_d))) \tag{8.3.4}$$

is the semi-copula associated with D.

Proof. It is easily shown that, for every $\mathbf{x} \in \mathbb{R}^d_+$, the survival function $\overline{F}_{h \circ \mathbb{P}}$ associated with $h \circ \mathbb{P}$ is given by

$$\overline{F}_{h \circ \mathbb{P}}(\mathbf{x}) = h\left[\mathbb{P}\left(]x_1, +\infty[\times \cdots \times]x_d, +\infty[\right)\right] = h \circ \overline{F}(\mathbf{x}).$$

Moreover, the marginal survival functions of $\overline{F}_{h \circ \mathbb{P}_{\overline{F}}}$ are given, for every $i \in \{1, \ldots, d\}$, by $h \circ \overline{F}_i$, where $\overline{F}_i$ is the marginal survival function of $\overline{F}$. In particular, if they are strictly decreasing and continuous, the semi-copula S associated with $\overline{F}_{h \circ \mathbb{P}_{\overline{F}}}$ is

$$
\begin{aligned}
S(u_1, \ldots, u_d) &= h\left[\overline{F}\left(\overline{F}_1^{-1}(h^{-1}(u_1)), \ldots, \overline{F}_d^{-1}(^{-1}(u_d))\right)\right] \\
&= \left[C(h^{-1}(u_1), \ldots, h^{-1}(u_d))\right],
\end{aligned}
$$

where C is the copula of $\overline{F}$. $\square$

As a by-product, Eq. (8.3.4) defines a way of transforming a copula C by means of an increasing bijection h into a semi-copula S. Transformations of this type are often called *distortions*. It may be of interest to check whether the distortion of a copula C is itself a copula; this will be clarified in the next section.

8.4 Transforms of semi-copulas

In this section, motivated by the appearance of semi-copulas in distorted probabilites, we study the behaviour of semi-copulas under a special class of transformations of type (8.3.4).

Denote by Θ the set of functions $h : \mathbb{I} \to \mathbb{I}$ that are continuous and strictly increasing with $h(1) = 1$ and by Θ_i the subset of Θ of those functions $h \in \Theta$ that are invertible, namely of those functions $h \in \Theta$ for which $h(0) - 0$. For $h \in \Theta_i$ the quasi inverse $h^{(-1)}$ coincides with the inverse h^{-1}. For a function $h \in \Theta \setminus \Theta_i$ the quasi-inverse $h^{(-1)} : \mathbb{I} \to \mathbb{I}$ is defined by

$$h^{(-1)}(t) := \begin{cases} h^{-1}(t), & t \in [h(0), 1], \\ 0, & t \in [0, h(0)]. \end{cases} \tag{8.4.1}$$

The following result, which we state in order to be able to refer to it, is easily proved.

Theorem 8.4.1. *For all h and g in Θ, one has*

(a) $h^{(-1)}$ *is continuous and strictly increasing in* $[h(0), 1]$;

(b) *for every* $t \in \mathbb{I}$, $h^{(-1)}(h(t)) = t$ *and* $h\left(h^{(-1)}(t)\right) = \max\{t, h(0)\}$;

(c) $(h \circ g)^{(-1)} = g^{(-1)} \circ h^{(-1)}$.

We may formally define the transformation of a semi-copula considered in (8.3.4).

Theorem 8.4.2. *For every $h \in \Theta$ and for every semi-copula $S \in \mathscr{S}_d$, the transform $S_h : \mathbb{I}^d \to \mathbb{I}$ defined, for all $\mathbf{u} = (u_1, \ldots, u_d) \in \mathbb{I}^d$, by*

$$S_h(\mathbf{u}) := h^{(-1)}\left(S(h(u_1), \ldots, h(u_d))\right) \tag{8.4.2}$$

is a semi-copula. Moreover, if S is continuous, so is its transform S_h.

Proof. For $j = 1, \ldots, d$, one has

$$\begin{aligned}
S_h(1, \ldots, 1, u_j, 1, \ldots, 1) &= h^{(-1)}\left(S(h(1), \ldots, h(1), h(u_j), h(1), \ldots, h(1))\right) \\
&= h^{(-1)}\left(S(1, \ldots, 1, h(u_j), 1, \ldots, 1)\right) \\
&= h^{(-1)} h(u_j) = u_j.
\end{aligned}$$

Moreover, since h is increasing, it is straightforward to show that S_h is increasing in each place. $\qquad\square$

A semi-copula of type (8.4.2) is called a *transformation* of S by means of h (or *distortion* of S via h). Formally, transformations of semi-copulas are defined via a mapping $\Psi : \mathscr{S}_d \times \Theta \to \mathscr{S}_d$ defined, for all $\mathbf{u} \in \mathbb{I}^d$, by

$$\Psi(S, h)(\mathbf{u}) := h^{(-1)}\left(S(h(u_1), \ldots, h(u_d))\right),$$

which will be denoted by

$$\Psi_h S := \Psi(S, h).$$

The set $\{\Psi_h : h \in \Theta\}$ is closed with respect to the composition operator $\circ$. Moreover, given h and g in Θ, one has, for every $S \in \mathscr{S}_d$ and for every $\mathbf{u} \in \mathbb{I}^d$,

$$\begin{aligned}
(\Psi_g \circ \Psi_h) S(\mathbf{u}) &= \Psi\left(\Psi(S, h), g\right)(\mathbf{u}) = g^{(-1)}\left(\Psi_h S(g(u_1), \ldots, g(u_d))\right) \\
&= g^{(-1)}\left(h^{(-1)} S\left((h \circ g)(u_1), \ldots, (h \circ g)(u_d)\right)\right) \\
&= (h \circ g)^{(-1)} S\left((h \circ g)(u_1), \ldots, (h \circ g)(u_d)\right) = \Psi_{h \circ g} S(\mathbf{u}).
\end{aligned}$$

The identity map in $\mathscr{S}_d$, which coincides with Ψ_{id_i}, is the neutral element of the composition operator $\circ$ in $\{\Psi_h : h \in \Theta\}$. Notice that only if h is in Θ_i does Ψ_h admit an inverse function, which is then given by $\Psi_h^{-1} = \Psi_{h^{-1}}$. In this case, the mapping $\Psi : \mathscr{S}_d \times \Theta_i \to \mathscr{S}_d$ is the action of the group Θ_i on $\mathscr{S}_d$. Moreover, since both h and $h^{(-1)}$ are increasing, one has, for every $h \in \Theta$,

$$\Psi_h M_d = M_d \qquad \text{and} \qquad \Psi_h Z_d = Z_d.$$

Finally, the transformation mapping preserves the pointwise order in the class of semi-copulas. In fact, if $S \prec S'$, then $\Psi_h S \prec \Psi_h S'$.

Example 8.4.3. For every $h \in \Theta$, $\Psi_h \Pi_2$ is an Archimedean and continuous t-norm; moreover, the operation Ψ gives rise to the whole family of continuous Archimedean t-norms (written with a multiplicative generator). It can be easily seen (see Section 8.2) that:

- $\{\Psi_h(\Pi_2) : h \in \Theta_i\}$ coincides with the class of strict, continuous and Archimedean t-norms;
- $\{\Psi_h(W_2) : h \in \Theta_i\}$ coincides with the class of nilpotent, continuous and Archimedean t-norms. ∎

Remark 8.4.4. For a given semi-copula S, the transforms S_h and S_g may be equal, $S_h = S_g$, even though the functions h and g are different, $h \neq g$. For instance, consider the copula W_2 and let h be the function defined on $\mathbb{I}$ by $h(t) := (t+1)/2$. Then $W_h = W_2$ and $W_{id_i} = W_2$, but $h \neq id$. ∎

Given a subset $\mathscr{E}$ of $\mathscr{S}_d$, it is of interest to consider whether $\mathscr{E}$ is closed under Ψ, i.e. $\Psi_h \mathscr{E} \subseteq \mathscr{D}$ for every $h \in \Theta$. Obviously, $\mathscr{S}_d$ and the subsets of symmetric and continuous semi-copulas are closed under Ψ. Another important closure result is provided by the next theorem.

Theorem 8.4.5. *The set of all triangular norms is closed under Ψ.*

Proof. It suffices to show that the function $T_h := \Psi_h T$, defined, for every $h \in \Theta$ and every t-norm T, by

$$T_h(u, v) := h^{(-1)}\left(T(h(u), h(v))\right) \qquad (u, v) \in \mathbb{I}^2 \,,$$

is associative. Set $\delta := h(0) \geq 0$. Then, if u, v and w all belong to $\mathbb{I}$, simple calculations lead to the following two expressions

$$T_h\left[T_h(u, v), w\right] = h^{(-1)} T\left[T(h(u), h(v)) \vee \delta, h(w)\right] \,,$$
$$T_h\left[u, T_h(v, w)\right] = h^{(-1)} T\left[h(u), T(h(v), h(w)) \vee \delta\right] \,.$$

If $T(h(u), h(v)) \leq \delta$, then

$$T_h\left[T_h(u, v), w\right] = h^{(-1)} T\left[\delta, h(w)\right] \leq h^{(-1)}(\delta) = 0 \,,$$

and either

$$\begin{aligned}
T_h\left[u, T_h(v, w)\right] &= h^{(-1)} T\left[h(u), T(h(v), h(w))\right] \\
&= h^{(-1)} T\left[T(h(u), h(v)), h(w)\right] \\
&\leq h^{(-1)} T(\delta, h(w)) \leq h^{(-1)}(\delta) = 0 \,,
\end{aligned}$$

or

$$T_h\left[u, T_h(v, w)\right] = h^{(-1)} T(h(u), \delta) \leq h^{(-1)}(\delta) = 0 \,.$$

Therefore, T is associative.

Analogous considerations hold if $T(h(u), h(v)) > \delta$. □

Notice that, in general, for a semi-copula S, if $\Psi_h S$ is symmetric (or associative), then S need not be symmetric or associative, as the following examples show.

Example 8.4.6. When h is not invertible, viz. when $h(0) \neq 0$, S need not be symmetric even if $\Psi_h S$ is symmetric. Take α and β in $]0, 1[$ with $\beta \geq 2\alpha$, and consider the partition of $\mathbb{I}$, by the two points $u = \alpha$ and $u = \beta$ and the shuffle of Min $C_{\alpha,\beta}$ given by

$$
C_{\alpha,\beta}(u, v) = \begin{cases} M_2(u + \alpha - \beta, v), & (u, v) \in [\beta - \alpha, \beta] \times [0, \alpha] , \\ M_2(u + v - \beta, v), & (u, v) \in ([0, \beta - \alpha] \times [0, \beta]) \\ & \quad \cup ([0, \beta] \times [\alpha, \beta]) , \\ M_2(u, v), & (u, v) \in ([\beta, 1] \times \mathbb{I}) \cup (\mathbb{I} \times [\beta, 1]) . \end{cases}
$$

Then $C_{\alpha,\beta}$ is not symmetric since

$$
C_{\alpha,\beta}\left(\beta - \frac{\alpha}{2}, \frac{\alpha}{2}\right) = M_2\left(\beta - \frac{\alpha}{2}, \frac{\alpha}{2}\right) = \frac{\alpha}{2} \neq 0 = C_{\alpha,\beta}\left(\frac{\alpha}{2}, \beta - \frac{\alpha}{2}\right) .
$$

On the other hand, for every $h \in \Theta_c$ with $h(0) = \beta$, one has $\Psi_h C_{\alpha,\beta} = M_2$, which, of course, is symmetric. ∎

Example 8.4.7. When h is not invertible, viz. when $h(0) \neq 0$, S need not be associative even if $\Psi_h S$ is associative. For instance, let $C \in \mathscr{C}_2$ be non-associative, take $\alpha \in]0, 1[$ and consider the ordinal sum $C_\alpha = \langle]0, \alpha[, C\rangle$, i.e.

$$
C_\alpha(u, v) = \begin{cases} \alpha\, C\left(\frac{u}{\alpha}, \frac{v}{\alpha}\right) , & (u, v) \in [0, \alpha]^2 , \\ M_2(u, v), & \text{elsewhere.} \end{cases}
$$

Since C is not associative there exists a triple $(u, v, w) \in \mathbb{I}^3$ such that

$$
C(C(u, v), w) \neq C(u, C(v, w)) .
$$

Then $u' = \alpha u$, $v' = \alpha v$ and $w' = \alpha w$ all belong to $[0, \alpha]$. Moreover $C_\alpha(u', v') = \alpha\, C(u, v) \in [0, \alpha]$, so that

$$
C_\alpha\left(C_\alpha(u', v'), w'\right) = \alpha\, C\left(C(u, v), w\right) .
$$

Similarly

$$
C_\alpha\left(u', C_\alpha(v', w')\right) = \alpha\, C(u, C(v, w)) ,
$$

which proves that C_α is not associative. However, for every $h \in \Theta_c$ with $h(0) = \alpha$, one has, for all u, v and w in $\mathbb{I}$,

$$
(\Psi_h C_\alpha)\left((\Psi_h C_\alpha)(u, v), w\right) = M_2\left(M_2(u, v), w\right) = M_2\left(u, M_2(v, w)\right)
$$
$$
= (\Psi_h C_\alpha)\left(u, (\Psi_h C_\alpha)(v, w)\right) ;
$$

therefore $\Psi_h C_\alpha$ is associative. ∎

Now, consider the transformation of a (quasi-)copula. First of all, notice that the class of (quasi-)copulas is not closed under the transformation Ψ_h, as the following example shows.

Example 8.4.8. Take $C = \Pi_2$; it was noticed above that $\Psi_h\Pi_2$ is an Archimedean and continuous t-norm for every $h \in \Theta$. Since an Archimedean t-norm is a copula if, and only if, its additive generator is convex (Theorem 8.2.15), it suffices to choose h in such a way that the corresponding additive generator $t \mapsto \psi(t) := -\ln h(t)$ is not convex; thus $\Psi_h\Pi_2$ is not a copula. For instance, let $h \in \Theta$ be defined by $h(t) := t^2$ for every $t \in \mathbb{I}$. Then

$$W_h(u,v) := h^{-1}\left(W_2(h(u), h(v))\right) = \sqrt{\max\{u^2 + v^2 - 1, 0\}},$$

namely

$$W_h(u,v) = \begin{cases} 0, & u^2 + v^2 \leq 1, \\ \sqrt{u^2 + v^2 - 1}, & u^2 + v^2 > 1. \end{cases}$$

The W_h-volume of the square $[6/10, 1]^2$ is

$$W_h(1,1) - W_h\left(1, \frac{6}{10}\right) - W_h\left(\frac{6}{10}, 1\right) + W_h\left(\frac{6}{10}, \frac{6}{10}\right) = -\frac{2}{10} < 0,$$

so that W_h is not even a quasi-copula. This shows that neither the family $\mathscr{C}_2$ of copulas nor that $\mathscr{Q}_2$ of quasi-copulas are closed under Ψ. ∎

When we pass to the transformation of quasi-copulas, the following result is easily derived from the very characterisation of these functions.

Theorem 8.4.9. *Let $Q \in \mathscr{Q}_d$ be a quasi-copula and let h be in Θ; then the following statements are equivalent*

(a) *the transform Q_h is a quasi-copula;*

(b) *for almost every $\mathbf{u} = (u_1, \ldots, u_d)$ in $\mathbb{I}^d$, and for $i = 1, \ldots, d$,*

$$h'(u_i)\, \partial_i Q(h(u_1), \ldots, h(u_d)) \leq h'\left(h^{(-1)}\left(Q(h(u_1), \ldots, h(u_d))\right)\right), \quad (8.4.3)$$

at all points where the partial derivatives $\partial_i Q$ $(i = 1, \ldots, d)$ exist.

Proof. For almost all $\mathbf{u}$ in $\mathbb{I}^d$ and for $i = 1, \ldots, d$, one has

$$\partial_i Q_h(\mathbf{u}) = \frac{h'(u_i)\, \partial_i Q(h(u_1), \ldots, h(u_d))}{h'\left(h^{(-1)}\left(Q(h(u_1), \ldots, h(u_d))\right)\right)}.$$

Now Q_h satisfies the boundary conditions and is increasing in each place; therefore, in view of Theorem 7.2.7, it is a quasi-copula if, and only if, for every $i \in \{1, \ldots, d\}$, $\partial_i Q_h \leq 1$, namely, if, and only if, condition (8.4.3) holds. □

In particular, it follows that

Corollary 8.4.10. *If Q is a quasi-copula, $Q \in \mathcal{Q}_2$, and if $h \in \Theta$ is concave, then also Q_h is a quasi-copula, $Q \in \mathcal{Q}_2$.*

Proof. For all u and v in $\mathbb{I}$, one has

$$u = h^{(-1)} \left(Q(h(u), 1) \right) \geq h^{(-1)} \left(Q(h(u), h(v)) \right);$$

since h' is decreasing a.e. on $\mathbb{I}$, and since $|\partial_i Q| \leq 1$ $(i = 1, 2)$,

$$h'(u) \left(\partial_i Q(h(u), h(v)) \right) \leq h'(u) \leq h' \left(h^{(-1)} \left(Q(h(u), h(v)) \right) \right) \qquad (i = 1, 2),$$

namely, condition (8.4.3). $\qquad\qquad\qquad\qquad\qquad\qquad\qquad\qquad\qquad\qquad\qquad\quad$ □

Now, consider the transform of a copula $C \in \mathcal{C}_d$,

$$C_h(\mathbf{u}) := h^{(-1)} \left(C(h(u_1), \ldots, h(u_d)) \right). \qquad\qquad (8.4.4)$$

We start with the two-dimensional case.

Theorem 8.4.11. *For $h \in \Theta$, the following statements are equivalent:*

(a) *h is concave;*

(b) *for every copula $C \in \mathcal{C}_2$, the transform (8.4.4) is a copula.*

Proof. (a) $\Longrightarrow$ (b) We only need to show that C_h is 2-increasing. To this end, let u_1, v_1, u_2 and v_2 be points of $\mathbb{I}$ such that $u_1 \leq u_2$ and $v_1 \leq v_2$. Then the points s_i $(i = 1, 2, 3, 4)$ defined by

$$\begin{aligned} s_1 &= C(h(u_1), h(v_1)), & s_2 &= C(h(u_1), h(v_2)), \\ s_3 &= C(h(u_2), h(v_1)), & s_4 &= C(h(u_2), h(v_2)), \end{aligned}$$

satisfy the inequalities

$$s_1 \leq \min\{s_2, s_3\} \leq \max\{s_2, s_3\} \leq s_4, \qquad \text{and} \qquad s_1 + s_4 \geq s_2 + s_3.$$

Thus, there exists $\alpha \in \mathbb{I}$ and $\tilde{s}_4 \leq s_4$ with $\tilde{s}_4 + s_1 = s_2 + s_3$ such that

$$\begin{aligned} h^{(-1)}(s_3) + h^{(-1)}(s_2) &= h^{(-1)} \left(\alpha s_1 + (1 - \alpha)\tilde{s}_4 \right) + h^{(-1)} \left((1 - \alpha)s_1 + \alpha\tilde{s}_4 \right) \\ &\leq \alpha h^{(-1)}(s_1) + (1 - \alpha)h^{(-1)}(\tilde{s}_4) + (1 - \alpha)h^{(-1)}(s_1) + \alpha h^{(-1)}(\tilde{s}_4) \\ &\leq h^{(-1)}(s_1) + h^{(-1)}(\tilde{s}_4) \leq h^{(-1)}(s_1) + h^{(-1)}(s_4), \end{aligned}$$

where the inequality follows from the fact that $h^{(-1)}$ is convex and increasing. Thus C_h is 2-increasing.

(b) $\Longrightarrow$ (a) It suffices to show that $h^{(-1)}$ is Jensen-convex, namely, for all s and t in $\mathbb{I}$,

$$h^{(-1)} \left(\frac{s + t}{2} \right) = \frac{h^{(-1)}(s) + h^{(-1)}(t)}{2}, \qquad\qquad (8.4.5)$$

because, then, $h^{(-1)}$ is convex and h is concave.

Without loss of generality, consider the copula W_2 and points s and t in $\mathbb{I}$ with $s \leq t$. If $(s+t)/2$ is in $[0, h(0)]$ then (8.4.5) is immediate. If $(s+t)/2$ is in $]h(0), 1]$, then one has

$$W_2\left(\frac{s+1}{2}, \frac{s+1}{2}\right) = s, \qquad \text{and} \qquad W_2\left(\frac{t+1}{2}, \frac{t+1}{2}\right) = t,$$

$$W_2\left(\frac{s+1}{2}, \frac{t+1}{2}\right) = \frac{s+t}{2} = W_2\left(\frac{t+1}{2}, \frac{s+1}{2}\right).$$

There are points u_1 and u_2 in $\mathbb{I}$ such that

$$h(u_1) = \frac{1+s}{2}, \qquad h(u_2) = \frac{1+t}{2}.$$

Since W_h is a copula, it is 2-increasing

$$W_h(u_1, u_1) - W_h(u_1, u_2) - W_h(u_2, u_1) + W_h(u_2, u_2) \geq 0;$$

as a consequence, one has

$$h^{(-1)}(s) - h^{(-1)}\left(\frac{s+t}{2}\right) - h^{(-1)}\left(\frac{s+t}{2}\right) + h^{(-1)}(t) \geq 0;$$

which is the desired conclusion. $\qquad\qquad\qquad\qquad\qquad\qquad\qquad\qquad\qquad\square$

Example 8.4.12. Let us denote by Θ_c the set of concave functions in Θ. For instance the following functions are in Θ_c:

(a) $h(t) := t^{1/\alpha}$ and $h^{-1}(t) = t^\alpha$, with $\alpha \geq 1$;

(b) $h(t) := \sin(\pi t/2)$ and $h^{-1}(t) = (2/\pi)\arcsin t$;

(c) $h(t) := (4/\pi)\arctan t$ and $h^{-1}(t) = \tan(\pi t/4)$.

Moreover, if h is in Θ_c, then both $t \mapsto h(t^\alpha)$ and $t \mapsto h^\alpha(t)$ are in Θ_c for every $\alpha \in]0, 1[$. A longer list of functions in Θ_c can be found in [Morillas, 2005, Table 1].

Furthermore, Θ_c (and its subset Θ_{ci} formed by the invertible concave functions) is convex, closed under composition and multiplication. In addition, if f belongs to Θ_{ci} then the function $\mathbb{I} \ni t \mapsto 1 - f^{-1}(1-t)$ also belongs to Θ_{ci}. $\qquad\blacksquare$

Example 8.4.13. In the case of power functions, an interesting probabilistic interpretation of the transform (8.4.4) can be given. In fact, if $h(t) = t^{1/n}$ for some $n \in \mathbb{N}$, then C_h is the copula associated with the componentwise maxima $X := \max\{X_1, \ldots, X_n\}$ and $Y = \max\{Y_1, \ldots, Y_n\}$ of a random sample $(X_1, Y_1), \ldots, (X_n, Y_n)$ from some arbitrary distribution function with underlying copula C. $\qquad\blacksquare$

Next we consider the transforms of d-copulas. As the following example shows the requirement that a function $h \in \Theta$ is concave is not enough to ensure that $\Psi_h C$ is a d-copula for $d \geq 3$.

Example 8.4.14. Consider the following mixture of 3-copulas

$$C(u_1, u_2, u_3) := \frac{1}{2} W_2 (u_1, M_2(u_2, u_3)) + \frac{1}{2} W_2 (u_2, M_2(u_1, u_3)) .$$

If $h(t) = t^{3/2}$, which belongs to Θ_{ci}, then

$$V_C ([27/64, 1] \times [27/64, 1] \times [8/27, 27/64]) \simeq -0.0194 < 0 ,$$

so that C is not a 3-copula.   ∎

In high dimensions, in order to preserve the copula property by distortion, a stronger notion of concavity for h is required (see, e.g., Morillas [2005]).

Theorem 8.4.15. *Let C be in $\mathscr{C}_d$ and let $h : \mathbb{I} \to \mathbb{I}$ be in Θ. If the pseudo-inverse $h^{(-1)}$ is d-monotone, then $\Psi_h C$ is a d-copula, i.e., $\Psi_h C \in \mathscr{C}_d$.*

Proof. Since $D := \Psi_h C$ satisfies the boundary conditions of a copula, it only remains to prove that it is d-increasing.

Let $\mathbf{a}$ and $\mathbf{b}$ be in $\mathbb{I}^d$ with $\mathbf{a} < \mathbf{b}$, and consider the box $B = [\mathbf{a}, \mathbf{b}]$. Since h is strictly increasing, and, hence, $h(a_j) < h(b_j)$ for every $j \in \{1, \ldots, d\}$, set

$$B' := [h(a_1), h(b_1)] \times \cdots \times [h(a_d), h(b_d)] .$$

Consider a sequence $(C_n)_{n \in \mathbb{N}}$ of d-copulas that converges uniformly to C, and such that, for every $n \in \mathbb{N}$, C_n has all the derivatives up to order d: to this end, it suffices to consider the Bernstein approximation of Section 4.1. Now, let P_j denote the set of partitions of $\{1, \ldots, d\}$ into j disjoint sets, let $S = \{j_1, \ldots, j_m\}$ denote one of the sets of the partition P_j and set

$$C_n^S := \frac{\partial^m}{\partial u_{j_1} \ldots \partial u_{j_m}} C_n ,$$

the mixed partial derivative of C_n with respect to the variable indexed in S. Now, it can be easily seen that the mixed partial derivatives of order d of $h^{(-1)} \circ C_n$ are given by

$$\frac{\partial^d}{\partial u_1 \ldots \partial u_d} \left(h^{(-1)} \circ C_n \right) = \sum_{j=1}^d \left((h^{(-1)})^{(j)} \circ C_n \right) \sum_{P \in P_j} \prod_{S \in P} C_n^S .$$

Since C_n is a copula and $h^{(-1)}$ is d-monotone, the r.h.s. in the previous equation is non-negative. As a consequence, $h^{(-1)} \circ C_n$ is d-increasing; in particular $V_{h^{(-1)} \circ C_n}(B') \geq 0$. But (C_n) converges uniformly to C, so that

$$V_{h^{(-1)} \circ C}(B') = \lim_{n \to +\infty} V_{h^{(-1)} \circ C_n}(B') \geq 0 ;$$

then

$$V_D(B) = V_{h^{(-1)} \circ C}(B') \geq 0 ,$$

which proves that D is d-increasing.   □

An interesting application of this distortion is presented by Valdez and Xiao [2011]: they show that distortions preserve the supermodular order among random vectors. Recall that, given two d-dimensional random vectors $\mathbf{X}$ and $\mathbf{Y}$ such that $\mathbb{E}[\varphi(\mathbf{X})] \leq \mathbb{E}[\varphi(\mathbf{Y})]$ for every supermodular function $\varphi : \mathbb{R}^d \to \mathbb{R}$, provided the expectations exist, then $\mathbf{X}$ is said to be smaller than $\mathbf{Y}$ in the supermodular order (denoted by $\mathbf{X} \leq_{sm} \mathbf{Y}$).

Theorem 8.4.16. *Suppose that the d-dimensional random vectors $\mathbf{X}$ and $\mathbf{Y}$, defined on a same probability space, have respective associated copulas $C_{\mathbf{X}}$ and $C_{\mathbf{Y}}$ with absolutely continuous univariate margins. Denote by $\mathbf{X}'$ and $\mathbf{Y}'$ the random vectors having the same univariate margins as $\mathbf{X}$ and $\mathbf{Y}$, respectively, and copulas $\Psi_h C_{\mathbf{X}}$ and $\Psi_h C_{\mathbf{Y}}$ for $h^{(-1)}$ absolutely monotone of order d. Then $\mathbf{X} \leq_{sm} \mathbf{Y}$ implies $\mathbf{X}' \leq_{sm} \mathbf{Y}'$.*

Proof. See [Valdez and Xiao, 2011, Corollary 4.8]. □

8.4.1 Transforms of bivariate semi-copulas

Here, we discuss in detail the transforms of bivariate semi-copulas. In particular, given the class of all transformations, we wonder whether, for instance, the class of continuous semi-copulas coincides with that of the distortions of copulas or quasi-copulas, a fact that might provide an important simplification of these concepts. However, as we show in the following, relationships of this type do not hold even in the bivariate case.

We start by a preliminary lemma.

Lemma 8.4.17. *Consider the following two conditions for a continuous semi-copula $S \in \mathscr{S}_2$:*

(a) *If $v' \in \,]0, 1[$ and two points u and u' with $u \neq u'$ exist such that $S(u, v') = S(u', v')$, then $S(u, v) = S(u', v)$ for every $v \in [0, v']$. Also $S(u', v) = S(u', v')$ implies $S(u, v) = S(u, v')$ for every $u \in [0, u']$.*

(b) *If u' and v are in $]0, 1[$, with $u' < v$, and $S(u', v) = u'$, then $S(u, v) = u$ for every $u \in \,]0, u'[$.*

Then condition (a) *implies condition* (b).

Proof. Assume, if possible, that S does not satisfy condition (b). Then assume, for instance, that there exist two points u' and v in $]0, 1[$, with $u' < v$, such that $S(u', v) = u'$ while $S(u, v) < u$ for at least a point $u \in \,]0, u'[$. Then

$$S(u', v) = u' = S(u', 1) \qquad \text{and} \qquad S(u, v) < u = S(u, 1),$$

so that condition (a) cannot hold. □

The following result gives a necessary condition that a semi-copula is the transform of a quasi-copula.

Theorem 8.4.18. *If the continuous semi-copula $S \in \mathscr{S}_2$ is the transform of a quasi-copula $Q \in \mathscr{Q}_2$ under the function $h \in \Theta_i$, then condition* (b) *of Lemma 8.4.17 holds.*

Proof. By assumption there exist a quasi-copula Q and a function $h \in \Theta$ such that $S = \Psi_h Q$; thus

$$S(u', v) = h^{(-1)}\left(Q(h(u'), h(v))\right),$$

so that

$$h\left(S(u', v)\right) = Q(h(u'), h(v)).$$

Since $S(u', v) = u'$, the 1-Lipschitz property of Q implies, for every $u \in [0, u']$,

$$
\begin{aligned}
h(u') - h(S(u, v)) &= h(S(u', v)) - h(S(u, v)) \\
&= Q\left(h(u'), h(v)\right) - Q\left(h(u), h(v)\right) \leq h(u') - h(u),
\end{aligned}
$$

which yields $h(S(u, v)) \geq h(u)$ and, hence, $S(u, v) \geq u$. Since $S(u, v) \leq \min\{u, v\}$ this proves the assertion. $\qquad \square$

This result will now be used to provide the example of a continuous semi-copula that cannot be represented as a transformation of any quasi-copula.

Example 8.4.19. Define a symmetric $S \in \mathscr{S}_2$ through its level curves as follows.

- For $u \in [0, 1/3]$, the level curves join $(u, 1)$ to $(3u/2, 3u/2)$;
- for $u \in [1/3, 2/3]$, the level curves join $(u, 1)$ to $[1/3 + u/2, 1/3 + u/2]$;
- for $u \in [2/3, 1]$, the level curves join $(u, 1)$ to (u, u).

Then

$$S\left(\frac{2}{3}, \frac{2}{3}\right) = \frac{2}{3} \quad \text{and} \quad S\left(\frac{4}{9}, \frac{2}{3}\right) = \frac{1}{3} < \frac{4}{9}.$$

Thus S does not fulfil the condition of Theorem 8.4.18; therefore it cannot be the image under Ψ of any quasi-copula. $\qquad \blacksquare$

As a by-product, the following relations hold among the sets of semi-copulas, copulas and quasi-copulas and their transforms for $d = 2$.

$$
\begin{array}{ccccccc}
\mathscr{C}_2 & \subset & \Psi\mathscr{C}_2 & \subset & \Psi\mathscr{S}_C & \subset & \Psi\mathscr{S}_2 \\
\cap & & \cap & & \| & & \| \\
\mathscr{Q}_2 & \subset & \Psi\mathscr{Q}_2 & \subset & \mathscr{S}_C & \subset & \mathscr{S}_2
\end{array}
$$

All the inclusions in the previous diagram are strict. The strict inclusion $\mathscr{C}_2 \subset \mathscr{Q}_2$ is well known and has been studied in Chapter 7; the relations $\mathscr{C}_2 \subset \Psi\mathscr{C}_2$ and $\mathscr{Q}_2 \subset \Psi\mathscr{Q}_2$ were established in Example 8.4.8. That the inclusion $\Psi\mathscr{Q}_2 \subset \mathscr{S}_C$ is strict has just been proved in Example 8.4.19.

Moreover, as will be shown shortly, $\Psi\mathscr{C}_2$ is not included in $\mathscr{Q}_2$.

But more is true, as proved in the following theorem.

Theorem 8.4.20. *For every bivariate copula* $C \in \mathcal{C}_2$, $C \neq M_2$, *there is a function* $h \in \Theta$ *such that* $\Psi_h\, C$ *is not a quasi-copula.*

Proof. Since C is different from M_2 there exists at least an element $u_0 \in\,]0, 1[$, such that $C(u_0, u_0) < u_0$. Two cases will be considered.

Case 1: u_0 is in $[1/2, 1[$. There exists $\delta > 0$ such that $C(u_0 + \delta, u_0 + \delta) = u_0$. Then choose h in Θ_i in such a manner that $h(\delta) = u_0$ and $h(u_0 + \delta) = u_0 + \delta$. Now

$$\Psi_h\, C(u_0 + \delta, u_0 + \delta) = h^{-1}\left(C\left(h(u_0 + \delta), h(u_0 + \delta)\right)\right) = h^{-1}(u_0) = \delta\,,$$

and

$$\Psi_h\, C(1, u_0 + \delta) = h^{-1}C(1, u_0 + \delta) = h^{-1}(u_0 + \delta) = u_0 + \delta\,.$$

Thus

$$\Psi_h\, C(1, u_0 + \delta) - \Psi_h\, C(u_0 + \delta, u_0 + \delta) = u_0 > 1 - u_0 - \delta\,,$$

since $u_0 \geq 1/2 > (1 - \delta)/2$. Therefore $\Psi_h\, C$ is not 1-Lipschitz, and, hence, not a quasi-copula.

Case 2: u_0 is in $]0, 1/2[$. There exist $\delta > 0$ such that $C(u_0 + \delta, u_0 + \delta) = u_0$. Clearly, $u_0 + \delta < 1/2$, so that $2\, u_0 + \delta < 2\, (u_0 + \delta) < 1$. Choose a function h in Θ_i satisfying the following constraints

$$h\left(u_0 + \frac{2}{3}\delta\right) = u_0 + \frac{2}{3}\delta,\ h(u_0 + \delta) = u_0 + \delta,\ h\left(\frac{\delta}{3}\right) = u_0\,,\ h(2\, u_0 + \delta) = \alpha,$$

where $\alpha \in\,]u_0 + \delta, 1[$ is such that $C(\alpha, u_0 + \delta) = u_0 + (2\,\delta)/3$. Then

$$\Psi_h\, C(u_0 + \delta, u_0 + \delta) = h^{-1}C(u_0 + \delta, u_0 + \delta) = h^{-1}(u_0) = \frac{1}{3}\delta\,,$$

$$\Psi_h\, C(2\, u_0 + \delta, u_0 + \delta) = h^{-1}C(\alpha, u_0 + \delta) = u_0 + \frac{2}{3}\delta\,.$$

Thus

$$\Psi_h\, C(2\, u_0 + \delta, u_0 + \delta) - \Psi_h\, C(u_0 + \delta, u_0 + \delta) = u_0 + \frac{1}{3}\delta > u_0\,,$$

so that $\Psi_h\, C$ is not 1-Lipschitz and, hence, not a quasi-copula. $\qquad\square$

Theorem 8.4.21. *If the continuous semi-copula* $S \in \mathcal{S}_2$ *is the transform of a copula* $C \in \mathcal{C}_2$ *under a function* h, $h \in \Theta_i$, *then condition* (a) *of Lemma* 8.4.17 *holds.*

Proof. Let $S = \Psi_h(C)$ be the transform of the copula C under some function $h \in \Theta_i$. Then

$$S(u, v) = h^{-1}\left(C(h(u), h(v))\right),$$

or, equivalently,

$$h\left(S(u, v)\right) = C(h(u), h(v))\,.$$

Take $u \le u'$ and $v \le v'$ and set

$$\alpha := S(u, v), \quad \beta := S(u', v), \quad \alpha' := S(u, v'), \quad \beta' := S(u', v'),$$

and

$$h(\alpha) = C(h(u), h(v)) =: a, \qquad h(\beta) = C(h(u'), h(v)) =: b,$$
$$h(\alpha') = C(h(u), h(v')) =: a', \qquad h(\beta') = C(h(u'), h(v')) =: b'.$$

Now the 2-increasing property of a copula yields $b' - a' \ge b - a$, which means

$$h(\beta') - h(\alpha') \ge h(\beta) - h(\alpha).$$

Thus, if $\beta' = \alpha'$, then $\beta = \alpha$, namely the first assertion. The proof of the second assertion is similar. $\qquad\square$

Corollary 8.4.22. *Let a symmetric quasi-copula $Q \in \mathcal{Q}_2$ satisfy, for u and v with $u < v$, the following condition*

$$Q(u, u) = a < b = Q(u, v) = Q(v, u) = Q(v, v);$$

then Q does not belong to $\Psi\,\mathscr{C}_2$.

Proof. The quasi-copula Q does not satisfy condition (a) of Lemma 8.4.17. $\qquad\square$

The quasi-copula of Example 7.4.2 satisfies the condition of the above corollary and, therefore, is not the image of any copula under any function h.

Historical remark 8.4.23. We conclude this section with a historical remark about the terminology used here. The term "semi-copula" appeared for the first time in Bassan and Spizzichino [2003], but it was not formally defined there. In fact, the authors wrote "we may occasionally refer to functions with the above properties as semi-copulae" (page 3). Instead, it was later discussed in depth by Bassan and Spizzichino [2005b] who used, however, a slightly different definition. Specifically, Bassan and Spizzichino [2005b] call "extended semicopula" a function defined as in Definition 8.1.1, while the term "semicopula" is used for a distortion of a bivariate exchangeable copula, namely to the class

$$\{\Psi_h C : h \in \Theta_i, \, C \in \mathscr{C}_2, \, C \text{ exchangeable}\}.$$

The terminology used in this book was introduced in [Durante and Sempi, 2005b] (see also [Durante et al., 2006b]) and is now generally adopted in the literature. $\qquad\blacksquare$

Further readings 8.4.24. The presentation of this section is based on the works by Durante and Sempi [2005a] and by Morillas [2005], but uses also the results by Alvoni et al. [2009]. More details on distortions of copulas can be also found in [Durante et al., 2010a; Valdez and Xiao, 2011]. For some applications of distortions, see also [Di Bernardino and Rullière, 2013]. $\qquad\blacksquare$

8.5 Semi-copulas and level curves

In this section, we aim to explain why semi-copulas emerge in the analysis of the level curves of a function $\overline{F}$ on $\mathbb{R}^d$. This interest is generally motivated by the possible applications of a level set representation of a function, which encompass Economics (e.g., the indifference curves for a utility function as considered by Cerqueti and Spizzichino [2013]) and various fields, including the recent development about level-set based risk measures [Embrechts and Puccetti, 2006; Nappo and Spizzichino, 2009; Salvadori et al., 2011; Cousin and Di Bernardino, 2013]. Moreover, as will be seen in the next section, the level curves of a survival function provide useful tools to explain concepts in reliability theory.

Hence, following the latter approach, in the rest of the chapter we consider a vector $\mathbf{X}$ (on the probability space $(\Omega, \mathscr{F}, \mathbb{P})$) of exchangeable, positive random variables (interpreted as lifetimes) with joint probability survival function $\overline{F}$ and marginal survival functions $\overline{F}_i = \overline{G}$ for $i = 1, \ldots, d$, which are continuous and strictly decreasing on $[0, +\infty[$.

The family of level curves of a survival function $\overline{F}$ will be described by $\mathscr{L}_{\overline{F}} = \{\ell_\alpha\}_{\alpha \in \mathbb{I}}$, where

$$\ell_\alpha = \{\mathbf{x} \in \mathbb{R}_+^d \mid \overline{F}(\mathbf{x}) = \alpha\}.$$

The class of all the survival functions with the same family of level curves $\mathscr{L}$ can be characterized as follows.

Theorem 8.5.1. *Let $\overline{F}_1$ and $\overline{F}_2$ be probability survival functions whose families of level curves are given, respectively, by $\mathscr{L}_{\overline{F}_1}$ and $\mathscr{L}_{\overline{F}_2}$. The following statements are equivalent:*

(a) $\mathscr{L}_{\overline{F}_1} = \mathscr{L}_{\overline{F}_2}$;

(b) *there exists an increasing bijection $h : \mathbb{I} \to \mathbb{I}$ such that $\overline{F}_2 = h \circ \overline{F}_1$.*

Proof. The proof is conducted in the case $d = 2$, the general case being analogous.

(a) $\implies$ (b) Note that, for every $\alpha \in \mathbb{I}$, there exist $\ell_\alpha \in \mathscr{L}_{\overline{F}_1}$ and just one $x \in \mathbb{R}_+$ such that $(x, x) \in \ell_\alpha$. Set $\beta := \overline{F}_2(x, x)$. Consider the function $h : \mathbb{I} \to \mathbb{I}$, defined by $h(\alpha) := \beta$. Now, one has $h(0) = 0$, because $\overline{F}_1(x, x) = 0$ only when $x = +\infty$ and, in this case, $\overline{F}_2(x, x) = 0$; $h(1) = 1$, because $\overline{F}_1(x, x) = 1$ when $x = 0$ and, in this case, $\overline{F}_2(x, x) = 0$. Moreover, h is strictly increasing, because, for $\alpha > \alpha'$, there exist $x, x' \in \mathbb{R}_+$ with $x < x'$ such that $\overline{F}_1(x, x) = \alpha > \overline{F}_1(x', x') = \alpha'$ and, analogously, $h(\alpha) = \overline{F}_2(x, x) > \overline{F}_2(x', x') = h(\alpha')$. Thus, h is a bijection.

(b) $\implies$ (a) For every $\alpha \in \mathbb{I}$,

$$\{(x, y) \in \mathbb{R}_+^2 \mid \overline{F}_2(x, y) = \alpha\} = \{(x, y) \in \mathbb{R}_+^2 \mid (h \circ \overline{F}_1)(x, y) = \alpha\}$$
$$= \{(x, y) \in \mathbb{R}_+^2 \mid \overline{F}_1(x, y) = h^{-1}(\alpha)\},$$

and, as a consequence, $\mathscr{L}_{\overline{F}_2} = \mathscr{L}_{\overline{F}_1}$. $\qquad\square$

In view of Theorem 8.5.1, the study of the families of the level curves of a survival function $\overline{F}$ can be developed in the more general framework of distorted probabilities and associated survival functions. In fact, if $\mathbb{P}_{F_1}$ is the probability measure induced by $F_1 := 1 - \overline{F}_1$, then $\overline{F}_2 = h \circ \overline{F}_1$ is the survival function of the distorted probability $h \circ \mathbb{P}_{F_1}$.

Let $\mathscr{L}$ be a fixed family of level curves for survival functions. It is important to notice at this point that any survival function $\overline{F}$ such that $\mathscr{L}_{\overline{F}} = \mathscr{L}$ possesses its own one-dimensional marginal and, corresponding to the latter, its survival copula. How can this class of survival functions be represented in a concise way?

Formally, in the class of survival functions one may define an equivalence relation R such that two survival functions are in the relation R if, and only if, they share the same family of level curves. Then, one may consider the equivalence class of all survival functions with respect to the relation R. It is therefore of natural interest to single out a representative of each equivalence class obtained from R. One possible method of achieving this amounts to fixing a marginal survival function $\overline{G}_0$, and then selecting just the survival function admitting $\overline{G}_0$ as its marginal. Starting from an arbitrary survival function $\overline{F}$, this can be done by considering, for every $\mathbf{x} \in \mathbb{R}^d$,

$$\overline{L}(\mathbf{x}) = \overline{G}_0 \left(\overline{G}^{-1}(\overline{F}(\mathbf{x})) \right) . \tag{8.5.1}$$

In fact, by Theorem 8.3.9, $\overline{L}$ is such that $\mathscr{L}_{\overline{L}} = \mathscr{L}_{\overline{F}}$ and it is immediately checked that all the marginals of $\overline{L}$ coincide with $\overline{G}_0$.

In this direction, Barlow and Spizzichino [1993] suggested considering $\overline{G}_0(x) = \exp(-x)$. In fact, this kind of survival function may be considered "indifferent" to ageing, in the sense that it has, at the same time, positive and negative ageing [Spizzichino, 2001]. Intuitively, this means that $\overline{L}$ is the object that allows expressing the behaviour of the level sets of a joint survival function by reducing it to a function having marginals with no univariate ageing.

Then, for any survival function $\overline{F}$, we put

$$\overline{L}_{\overline{F}}(\mathbf{x}) = \exp \left(-\overline{G}^{-1}(\overline{F}(\mathbf{x})) \right) . \tag{8.5.2}$$

The following result holds.

Theorem 8.5.2. *If two survival functions $\overline{F}_1$ and $\overline{F}_2$ have the same level curves, then $\overline{L}_{\overline{F}_1} = \overline{L}_{\overline{F}_2}$.*

Proof. In view of Theorem 8.5.1, if $\overline{F}_1$ and $\overline{F}_2$ have the same level curves, then there exists $h : \mathbb{I} \to \mathbb{I}$ such that $\overline{F}_2 = h \circ \overline{F}_1$. In particular, the marginal survival function $\overline{G}_2$ of $\overline{F}_2$ is given by $h \circ \overline{G}_1$. Thus

$$\overline{L}_{\overline{F}_2}(\mathbf{x}) = \exp \left(-\overline{G}_2^{-1}(\overline{F}_2(\mathbf{x})) \right) = \exp \left(-\overline{G}_1^{-1} \circ h^{-1}(\overline{F}_2(\mathbf{x})) \right)$$
$$= \exp \left(-\overline{G}_1^{-1}(\overline{F}_1(\mathbf{x})) \right) = \overline{L}_{\overline{F}_1}(\mathbf{x}),$$

which is the desired assertion. $\square$

Thus the survival function $\overline{L}$ of eq. (8.5.2) is a representative element of all the survival functions sharing the same family of level curves of $\overline{F}$. In some sense, it expresses all the properties of a survival function that are related to the level curves, regardless the marginal behaviour. Therefore, mimicking the copula approach for d.f.'s, one may represent $\overline{L}$ via a suitable function on the domain $\mathbb{I}^d$. This is the content of the next result, which is based on Theorem 8.3.9.

Theorem 8.5.3. *The survival semi-copula associated with the survival function $\overline{L}$ of* (8.5.2) *is given, for every* $\mathbf{u} \in \mathbb{I}^d$, *by*

$$B(\mathbf{u}) = \exp\left(-\overline{G}^{-1}\left(\overline{F}(-\ln(u_1), \ldots, -\ln(u_d))\right)\right) . \qquad (8.5.3)$$

The semi-copula given by (8.5.3) is called the *(multivariate) ageing function* of $\overline{F}$. It can also be expressed in the following equivalent way:

$$B_{\overline{F}}(\mathbf{u}) = \exp\left(-\overline{G}^{-1}\left(C(\overline{G}(-\ln(u_1)), \ldots, \overline{G}(-\ln(u_d)))\right)\right) , \qquad (8.5.4)$$

where C is the survival copula of $\overline{F}$. Thus B is obtained by means of a distortion of the copula C of $\overline{F}$ by means of the bijective transformation $\overline{G}(-\ln u)$. As a matter of fact, B is a continuous semi-copula, but not necessarily a copula.

The ageing function $B_{\overline{F}}$ is a semi-copula representing all the distorted survival functions having the same family of level curves and the survival semi-copula of one of them (namely $\overline{L}$, the one with standard exponential marginals). Therefore, $B_{\overline{F}}$ is the natural tool for studying the properties of the family of the level curves of a survival function and, hence, as will be seen in the next section, it allows to define certain ageing properties of $\overline{F}$. Notice that any copula C is a multivariate ageing function, since it can be obtained as the multivariate ageing function of a survival function $\overline{F}$ having copula C and univariate survival marginal $\overline{G}(t) = \exp(-t)$.

Example 8.5.4. Let $\overline{F}$ be a time-transformed exponential (TTE) model as described by Bassan and Spizzichino [2005b]. Then, it is known that the copula C of $\overline{F}$ is a (strict) Archimedean copula additively generated by a convex and decreasing function $\varphi : \mathbb{I} \to [0, +\infty]$ with $\varphi(1) = 0$ and $\varphi(0) = +\infty$, i.e., $C(u, v) = \varphi^{-1}(\varphi(u) + \varphi(v))$. Moreover, the related B is a strict triangular norm additively generated by $t = \varphi \circ \overline{G} \circ (-\ln)$. ∎

Remark 8.5.5. From the results of the last part of Section 8.4, the class of multivariate ageing functions cannot be expected to coincide with the class of quasi-copulas or continuous semi-copulas. ∎

Further readings 8.5.6. Further motivations and results about the multivariate ageing function and its use in Reliability Theory can be found in a series of papers by Bassan and Spizzichino [2001, 2003, 2005b]; Foschi and Spizzichino [2008]; Spizzichino [2001]. ∎

8.6   Multivariate ageing notions of NBU and IFR

The ageing function $B_{\overline{F}}$ of (8.5.3) may be used in order to define multivariate ageing notions by means of the following scheme (see [Bassan and Spizzichino, 2001, 2005b] for the bivariate case and [Durante et al., 2010b] for the general case):

(i)   Consider a univariate ageing notion P.

(ii) Take the joint survival function $\overline{F}$ of d independent and identically distributed (i.i.d.) lifetimes and prove results of the following type: each lifetime has the property P if, and only if, $B_{\overline{F}}$ has the property $\widetilde{P}$.

(iii) Define a multivariate ageing notion as follows: any exchangeable survival function $\overline{F}$ is multivariate-P if $B_{\overline{F}}$ has the property $\widetilde{P}$.

The rationale behind this procedure is grounded on the seminal work by Barlow and Mendel [1992] and Barlow and Spizzichino [1993], which have underlined the use of level sets to define suitable notions of multivariate ageing, related to Schur-concave functions.

Specifically, we concentrate our attention on notions that generalise the univariate concepts of *New Better than Used* (NBU) and *Increasing Failure Rate* (IFR), which are recalled below.

Definition 8.6.1. Let X be a lifetime on the probability space $(\Omega, \mathscr{F}, \mathbb{P})$ with survival function $\overline{G}$.

- $\overline{G}$ is NBU (i.e., it is *New Better than Used*) if $\mathbb{P}(X - s > t \mid X > s) \leq \mathbb{P}(X > t)$ for all $s, t \geq 0$, or, equivalently, $\overline{G}(x + y) \leq \overline{G}(x)\overline{G}(y)$ for all $x, y \in \mathbb{R}$.

- $\overline{G}$ is IFR (i.e., it has *Increasing Failure Rate*) if $s \mapsto \mathbb{P}(X - s > t \mid X > s)$ is decreasing for every $t \geq 0$, or, equivalently, $\overline{F}$ is log-concave. $\Diamond$

As will be shown, the multivariate notions to be introduced are satisfied by random vectors whose components are conditionally independent and identically distributed having an NBU (respectively, IFR) univariate conditional survival function. This circumstance has been considered as a natural requirement for Bayesian notions of multivariate ageing (see, e.g., [Barlow and Mendel, 1992]). Furthermore, these notions may also be interpreted in terms of comparison among conditional survival functions of residual lifetimes, given a same history of observed survivals.

In order to simplify notations we shall often formulate our results referring to one of the following assumptions:

Assumption 1 (exchangeable case): Let $\mathbf{X} = (X_1, \ldots, X_d)$ $(d \geq 2)$ be an exchangeable random vector of continuous lifetimes with joint survival function $\overline{F} : \mathbb{R}^d \to \mathbb{I}$ and identical univariate survival marginals equal to $\overline{G}$. We suppose that $\overline{G}$ is strictly decreasing on $\mathbb{R}$ with $\overline{G}(0) = 1$ and $\overline{G}(+\infty) = 0$.

Assumption 2 (i.i.d. case): Under Assumption 1, we suppose in addition that the components of $\mathbf{X}$ are independent.

We shall denote by B_Π the multivariate ageing function corresponding to the copula Π_d, namely

$$B_\Pi(\mathbf{u}) = \exp\left(-\overline{G}^{-1}\left(\overline{G}(-\ln(u_1))\cdots\overline{G}(-\ln(u_d))\right)\right). \qquad (8.6.1)$$

Within the family of the multivariate ageing functions, we define the following classes that may be used to express multivariate ageing notions.

Definition 8.6.2. Let B be a multivariate ageing function. It will be said that:

(A1) B belongs to $\mathscr{A}_1^+$ if, and only if, for every $\mathbf{u} \in \mathbb{I}^d$,

$$B(u_1, \ldots, u_d) \geq \Pi_d(u_1, \ldots, u_d). \qquad (8.6.2)$$

(A2) B belongs to $\mathscr{A}_2^+$ if, and only if, for all $i, j \in \{1, \ldots, d\}$, $i \neq j$, and for every $\mathbf{u} \in \mathbb{I}^d$,

$$B(u_1, \ldots, u_{i-1}, u_i, u_{i+1}, \ldots, u_{j-1}, u_j, u_{j+1}, \ldots, u_d)$$
$$\geq B(u_1, \ldots, u_{i-1}, u_i \cdot u_j, u_{i+1}, \ldots, u_{j-1}, 1, u_{j+1}, \ldots, u_d). \quad (8.6.3)$$

(A3) B belongs to $\mathscr{A}_3^+$ if, and only if, for all $i, j \in \{1, \ldots, d\}$, $i \neq j$, for all $u_i, u_j \in \mathbb{I}$, $u_i \geq u_j$, and for every $s \in]0, 1[$,

$$B(u_1, \ldots, u_i s, \ldots, u_j, \ldots, u_d) \geq B(u_1, \ldots, u_i, \ldots, u_j s, \ldots, u_d). \quad (8.6.4)$$

The corresponding classes $\mathscr{A}_i^-$ ($i = 1, 2, 3$) are defined by reversing the inequality signs in (8.6.2), (8.6.3) and (8.6.4), respectively. $\qquad \diamond$

Property (8.6.2) is a pointwise comparison between the multivariate ageing function B and the copula Π_d. The properties expressed by equations (8.6.3) and (8.6.4) are essentially inequalities related to the bivariate sections of B. In particular, (8.6.4) translates *supermigrativity* (compare [Durante and Ghiselli-Ricci, 2009, 2012]) of all the bivariate sections of B, while (8.6.3) is one of its weaker forms, obtained by letting $u_i = 1$ and $s = \frac{1}{u_j}$ in (8.6.4). Therefore, $\mathscr{A}_3^+ \subseteq \mathscr{A}_1^2$; but the converse inclusion is not true, as shown in Example 8.6.5 below.

Furthermore, $\mathscr{A}_2^+ \subseteq \mathscr{A}_1^+$. In fact, by iteratively applying (8.6.3), one obtains, for every $\mathbf{u} \in \mathbb{I}^d$,

$$B(u_1, \ldots, u_i, \ldots, u_j, \ldots, u_k, \ldots, u_d) \geq B(u_1, \ldots, u_i u_j, \ldots, 1, \ldots, u_k, \ldots, u_d)$$
$$\geq B(u_1, \ldots, u_i u_j u_k, \ldots, 1, \ldots, 1, \ldots, u_d)$$
$$\cdots$$
$$\geq B(1, \ldots, u_1 \cdots u_d, \ldots, 1) = u_1 \cdots u_d.$$

Since a multivariate ageing function has uniform margins, $B \in \mathscr{A}_2^+$ is equivalent to $B \in \mathscr{A}_1^+$ for the case $d = 2$. However, in the d-dimensional case, $d \geq 3$, $\mathscr{A}_2^+$ is strictly included in $\mathscr{A}_1^+$, as will be shown in Example 8.6.4 below.

In the following example, we consider the case of the so-called TTE models (see [Barlow and Spizzichino, 1993; Spizzichino, 2001]), whose multivariate survival functions admit an Archimedean copula.

Example 8.6.3. Let B be a multivariate ageing function that can be written in the form

$$B(\mathbf{u}) = \psi^{-1}\left(\sum_{i=1}^{d} \psi(u_i) \right) \qquad (8.6.5)$$

with a strictly decreasing additive generator $\psi : \mathbb{I} \to \mathbb{R}$ such that $\psi(0) = +\infty$ and $\psi(1) = 0$. Then B belongs to the class of the d-dimensional strict triangular norms. In particular, B is also a copula when ψ^{-1} is d-completely monotone (see Theorem 6.5.7). Now, for a semi-copula B of type (8.6.5) the following statements can be proved:

(i) B is in $\mathscr{A}_1^+$ if, and only if, B is in $\mathscr{A}_2^+$, and this happens when $\psi(uv) \leq \psi(u) + \psi(v)$ for all u and v in $\mathbb{I}$;

(ii) B is in $\mathscr{A}_3^+$ if, and only if, ψ^{-1} is log-convex (see [Bassan and Spizzichino, 2005b; Durante and Ghiselli-Ricci, 2009]).

Notice that the multivariate ageing functions B_{Π_d} of (8.6.1) are of the form (8.6.5) with $\psi(t) = -\ln(\overline{G}(-\ln(t)))$. ∎

The following examples clarify the relations among the classes introduced above.

Example 8.6.4. Let $f : \mathbb{I} \to \mathbb{I}$ be the function given by

$$f(t) = \begin{cases} et, & t \in \left[0, e^{-2}\right], \\ e^{-1}, & t \in \left]e^{-2}, e^{-1}\right], \\ t, & t \in \left]e^{-1}, 1\right]. \end{cases}$$

Let $C : \mathbb{I}^3 \to \mathbb{I}$ be given by $C(u_1, u_2, u_3) = u_{(1)} f(u_{(2)}) f(u_{(3)})$, where $u_{(1)}, u_{(2)}, u_{(3)}$ denote the components of $\mathbf{u}$ rearranged in increasing order. Since $f(1) = 1$, f is increasing, and $f(t)/t$ is decreasing on $]0, 1]$, it follows that C is a copula (see Example 1.3.7). It can be proved that C belongs to $\mathscr{A}_1^+$. However, C does not belong to $\mathscr{A}_2^+$. In fact, one has

$$C(u_1, u_2, u_3) = e^{-4} < e^{-7/2} = C(u_1, u_2 u_3, 1),$$

by taking $u_1 = e^{-5/2}$, $u_2 = e^{-3/2}$ and $u_3 = e^{-1/2}$. ∎

Example 8.6.5. Let B be the multivariate ageing function of type (8.6.5), where $\psi : \mathbb{I} \to \mathbb{R}$ is given by

$$\psi(t) = \begin{cases} -\ln t, & t \in \left]0, e^{-2-\varepsilon}\right], \\ -\frac{\varepsilon}{1+\varepsilon}(\ln t + 1) + 2, & t \in \left]e^{-2-\varepsilon}, e^{-1}\right], \\ -2\ln t, & t \in \left]e^{-1}, 1\right], \end{cases}$$

with $\varepsilon \in \,]0, 1[$. Now, let us consider $g : \mathbb{R} \to \mathbb{R}$, $g(t) = \psi(\exp(-t))$, given by

$$g(t) = \begin{cases} 2t, & t \in [0, 1], \\ \frac{\varepsilon}{1+\varepsilon}(t - 1) + 2, & t \in \,]1, 2+\varepsilon], \\ t, & t \in \,]2 + \varepsilon, +\infty[. \end{cases}$$

Now, g is not concave and, hence, ψ^{-1} is not log-convex. Thus, in view of Example 8.6.3, B does not belong to $\mathscr{A}_3^+$. However, it can be shown that $\psi(uv) \leq \psi(u)+\psi(v)$ for all u and v in $\mathbb{I}$. From Example 8.6.3, it follows that B is in $\mathscr{A}_2^+$.   ■

The families $\mathscr{A}_1^+, \mathscr{A}_2^+, \mathscr{A}_3^+$ will be used in order to define notions of positive ageing in terms of the multivariate ageing function B. Notice that, since negative properties can be introduced and studied in a similar way, they will not be considered in detail.

Theorem 8.6.6. *When Assumption 2 holds the following statements are equivalent:*

(a) $\overline{G}$ *is NBU;*

(b) B_Π *is in* $\mathscr{A}_1^+$;

(c) B_Π *is in* $\mathscr{A}_2^+$.

Proof. (a) $\Longleftrightarrow$ (b) Let $\overline{G}$ be NBU. It is easily proved by induction that, for all x and y in $\mathbb{R}$, $\overline{G}(x+y) \leq \overline{G}(x)\overline{G}(y)$ is equivalent to

$$\overline{G}\left(\sum_{i=1}^{d} x_i \right) \leq \prod_{i=1}^{d} \overline{G}(x_i),$$

for every $\mathbf{x} \in \mathbb{R}^d$. Setting $x_i = -\ln(u_i)$, one has, for every $\mathbf{u} \in \mathbb{I}^d$,

$$\overline{G}(-\ln(u_1 \cdots u_d)) \leq \overline{G}(-\ln(u_1)) \cdots \overline{G}(-\ln(u_d)). \qquad (8.6.6)$$

Thus, if one applies $\exp \circ (-\overline{G}^{-1})$ to both sides of the previous inequality, it follows in a straightforward manner that $B_\Pi \geq \Pi_d$.

(a) $\Longleftrightarrow$ (c) Since $\overline{G}$ is NBU, $\overline{G}(-\ln(u_i u_j)) \leq \overline{G}(-\ln(u_i))\overline{G}(-\ln(u_j))$ holds for all u_i and $u_j \in \mathbb{I}$. By multiplying both sides of this inequality by $\prod_{k \in \mathscr{I}} \overline{G}(-\ln(u_k))$, where $\mathscr{I} = \{1, \ldots, d\} \setminus \{i, j\}$ and $u_k \in \mathbb{I}$ for every $k \in \mathscr{I}$, and applying the function $\exp \circ (-\overline{G}^{-1})$ to both sides, one obtains

$$B_\Pi(u_1, \ldots, u_i, \ldots, u_j, \ldots, u_d) \geq B_\Pi(u_1, \ldots, u_i u_j, \ldots, 1, \ldots, u_d),$$

viz. $B_\Pi \in \mathscr{A}_2^+$.   □

Therefore, we can write $\mathscr{A}_1^+ \cap \{B_\Pi : \overline{G} \text{ is NBU}\} = \mathscr{A}_2^+ \cap \{B_\Pi : \overline{G} \text{ is NBU}\}$. Notice that, in general, $\mathscr{A}_1^+ \neq \mathscr{A}_2^+$.

Theorem 8.6.7. *When Assumption 2 holds, the following statements are equivalent:*

(a) $\overline{G}$ *is IFR;*

(b) $B_\Pi \in \mathscr{A}_3^+$.

Proof. Let $\overline{G}$ be IFR. This is equivalent to

$$\frac{\overline{G}(x_i + \sigma)}{\overline{G}(x_i)} \geq \frac{\overline{G}(x_j + \sigma)}{\overline{G}(x_j)},$$

for all x_i and x_j in $\mathbb{R}_+$, with $x_i \leq x_j$ and $\sigma \geq 0$. By setting $x_i = -\ln u_i$, $x_j = -\ln u_j$ and $\sigma = -\ln s$, one obtains

$$\overline{G}(-\ln(u_i s))\,\overline{G}(-\ln(u_j)) \geq \overline{G}(-\ln(u_j s))\,\overline{G}(-\ln u_i)\,,$$

for all u_i and u_j in $]0,1]$, with $u_i \geq u_j$ and $s \in\,]0,1[$. By multiplying both sides of the inequality by $\prod_{k \in \mathscr{I}} \overline{G}(-\ln(u_k))$, where $\mathscr{I} = \{1,\ldots,d\} \setminus \{i,j\}$ and $u_k \in \mathbb{I}$ for every $k \in \mathscr{I}$, and applying the function $\exp \circ (-\overline{G}^{-1})$ to both sides, one obtains

$$B_\Pi(u_1,\ldots,u_i s,\ldots,u_j,\ldots,u_d) \geq B_\Pi(u_1,\ldots,u_i,\ldots,u_j s,\ldots,u_d)\,,$$

namely $B_\Pi \in \mathscr{A}_1^3$. $\square$

The previous result is actually a reformulation in terms of the multivariate ageing function of well-known results concerning the joint survival function $\overline{F} = \overline{F}_{\Pi,\overline{G}}$ of independent and identically distributed lifetimes that are IFR. As noted several times in the literature (for example, in [Barlow and Mendel, 1992; Barlow and Spizzichino, 1993; Spizzichino, 2001]), such a $\overline{F}$ is *Schur-concave*, i.e., for every $s \geq 0$ and for all i and j in $\{1,\ldots,d\}$, $i < j$, the mapping

$$x_i \mapsto \overline{F}(x_1,\ldots,x_{i-1},x_i,x_{i+1}\ldots,x_{j-1},s-x_i,x_{j+1},\ldots,x_d)$$

is decreasing on $\left[\frac{s}{2},+\infty\right]$ (see [Marshall and Olkin, 1979]). This is equivalent to

$$\overline{F}(x_1,\ldots,x_i+\tau,\ldots,x_j-\tau,\ldots,x_d) \geq \overline{F}(x_1,\ldots,x_i,\ldots,x_j,\ldots,x_d) \quad (8.6.7)$$

for all $i,j \in \{1,\ldots,d\}$, $i < j$, for every $\mathbf{x} \in \mathbb{R}^d$ such that $x_i \leq x_j$ and for every $\tau \in [0,x_j-x_i]$.

Now, by using Propositions 8.6.6 and 8.6.7 and the scheme presented before, the following notions of multivariate ageing for an exchangeable survival function $\overline{F}$ can be introduced.

Definition 8.6.8. When Assumption 1 holds, one says that:

- $\overline{F}$ is *B-multivariate-NBU of the first type* (*B*-MNBU1) if, and only if, $B_{\overline{F}}$ is in $\mathscr{A}_1^+$;
- $\overline{F}$ is *B-multivariate-NBU of the second type* (*B*-MNBU2) if, and only if, $B_{\overline{F}}$ is in $\mathscr{A}_2^+$;
- $\overline{F}$ is *B-multivariate-IFR* (*B*-MIFR) if, and only if, $B_{\overline{F}}$ is in $\mathscr{A}_3^+$. $\diamond$

In order to avoid confusion with other multivariate notions of ageing introduced in the literature, we used the prefix B for the notions introduced above. This also underlines the fact that all these notions are expressed in terms of the multivariate ageing functions B of $\overline{F}$. Now, we should like to investigate some properties of these notions.

First, notice that any k-dimensional marginal of $\overline{F}$ ($2 \leq k \leq d-1$) has the same multivariate ageing property of $\overline{F}$. This point is formalised in the following result.

Theorem 8.6.9. *Suppose that Assumption 1 holds. For every* $k \in \{2, \ldots, d\}$, *let* $\overline{F}^{(k)}$ *be the* k-*dimensional marginal of* $\overline{F}$. *If* $\overline{F}$ *is B-MNBU1 (respectively, B-MNBU2 or B-MIFR), then* $\overline{F}^{(k)}$ *is B-MNBU1 (respectively, B-MNBU2 or B-MIFR).*

Proof. If $\overline{F}^{(k)} : \mathbb{R}^k \to \mathbb{I}$ is the k-dimensional margin of $\overline{F}$ ($2 \leq k \leq d$), given by

$$\overline{F}^{(k)}(x_1, \ldots, x_k) = \overline{F}(x_1, \ldots, x_k, 0, \ldots, 0),$$

then it follows that

$$B_{\overline{F}^{(k)}}(u_1, \ldots, u_k) = B_{\overline{F}}(u_1, \ldots, u_k, 1, \ldots, 1).$$

Easy calculations show that $B_{\overline{F}^{(k)}}$ is in $\mathscr{A}_1^+$ (respectively, $\mathscr{A}_2^+$ or $\mathscr{A}_3^+$), when $B_{\overline{F}}$ is in $\mathscr{A}_1^+$ (respectively, $\mathscr{A}_2^+$ or $\mathscr{A}_3^+$), which is the desired assertion. $\square$

The previous definitions of multivariate ageing admit a probabilistic interpretation in terms of conditional survival probabilities for residual lifetimes. Before stating them, we clarify the notation. For every $\mathbf{x} \in \mathbb{R}^d$ we denote by $\hat{\mathbf{x}}_i$ the vector of $\mathbb{R}^{d-1}$ obtained by depriving $\mathbf{x}$ of its i-th component, $\hat{\mathbf{x}}_i = (x_1, \ldots, x_{i-1}, x_{i+1}, \ldots, x_d)$. A similar convention will apply for random vectors.

Theorem 8.6.10. *If Assumption 1 holds, then*

(a) $\overline{F}$ *is B-MNBU1 if, and only if, for all* $i \in \{1, \ldots, d\}$, $\mathbf{x} \in \mathbb{R}^d$ *and* $\tau > 0$,

$$\mathbb{P}\left(X_1 > x_1, \ldots, X_i > x_i + \tau, \ldots X_d > x_d \mid X_i > x_i\right)$$
$$\geq \mathbb{P}\left(X_i > x_1 + \cdots + x_i + \cdots + x_d + \tau \mid X_i > x_i\right). \quad (8.6.8)$$

(b) $\overline{F}$ *is B-MNBU2 if, and only if, for all* $i, j \in \{1, \ldots, d\}$, $i \neq j$, $\hat{\mathbf{x}}_j \in \mathbb{R}^{d-1}$ *and* $\tau > 0$,

$$\mathbb{P}\left(X_j > \tau \mid \hat{\mathbf{X}}_j > \hat{\mathbf{x}}_j\right) \geq \mathbb{P}\left(X_i > \tau + x_i \mid \hat{\mathbf{X}}_j > \hat{\mathbf{x}}_j\right). \quad (8.6.9)$$

(c) $\overline{F}$ *is B-MIFR if, and only if, for all* $i, j \in \{1, \ldots, d\}$, *for every* $\mathbf{x} \in \mathbb{R}^d$ *such that* $x_i \leq x_j$, *and for every* $\tau > 0$,

$$\mathbb{P}\left(X_i > x_i + \tau \mid \mathbf{X} > \mathbf{x}\right) \geq \mathbb{P}\left(X_j > x_j + \tau \mid \mathbf{X} > \mathbf{x}\right). \quad (8.6.10)$$

Proof. (a) By definition, $\overline{F}$ is B-MNBU1 if, and only if, for every $\mathbf{u} \in \mathbb{I}^d$,

$$\exp\left(\overline{G}^{-1}\left(\overline{F}(-\ln u_1, \ldots, -\ln u_d)\right)\right) \geq u_1 \cdots u_d.$$

Thus, for every $\mathbf{x} \in \mathbb{R}^d$, one has

$$\overline{F}(x_1, \ldots, x_d) \geq \overline{G}(x_1 + \cdots + x_d), \quad (8.6.11)$$

which is equivalent to eq. (8.6.8).

(b) Since $\overline{F}$ is B-MNBU2, for all $i, j \in \{1, \ldots, d\}$, with $i \neq j$, and for every $\mathbf{u} \in \mathbb{I}^d$,

$$\exp\left(\overline{G}^{-1}\left(\overline{F}(-\ln u_1, \ldots, -\ln u_i, \ldots, -\ln u_j, \ldots, -\ln u_d))\right)\right)$$

$$\geq \exp\left(\overline{G}^{-1}\left(\overline{F}(-\ln u_1, \ldots, -\ln u_i u_j, \ldots, 1, \ldots, -\ln u_d))\right)\right),$$

which is equivalent to

$$\overline{F}(x_1, \ldots, x_i, \ldots, \tau, \ldots, x_d) \geq \overline{F}(x_1, \ldots, \tau + x_i, \ldots, 0, \ldots, x_d) \qquad (8.6.12)$$

for all $\hat{\mathbf{x}}_i \in \mathbb{R}^{d-1}$ and $\tau > 0$. This last condition can be expressed as

$$\mathbb{P}(X_j > \tau \mid X_1 > x_1, \ldots, X_i > x_i, \ldots, X_j > 0, \ldots, X_d > x_d)$$

$$\geq \mathbb{P}(X_i > \tau + x_i \mid X_1 > x_1, \ldots, X_i > x_i, \ldots, X_j > 0, \ldots, X_d > x_d),$$

which is the desired assertion.

(c) One has just to prove that $\overline{F}$ is B-MIFR if, and only if, $\overline{F}$ is Schur-concave. Then, the assertion will follow, since the Schur-concavity of $\overline{F}$ is equivalent to the fact that eq. (8.6.10) holds (see [Spizzichino, 2001, Proposition 4.15]).

Now, the equivalence between $\overline{F}$ being B-MIFR and $\overline{F}$ being Schur-concave follows by extending Lemma 4.2 by Bassan and Spizzichino [2005b] from the bivariate to the d-dimensional case. In detail, $\overline{F}$ is Schur-concave if, and only if, for all $i, j \in \{1, \ldots, d\}$, $(i < j)$, for every $\mathbf{x} \in \mathbb{R}^d$ such that $x_i \leq x_j$ and for every $\tau \in [0, (x_j - x_i)/2]$, inequality (8.6.7) holds. In terms of $B_{\overline{F}}$, this is equivalent to

$$B_{\overline{F}}(e^{-x_1}, \ldots, e^{-x_{i-1}}, e^{-x_i-\tau}, e^{-x_{i+1}}, \ldots, e^{-x_{j-1}}, e^{-x_j+\tau}, e^{-x_{j+1}}, \ldots, e^{-x_d})$$

$$\geq B_{\overline{F}}(e^{-x_1}, \ldots, e^{-x_{i-1}}, e^{-x_i}, e^{-x_{i+1}}, \ldots, e^{-x_{j-1}}, e^{-x_j}, e^{-x_{j+1}}, \ldots, e^{-x_d}). \qquad (8.6.13)$$

In other words,

$$B_{\overline{F}}\left(u_1, \ldots, u_i s, \ldots, \frac{u_j}{s}, \ldots, u_d\right) \geq B_{\overline{F}}(u_1, \ldots, u_d), \qquad (8.6.14)$$

for all $i, j \in \{1, \ldots, d\}$, $i < j$, for every $\mathbf{u} \in]0, 1]^d$ such that $u_i \geq u_j$ and for every $s \in [u_j/u_i, 1]$, which is an equivalent way of expressing the fact that $B_{\overline{F}}$ is in $\mathscr{A}_3^+$. □

Note that conditions (8.6.9) and (8.6.10) can be expressed as a comparison between residual lifetimes, conditionally on a same history. Specifically, the condition "$\overline{F}$ is B-MNBU2" is equivalent to

$$\left[X_i \mid \hat{\mathbf{X}}_i > \hat{\mathbf{x}}_i\right] \geq_{st} \left[X_j - x_j \mid \hat{\mathbf{X}}_i > \hat{\mathbf{x}}_i\right], \qquad (8.6.15)$$

for all $i, j \in \{1, \ldots, d\}$, $i \neq j$, and for every $\mathbf{x} \in \mathbb{R}^d$, where $\geq_{st}$ denotes the univariate usual stochastic order (see [Shaked and Shanthikumar, 2007]). Instead, the fact that $\overline{F}$ is B-MIFR may be expressed as

$$[X_i - x_i \mid \mathbf{X} > \mathbf{x}] \geq_{st} [X_j - x_j \mid \mathbf{X} > \mathbf{x}], \qquad (8.6.16)$$

for all $i, j \in \{1, \ldots, d\}$, for every $\mathbf{x} \in \mathbb{R}^d$ such that $x_i \leq x_j$.

Thanks to the probabilistic interpretations given by (8.6.15) and (8.6.16), an interesting link between B-MNBU2 and B-MIFR can be proved. Consider the vector of the residual lifetimes of $\mathbf{X}$ at time $t > 0$, $\mathbf{X}_t = [\mathbf{X} - \mathbf{t} \mid \mathbf{X} > \mathbf{t}]$, where $\mathbf{t} = (t, \ldots, t)$. Let us denote by $\overline{F}_t : \mathbb{R}^d \to \mathbb{I}$ the joint survival function of $\mathbf{X}_t$ and by $B_{\overline{F}_t}$ the corresponding multivariate ageing function. By extending results related to the bivariate case (see [Bassan and Spizzichino, 2003, 2005a,b; Foschi and Spizzichino, 2008]), the following can be proved.

Theorem 8.6.11. *When Assumption 1 holds, $\overline{F}_t$ is B-MNBU2 for every $t \geq 0$ if, and only if, $\overline{F}$ is B-MIFR.*

Proof. $\overline{F}_t$ is B-MNBU2 for every $t \geq 0$ if, and only if,

$$\overline{F}(x_1 + t, \ldots, x_i + t, \ldots, x_j + t, \ldots, x_d + t)$$
$$\geq \overline{F}(x_1 + t, \ldots, t, \ldots, x_j + x_i + t, \ldots, x_d + t),$$

for all $t \geq 0$, $\mathbf{x} \in \mathbb{R}^d$ and $i, j \in \{1, \ldots, d\}$, which is equivalent to the fact that $\overline{F}$ is Schur-concave. $\qquad\square$

Note that if $\overline{F}$ is B-MNBU2, then $\overline{F}_t$ may not be B-MNBU2 for some $t > 0$ (see [Foschi, 2013] for an example in the bivariate case). However, for the notion of B-MIFR, one can prove the following result.

Corollary 8.6.12. *When Assumption 1 holds, if $\overline{F}$ is B-MIFR, then $\overline{F}_t$ is B-MIFR for every $t \geq 0$.*

Proof. From Theorem 8.6.11, if $\overline{F}$ is B-MIFR, then $\overline{F}_{t+s}$ is B-MNBU2 for every $t, s \geq 0$. As a consequence, $\overline{F}_t$ is B-MIFR for every $t \geq 0$. $\qquad\square$

As for inequality (8.6.8), it is not clear whether it can also be expressed as a comparison of lifetimes conditionally on the same history, in a similar way to the inequalities in (8.6.15) and (8.6.16). However, it is possible to give it an intuitive interpretation in reliability terms, similarly to what was done in Example 4.2 of [Bassan and Spizzichino, 2005b] for the case $d = 2$.

Remark 8.6.13. Inequality (8.6.9) implies inequality (8.6.8); this can be seen by using the multivariate ageing function $B_{\overline{F}}$ and the definitions of B-MNBU1 and B-MNBU2. Actually, as shown in Example (8.6.3), inequalities (8.6.9) and (8.6.8) coincide for TTE models, but not in general. Consider, for instance, a multivariate survival function $\overline{F}$ whose marginals are exponential and whose copula is that of Example 8.6.4. $\qquad\blacksquare$

The notions of multivariate ageing introduced in Definition 8.6.8 are preserved under mixtures, as specified by the following theorem.

Theorem 8.6.14. *Let* $(\overline{F}_\theta)_{\theta \in \Theta}$ *be a family of survival functions satisfying Assumption 1. Let* μ *be a distribution on* Θ. *Let* $\overline{F}$ *be the mixture of* $(\overline{F}_\theta)_{\theta \in \Theta}$ *with respect to* μ, *given, for every* $\mathbf{x} \in \mathbb{R}^d$, *by*

$$\overline{F}(\mathbf{x}) = \int_\Theta \overline{F}_\theta(\mathbf{x}) \, d\mu(\theta).$$

The following statements hold:

(a) *if* $\overline{F}_\theta$ *is B-MNBU1 for every* $\theta \in \Theta$, *then* $\overline{F}$ *is B-MNBU1;*

(b) *if* $\overline{F}_\theta$ *is B-MNBU2 for every* $\theta \in \Theta$, *then* $\overline{F}$ *is B-MNBU2;*

(c) *if* $\overline{F}_\theta$ *is B-MIFR for every* $\theta \in \Theta$, *then* $\overline{F}$ *is B-MIFR.*

Proof. Part (a) follows by considering that every $\overline{F}_\theta$ satisfies (8.6.11) and hence the mixture $\overline{F}$ satisfies (8.6.11), which is an equivalent formulation of the B-MNBU1 property for $\overline{F}$. Analogously, part (b) easily follows from the fact that every $\overline{F}_\theta$ satisfies (8.6.12).

Finally, if every $\overline{F}_\theta$ is B-MIFR, then it is Schur-concave. As a consequence, the mixture $\overline{F}$ is also Schur-concave and therefore B-MIFR (see [Marshall and Olkin, 1979]). □

Consequently, the following interesting result can be easily derived.

Theorem 8.6.15. *Let Assumption 1 hold and suppose that* $\overline{F}$ *is the survival function of conditionally i.i.d. lifetimes given a common factor* Θ *with prior distribution* μ. *Moreover, suppose that* $\overline{G}(\cdot \mid \theta)$ *is NBU (respectively, IFR). Then* $\overline{F}$ *is B-MNBU2 (respectively, B-MIFR).*

Thus, the given definitions of multivariate ageing exhibit an interesting property: mixtures of independent and identically distributed lifetimes that are NBU (respectively, IFR) conditionally on the same factor Θ are also multivariate NBU (respectively, IFR).

Finally, we should like to discuss a possible application of the previous results to the construction of multivariate stochastic models. To this end, we present the following theorem that extends results by Bassan and Spizzichino [2005b] to the multivariate case. It provides a link between the survival copula C of a multivariate survival function $\overline{F}$ and the ageing properties of $\overline{F}$ itself.

Theorem 8.6.16. *If Assumption 1 is satisfied, then the following statements hold:*

(a) *if* C *is in* $\mathscr{A}_1^+$ *and* $\overline{G}$ *is NBU, then* $\overline{F}$ *is B-MNBU1;*

(b) *if* C *is in* $\mathscr{A}_2^+$ *and* $\overline{G}$ *is NBU, then* $\overline{F}$ *is B-MNBU2;*

(c) *if* C *is in* $\mathscr{A}_3^+$ *and* $\overline{G}$ *is IFR, then* $\overline{F}$ *is B-MIFR.*

Proof. (a) Let C be in $\mathscr{A}_1^+$. Then, for every $\mathbf{u} \in \mathbb{I}^d$,

$$\exp\left(-\overline{G}^{-1}\left(C(\overline{G}(-\ln u_1), \ldots, \overline{G}(-\ln u_d)))\right)\right) \geq B_\Pi(u_1, \ldots, u_d).$$

By considering Theorem 8.6.6(a), it follows that B belongs to $\mathscr{A}_1^+$.

(b) Let C be in $\mathscr{A}_2^+$. Then, for every $\mathbf{u} \in \mathbb{I}^d$,

$$\overline{G}\left(-\ln\left(B\left(e^{-\overline{G}^{-1}(u_1)}, \ldots, e^{-\overline{G}^{-1}(u_i)}, \ldots, e^{-\overline{G}^{-1}(u_j)}, \ldots, e^{-\overline{G}^{-1}(u_d)}\right)\right)\right)$$

$$\geq \overline{G}\left(-\ln\left(B\left(e^{-\overline{G}^{-1}(u_1)}, \ldots, e^{-\overline{G}^{-1}(u_i u_j)}, \ldots, 1, \ldots, e^{-\overline{G}^{-1}(u_d)}\right)\right)\right)$$

$$\geq \overline{G}\left(-\ln\left(B\left(e^{-\overline{G}^{-1}(u_1)}, \ldots, e^{-(\overline{G}^{-1}(u_i)+\overline{G}^{-1}(u_j))}, \ldots, 1, \ldots, e^{-\overline{G}^{-1}(u_d)}\right)\right)\right),$$

where the last inequality follows from the fact that $\overline{G}$ is NBU.

Setting $x_i = e^{-\overline{G}^{-1}(u_i)}$, it follows that B belongs to $\mathscr{A}_2^+$, which is the desired assertion.

(c) Since C is in $\mathscr{A}_3^+$ for every $\mathbf{u} \in \mathbb{I}^d$ such that $u_i \geq u_j$ and for every $s \in \,]0,1[$,

$$C(u_1, \ldots, u_i\,s, \ldots, u_j, \ldots, u_d) \geq C(u_1, \ldots, u_i, \ldots, u_j\,s, \ldots, u_d).$$

In particular, for every $0 < s_j \leq s_i < 1$,

$$C(u_1, \ldots, u_i\,s_i, \ldots, u_j, \ldots, u_d) \geq C(u_1, \ldots, u_i, \ldots, u_j\,s_j, \ldots, u_d). \quad (8.6.17)$$

Now, for every $k \in \{1, \ldots, d\}$, set

$$\alpha_k = \overline{G}^{-1}(u_k), \quad s_i = \frac{\overline{G}(\alpha_i + \sigma)}{\overline{G}(\alpha_i)}, \quad s_j = \frac{\overline{G}(\alpha_j + \sigma)}{\overline{G}(\alpha_j)},$$

where $\sigma = \overline{G}^{-1}(u_i\,s_i) - \overline{G}^{-1}(u_i) = \overline{G}^{-1}(u_j\,s_j) - \overline{G}^{-1}(u_j)$. Since $\overline{G}$ is IFR, $s_i \geq s_j$. Moreover, (8.6.17) yields

$$C\left(\overline{G}(\alpha_1), \ldots, \overline{G}(\alpha_i + \sigma), \ldots, \overline{G}(\alpha_j), \ldots, \overline{G}(\alpha_d)\right)$$

$$\geq C\left(\overline{G}(\alpha_1), \ldots, \overline{G}(\alpha_i), \ldots, \overline{G}(\alpha_j + \sigma), \ldots, \overline{G}(\alpha_d)\right).$$

By applying to both sides of this inequality the transformation $\exp \circ (-\overline{G}^{-1})$, one has

$$B(x_1, \ldots, x_i\,s', \ldots, x_j, \ldots, x_d) \geq B(x_1, \ldots, x_i, \ldots, x_j\,s', \ldots, x_d),$$

for every $\mathbf{x} \in \mathbb{I}^d$ such that $x_i \geq x_j$ and for every $s' \in \,]0,1[$, namely $B_{\overline{F}} \in \mathscr{A}_3^+$. $\square$

Remark 8.6.17. Let C be a copula. As already noted, either condition $C \in \mathscr{A}_2^+$ or $C \in \mathscr{A}_3^+$ implies $C \in \mathscr{A}_1^+$, which is considered as a notion of multivariate positive dependence. Thus, roughly speaking, Theorem 8.6.16 suggests that positive univariate ageing and (a kind of) positive dependence play in favour of positive multivariate ageing. However, note that positive multivariate ageing can coexist with several forms of dependence and univariate ageing: this fact was already stressed, for example, by Bassan and Spizzichino [2005b] (see also [Pellerey, 2008]). ∎

Interestingly (at least for statistical purposes), Theorem 8.6.16 can be used when one wishes to construct, for components judged to be similar, a multivariate survival model that satisfies one of the ageing conditions. In fact, by using Sklar's theorem, such a model can be constructed just by conveniently choosing univariate survival functions $\overline{G}$ (e.g., satisfying NBU or IFR property) and a suitable copula C (belonging to one of the classes $\mathscr{A}_i^+$); then they are joined in order to obtain the multivariate survival function $\overline{F} = C(\overline{G}, \ldots, \overline{G})$. This procedure hence provides sufficient conditions for multivariate ageing in terms of univariate ageing and stochastic dependence.

Bibliography

Aas, K., Czado, C., Frigessi, A., and Bakken, H. (2009). Pair-copula constructions of multiple dependence. *Insurance Math. Econom.*, 44(2):182–198.

Abel, N. H. (1826). Untersuchungen der Functionen zweier unabghängig verändlichen Grössen x und y wie $f(x, y)$ welche die Eigenschaft haben, dass $f(z, (f(x, y))$ eine symmetrische Function von x, y und z ist. *J. Reine Angew. Math.*, 1:11–15. Also in *Oeuvres complètes de N.H. Abel*, Vol. I, Christiana, 1881, pp. 61–65.

Aczél, J. (1949). Sur les opérations définies pour nombres réels. *Bull. Soc. Math. France*, 76:59–64.

Aigner, M. and Ziegler, G. M. (1999). *Proofs from THE BOOK*. Springer, Berlin.

Albanese, A. and Sempi, C. (2004). Countably generated idempotent copulæ. In López-Díaz, M., Gil, M. A., Grzegorzewski, P., Hryniewicz, O., and Lawry, J., editors, *Soft methodology and random information systems*, pages 197–204. Springer, Berlin/Heidelberg.

Ali, M. M., Mikhail, N. N., and Haq, M. S. (1978). A class of bivariate distributions including the bivariate logistic. *J. Multivariate Anal.*, 8(3):405–412.

Alsina, C., Frank, M. J., and Schweizer, B. (2006). *Associative functions. Triangular norms and copulas*. World Scientific, Singapore.

Alsina, C., Nelsen, R. B., and Schweizer, B. (1993). On the characterization of a class of binary operations on distribution functions. *Statist. Probab. Lett.*, 17:85–89.

Alsina, C. and Schweizer, B. (1988). Mixtures are not derivable. *Found. Phys. Lett.*, 1:171–174.

Alvoni, E., Papini, P. L., and Spizzichino, F. (2009). On a class of transformations of copulas and quasi–copulas. *Fuzzy Sets and Systems*, 160:334–343.

Amblard, C. and Girard, S. (2001). Une famille semi-paramétrique de copules symétriques bivariées. *C. R. Acad. Sci. Paris Sér. I Math.*, 333:129–132.

Amblard, C. and Girard, S. (2002). Symmetry and dependence properties within a semiparametric family of bivariate copulas. *J. Nonparametr. Stat.*, 14:715–727.

Amblard, C. and Girard, S. (2009). A new symmetric extension of FGM copulas. *Metrika*, 70:1–17.

Angus, J. E. (1994). The probability integral transform and related results. *SIAM Rev.*, 36(4):652–654.

Arnold, B. C., Balakrishnan, N., and Nagaraja, H. N. (1992). *A first course in order statistics*. Wiley, New York. Reprinted, SIAM, Philadelphia, 2008.

Ash, R. B. (2000). *Real analysis and probability*. Harcourt/Academic Press, Burlington, MA. 2nd. ed., with contributions by Catherine Doléans–Dade.

Barlow, R. E. and Mendel, M. B. (1992). de Finetti–type representations for life distributions. *J. Amer. Statist. Assoc.*, 87:1116–1122.

Barlow, R. E. and Spizzichino, F. (1993). Schur–concave survival functions and survival analysis. *J. Comput. Appl. Math.*, 46:437–447.

Barnsley, M. F. (1988). *Fractals everywhere*. Academic Press, Boston.

Bassan, B. and Spizzichino, F. (2001). Dependence and multivariate aging: the role of level sets of the survival function. In Hayakawa, Y., Irony, T., and Xie, M., editors, *System and Bayesian reliability*, volume 5 of *Series on Quality, Reliability and Engineering Statistics*, pages 229–242. World Scientific, Singapore.

Bassan, B. and Spizzichino, F. (2003). On some properties of dependence and aging for residual ifetimes in the exchangeable case. In Lindqvist, B. H. and Doksum, K. A., editors, *Mathematical and statistical methods in reliability*, volume 7 of *Series on Quality, Reliability and Engineering Statistics*, pages 235–249. World Scientific, Singapore.

Bassan, B. and Spizzichino, F. (2005a). Bivariate survival models with Clayton aging functions. *Insurance Math. Econom.*, 37:6–12.

Bassan, B. and Spizzichino, F. (2005b). Relations among univariate aging, bivariate aging and dependence for exchangeable lifetimes. *J. Multivariate Anal.*, 93:313–339.

Bauer, H. (1996). *Probability theory*. de Gruyter, Berlin/New York.

Bauer, H. (2001). *Measure and integration theory*. de Gruyter, Berlin.

Beare, B. K. (2010). Copulas and temporal dependence. *Econometrica*, 78:395–410.

Bedford, T. and Cooke, R. M. (2002). Vines–a new graphical model for dependent random variables. *Ann. Statist.*, 30(4):1031–1068.

Beirlant, J., Goegebeur, Y., Teugels, J., and Segers, J. (2004). *Statistics of extremes*. Wiley Series in Probability and Statistics. John Wiley & Sons Ltd., Chichester.

Beneš, V. and Štěpán, J. (1991). Extremal solutions in the marginal problem. In Dall'Aglio, G., Kotz, S., and Salinetti, G., editors, *Advances in probability distributions with given marginals (Rome, 1990)*, pages 189–206. Kluwer, Dordrecht.

Beneš, V. and Štěpán, J., editors (1997). *Distributions with given marginal and moment problems*. Kluwer, Dordrecht.

Bernard, C. and Czado, C. (2015). Conditional quantiles and tail dependence. *J. Multivariate Anal.*, page in press.

Bertino, S. (1977). Sulla dissomiglianza tra mutabili cicliche. *Metron*, 35:53–88. In Italian.

Bielecki, T. R., Jakubowski, J., and Nieweglowski, M. (2010). Dynamic modeling of dependence in finance via copulae between stochastic processes. In Jaworski, P., Durante, F., Härdle, W. K., and Rychlik, T., editors, *Copula theory and its applications*, volume 198 of *Lecture Notes in Statistics - Proceedings*, pages 33–76. Springer, Berlin/Heidelberg.

Billingsley, P. (1968). *Convergence of probability measures*. Wiley, New York.

Billingsley, P. (1979). *Probability and measure*. Wiley, New York. 3rd ed., 1995.

Birkhoff, G. (1940). *Lattice theory*, volume 25 of *AMS Colloquium Publications*. American Mathematical Society, Providence, RI. 3rd ed., 1967.

Birkhoff, G. (1946). Tres observaciones sobre el algebra lineal. *Univ. Nac. Tucumán, Rev. Ser. A*, 5:147–151.

Blomqvist, N. (1950). On a measure of dependence between two random variables. *Ann. Math. Statistics*, 21:593–600.

Bonferroni, C. E. (1936). Teoria statistica delle classi e calcolo delle probabilità. *Pubbl. R. Ist. Super. Sci. Econom. Commerciali di Firenze*, 8:1–62. In Italian.

Brezis, H. (1983). *Analyse fonctionnelle. Théorie et applications*. Masson, Paris. English version: *Functional analysis, Sobolev spaces and partial differential equations*, Springer, New York, 2010.

Brigo, D., Pallavicini, A., and Torresetti, R. (2010). *Credit models and the crisis: a journey into CDOs, copulas, correlations and dynamic models*. Wiley Finance, Chichester.

Brouwer, L. E. J. (1909). Die Theorie der endlichen kontinuierlichen Gruppen unabhängig von der Axiomen von Lie. *Math. Ann.*, 67:127–136.

Brown, J. R. (1965). Doubly stochastic measures and Markov operators. *Michigan Math. J.*, 12:367–375.

Brown, J. R. (1966). Approximation theorems for Markov operators. *Pacific J. Math.*, 16:13–23.

Bruckner, A. M. (1971). Differentiation of integrals. *Amer. Math. Monthly*, 78:ii+1–51.

Bücher, A., Segers, J., and Volgushev, S. (2014). When uniform weak convergence fails: empirical processes for dependence functions and residuals via epi- and hypographs. *Ann. Stat.*, 42(4):1598–1634.

Burchard, A. and Hajaiej, H. (2006). Rearrangement inequalities for functionals with monotone integrands. *J. Funct. Anal.*, 233:561–582.

Busemann, H. and Feller, W. (1934). Zur Differentiation der Lebegueschen Integrale. *Fund. Math.*, 22:226–256.

Capéraà, P., Fougères, A.-L., and Genest, C. (1997). A stochastic order based on a decomposition of Kendall's tau. In Beneš, V. and Štěpán, J., editors, *Distributions with given marginal and moment problems*, pages 81–86. Kluwer, Dordrecht.

Capéraà, P., Fougères, A.-L., and Genest, C. (2000). Bivariate distribution with given extreme value attractor. *J. Multivariate Anal.*, 72:30–49.

Carley, H. and Taylor, M. (2003). A new proof of Sklar's theorem. In Cuadras, C. M., Fortiana, J., and Rodríguez Lallena, J., editors, *Distributions with given marginals and Statistical Modelling*, pages 29–34. Kluwer, Dordrecht.

Cartan, E. (1930). La théorie des groupes finis et continus at l'analyse situs. *Mem. Sci. Math.*, 42:1–61.

Cerqueti, R. and Spizzichino, F. (2013). Extension of dependence properties to semi–copulas and applications to the mean–variance model. *Fuzzy Sets and Systems*, 220:99–108.

Charpentier, A., Fougègers, A.-L., Genest, C., and Nešlehová, J. (2014). Multivariate Archimax copulas. *J. Multivariate Anal.*, 126:118–136.

Charpentier, A. and Juri, A. (2006). Limiting dependence structures for tail events, with applications to credit derivatives. *J. Appl. Probab.*, 43(2):563–586.

Charpentier, A. and Segers, J. (2007). Lower tail dependence for Archimedean copulas: characterizations and pitfalls. *Insurance Math. Econom.*, 40(3):525–532.

Charpentier, A. and Segers, J. (2009). Tails of multivariate Archimedean copulas. *J. Multivariate Anal.*, 100(7):1521–1537.

Cherubini, U., Luciano, E., and Vecchiato, W. (2004). *Copula methods in finance*. Wiley, New York.

Cherubini, U., Mulinacci, S., and Durante, F., editors (2015). *Marshall–Olkin distributions – Advances in theory and practice*, Springer Proceedings in Mathematics & Statistics. Springer.

Cherubini, U., Mulinacci, S., Gobbi, F., and Romagnoli, S. (2011a). *Dynamic copula methods in finance*. Wiley, New York.

Cherubini, U., Mulinacci, S., and Romagnoli, S. (2011b). A copula–based model of speculative price dynamics in discrete time. *J. Multivariate Anal.*, 102:1047–1063.

Cherubini, U., Mulinacci, S., and Romagnoli, S. (2011c). On the distribution of the (un)bounded sum of random variables. *Insurance: Math. and Econom.*, 48:56–63.

Choquet, G. (1953–1954). Théorie des capacités. *Ann. Inst. Fourier Grenoble*, 5:131–195.

Choroś, B., Ibragimov, R., and Permiakova, E. (2010). Copula estimation. In Jaworski, P., Durante, F., Härdle, W. K., and Rychlik, T., editors, *Copula theory and its applications*, volume 198 of *Lecture Notes in Statistics – Proceedings*, pages 77–91. Springer, Berlin/Heidelberg.

Clayton, D. G. (1978). A model for association in bivariate life tables and its application in epidemiological studies in familial tendency in chronic desease indicidence. *Biometrika*, 65:141–151.

Clemen, R. T. and Reilly, T. (1999). Correlations and copulas for decision and risk analysis. *Management Science*, 45(2):208–224.

Clifford, A. H. (1954). Naturally totally ordered commutative semigroups. *American J. Math.*, 76:631–646.

Climescu, A. C. (1946). Sur l'équation fonctionnelle de l'associativité. *Bul. Politehn. Gh. Asachi. Iaşi*, 1:211–224.

Cook, R. D. and Johnson, M. E. (1981). A family of distributions for modelling nonelliptically symmetric multivariate data. *J. Roy. Statist. Soc. Ser. B*, 43:210–218.

Cornfeld, I. P., Fomin, S. V., and Sinaĭ, Y. G. (1982). *Ergodic theory*, volume 245 of *Grundlehren der Mathematischen Wissenschaften*. Springer, New York.

Cousin, A. and Di Bernardino, E. (2013). On multivariate extensions of Value–at–Risk. *J. Multivariate Anal.*, 119:32–46.

Cuadras, C. M. (2009). Constructing copula functions with weighted geometric means. *J. Statist. Plan. Infer.*, 139(11):3766–3772.

Cuadras, C. M. and Augé, J. (1981). A continuous general multivariate distribution and its properties. *Comm. Statist. A–Theory Methods*, 10:339–353.

Cuadras, C. M., Fortiana, J., and Rodríguez Lallena, J. A., editors (2002). *Distributions with given marginals and Statistical Modelling*. Kluwer, Dordrecht.

Cuculescu, I. and Theodorescu, R. (2001). Copulas: diagonals, tracks. *Rev. Roumaine Math. Pures Appl.*, 46:731–742.

Czado, C. (2010). Pair-copula constructions. In Jaworski, P., Durante, F., Härdle, W. K., and Rychlik, T., editors, *Copula theory and its applications*, volume 198 of *Lecture Notes in Statistics – Proceedings*, pages 3–31. Springer, Berlin/Heidelberg.

Czado, C. and Brechmann, E. C. (2013). Selection of vine copulas. In Jaworski, P., Durante, F., and Härdle, W. K., editors, *Copulae in mathematical and quantitative finance*, volume 213 of *Lecture Notes in Statistics*, pages 17–37. Springer, Berlin/Heidelberg.

Dall'Aglio, G. (1959). Sulla compatibilità delle funzioni di ripartizione doppia. *Rend. Mat. e Appl. (5)*, 18:385–413.

Dall'Aglio, G. (1972). Fréchet classes and compatibility of distribution functions. *Symposia Math.*, 9:131–150.

Dall'Aglio, G. (1991). Fréchet classes: the beginnings. In Dall'Aglio, G., Kotz, S., and Salinetti, G., editors, *Probability distributions with given marginals*, pages 1–12. Kluwer, Dordrecht.

Dall'Aglio, G., Kotz, S., and Salinetti, G., editors (1991). *Probability distributions with given marginals*. Kluwer, Dordrecht.

Darsow, W. F., Nguyen, B. E., and Olsen, T. (1992). Copulas and Markov processes. *Illinois J. Math.*, 36:600–642.

Darsow, W. F. and Olsen, E. T. (1995). Norms for copulas. *Int. J. Math. Math. Sci.*, 18:417–436.

Darsow, W. F. and Olsen, E. T. (2010). Characterization of idempotent copulas. *Note Mat.*, 30:147–177.

Davey, B. A. and Priestley, H. A. (2002). *Introduction to lattices and order*. Cambridge University Press.

David, H. A. (1981). *Order statistics*. Wiley, New York.

de Amo, E., Díaz Carrillo, M., and Fernández Sánchez, J. (2011). Measure–preserving functions and the independence copula. *Mediterr. J. Math.*, 8:445–464.

de Amo, E., Díaz Carrillo, M., and Fernández Sánchez, J. (2012). Characterization of all copulas associated with non-continuous random variables. *Fuzzy Sets and Systems*, 191:103–112.

De Baets, B. and De Meyer, H. (2007). Orthogonal grid constructions of copulas. *IEEE Trans. Fuzzy Syst.*, 15:1053–1062.

De Baets, B., De Meyer, H., and Mesiar, R. (2010). Lipschitz continuity of copulas w.r.t. l_p–norms. *Nonlinear Anal.*, 72:3722–3731.

De Baets, B., De Meyer, H., and Úbeda Flores, M. (2007). Extremes of the mass distribution associated with a trivariate quasi–copula. *C. R. Math. Acad. Sci. Paris*, 344:587–590.

de Finetti, B. (1937). A proposito di "correlazione". *Supplemento Statistico ai Nuovi problemi di Politica Storia ed Economia*, 3:41–57. (in Italian); also reproduced in Bruno de Finetti "Opere Scelte", pp. 317–333, Cremonese, Roma, 2006; English translation, *J. Électron. Hist. Probab. Stat.* 4: Article 4, 15 pages, 2008.

de Haan, L. (2006). Discussion of: "Copulas: tales and facts" by T. Mikosch, *Extremes* 9: 3-20 (2006). *Extremes*, 9:21–22.

de Haan, L. and Resnick, S. I. (1977). Limit theory for multivariate sample extremes. *Z. Wahrscheinlichkeitstheorie und Verw. Gebiete*, 40:317–337.

de la Peña, V. H., Ibragimov, R., and Sharakhmetov, S. (2006). Characterizations of joint distributions, copulas, information dependence and decoupling, with applications to time series. In *2nd Lehman Symposium — Optimality*, volume 49 of *Lecture Notes-Monograph Series*, pages 183–209. Institute of Mathematical Statistics, Beachwood, OH.

Deheuvels, P. (1978). Caractérisation complète des lois extrêmes multivariées et de la convergence des types extrêmes. *Publ. Inst. Stat. Univ. Paris*, 23:1–36.

Deheuvels, P. (1979). Propriétés d'existence et propriétés topologiques des fonctions de dépendance avec applications à la convergence des types pour des lois multivariées. *C. R. Acad. Sci. Paris Sér. A-B*, 288(2):A145–A148.

Deheuvels, P. (2009). A multivariate Bahadur–Kiefer representation for the empirical copula process. *J. Math. Sci. (N.Y.)*, 163:382–398.

Demarta, S. and McNeil, A. J. (2005). The t copula and related copulas. *Int. Stat. Rev.*, 73(1):111–129.

Denneberg, D. (1994). *Non–additive measure and integral.* Kluwer, Dordrecht.

DeVore, R. A. and Lorentz, G. G. (1993). *Constructive approximation*, volume 303 of *Grundlehren der mathematischen Wissenschaften.* Springer, Berlin.

Dhaene, J., Denuit, M., Goovaerts, M. J., Kaas, R., and Vyncke, D. (2002a). The concept of comonotonicity in actuarial science and finance: applications. *Insurance Math. Econom.*, 31:133–161.

Dhaene, J., Denuit, M., Goovaerts, M. J., Kaas, R., and Vyncke, D. (2002b). The concept of comonotonicity in actuarial science and finance: theory. *Insurance Math. Econom.*, 31:3–33.

Di Bernardino, E. and Prieur, C. (2014). Estimation of multivariate conditional-tail-expectation using Kendall's process. *Journal of Nonparametric Statistics*, 26(2):241–267.

Di Bernardino, E. and Rullière, D. (2013). On certain transformations of Archimedean copulas: application to the non-parametric estimation of their generators. *Dependence Modeling*, 1:1–36.

Dolati, A. and Úbeda Flores, M. (2006). On measures on multivariate concordance. *JPSS J. Probab. Stat. Sci.*, 4:147–163.

Dolati, A. and Úbeda Flores, M. (2009). Constructing copulas by means of pairs of order statistics. *Kybernetika (Prague)*, 45:992–1002.

Donnelly, C. and Embrechts, P. (2010). The devil is in the tails: actuarial mathematics and the subprime mortgage crisis. *Astin Bull.*, 40:1–33.

Doob, J. L. (1953). *Stochastic processes.* Wiley, New York.

Dou, X., Kuriki, S., Lin, G. D., and Richards, D. (2015). EM algorithms for estimating the Bernstein copula. *Comput. Statist. Data Anal.*, in press.

Drouet-Mari, D. and Kotz, S. (2001). *Correlation and dependence.* Imperial College Press, London.

Dudley, R. (1989). *Real analysis and probability.* Wadsworth & Brooks/Cole, Pacific Grove, CA.

Dunford, N. and Schwartz, J. T. (1958). *Linear operators. Part I: General theory.* Wiley.

Durante, F. (2006). A new class of symmetric bivariate copulas. *J. Nonparametr. Stat.*, 18:499–510.

Durante, F. (2009). Construction of non-exchangeable bivariate distribution functions. *Statist. Papers*, 50(2):383–391.

Durante, F. and Fernández-Sánchez, J. (2010). Multivariate shuffles and approximation of copulas. *Statist. Probab. Lett.*, 80:1827–1834.

Durante, F., Fernández-Sánchez, J., and Pappadà, R. (2015a). Copulas, diagonals and tail dependence. *Fuzzy Sets and Systems*, 264:22–41.

Durante, F., Fernández-Sánchez, J., Quesada-Molina, J. J., and Úbeda-Flores, M. (2015b). Convergence results for patchwork copulas. *European J. Oper. Res.*, submitted.

Durante, F., Fernández-Sánchez, J., and Sempi, C. (2012). Sklar's theorem via regularization techniques. *Nonlinear Anal.*, 75:769–774.

Durante, F., Fernández-Sánchez, J., and Sempi, C. (2013a). Multivariate patchwork copulas: a unified approach with applications to partial comonotonicity. *Insurance Math. Econom.*, 53:897–905.

Durante, F., Fernández-Sánchez, J., and Sempi, C. (2013b). A note on the notion of singular copula. *Fuzzy Sets and Systems*, 211(1):120–122.

Durante, F., Fernández-Sánchez, J., and Trutschnig, W. (2014a). Multivariate copulas with hairpin support. *J. Multivariate Anal.*, 130:323–334.

Durante, F., Fernández Sánchez, J., and Úbeda Flores, M. (2013c). Bivariate copulas generated by perturbations. *Fuzzy Sets and Systems*, 228:137–144.

Durante, F., Foschi, R., and Sarkoci, P. (2010a). Distorted copulas: constructions and tail dependence. *Comm. Statist. Theory Methods*, 39:2288–2301.

Durante, F., Foschi, R., and Spizzichino, F. (2010b). Aging function and multivariate notions of NBU and IFR. *Probab. Engrg. Inform. Sci.*, 24:263–278.

Durante, F. and Ghiselli-Ricci, R. (2009). Supermigrative semi–copulas and triangular norms. *Inform. Sci.*, 179:2689–2694.

Durante, F. and Ghiselli-Ricci, R. (2012). Supermigrative copulas and positive dependence. *AStA Adv. Stat. Anal.*, 6:327–342.

Durante, F. and Jaworski, P. (2010). A new characterization of bivariate copulas. *Comm. Statist. Theory Methods*, 39:2901–2912.

Durante, F. and Jaworski, P. (2012). Invariant dependence structure under univariate truncation. *Statistics*, 46(2):263–277.

Durante, F., Jaworski, P., and Mesiar, R. (2011). Invariant dependence structures and Archimedean copulas. *Stat. Probab. Lett.*, 81:1995–2003.

Durante, F., Klement, E. P., Mesiar, R., and Sempi, C. (2007a). Conjunctors and their residual implicators: characterizations and construction methods. *Mediterr. J. Math*, 4:343–356.

Durante, F., Klement, E. P., and Quesada-Molina, J. J. (2008a). Bounds for trivariate copulas with given bivariate marginals. *J. Inequal. Appl.*, 2008:1–9.

Durante, F., Klement, E. P., Quesada-Molina, J. J., and Sarkoci, P. (2007b). Remarks on two product-like constructions for copulas. *Kybernetika (Prague)*, 43:235–244.

Durante, F., Klement, E. P., Sempi, C., and Úbeda Flores, M. (2010c). Measures of non-exchangeability for bivariate random vectors. *Statist. Papers*, 51:687–699.

Durante, F., Kolesárová, A., Mesiar, R., and Sempi, C. (2008b). Semilinear copulas. *Fuzzy Sets and Systems*, 159:63–76.

Durante, F., Mesiar, R., and Papini, P. L. (2008c). The lattice-theoretic structure of the sets of triangular norms and semi–copulas. *Nonlinear Anal.*, 69:46–52.

Durante, F., Mesiar, R., and Sempi, C. (2006a). On a family of copulas constructed from the diagonal section. *Soft Computing*, 10:490–494.

Durante, F., Nelsen, R. B., Quesada-Molina, J. J., and Úbeda-Flores, M. (2014b). Pairwise and global dependence in trivariate copula models. In Laurent, A., Strauss, O., Bouchon-Meunier, B., and Yager, R. R., editors, *Information processing and management of uncertainty in knowledge-based systems*, volume 444 of *Communications in Computer and Information Science*, pages 243–251. Springer International Publishing.

Durante, F. and Okhrin, O. (2015). Estimation procedures for exchangeable Marshall copulas with hydrological application. *Stoch. Environ. Res Risk Asess.*, 29(1):205–226.

Durante, F., Quesada Molina, J. J., and Sempi, C. (2006b). Semicopulas: characterizations and applicability. *Kybernetika (Prague)*, 42:287–302.

Durante, F., Quesada Molina, J. J., and Úbeda Flores, M. (2007c). On a family of multivariate copulas for aggregation processes. *Inf. Sci.*, 177:5715–5724.

Durante, F. and Salvadori, G. (2010). On the construction of multivariate extreme value models via copulas. *Environmetrics*, 21(2):143–161.

Durante, F., Saminger-Platz, S., and Sarkoci, P. (2009a). Rectangular patchwork for bivariate copulas and tail dependence. *Comm. Statist. Theory Methods*, 38:2515–2527.

Durante, F., Sarkoci, P., and Sempi, C. (2009b). Shuffles of copulas. *J. Math. Anal. Appl.*, 352:914–921.

Durante, F. and Sempi, C. (2005a). Copula and semi–copula transforms. *Int. J. Math. Math. Sci.*, 4:645–655.

Durante, F. and Sempi, C. (2005b). Semicopulæ. *Kybernetika (Prague)*, 41:315–328.

Durante, F. and Spizzichino, F. (2010). Semi–copulas, capacities and families of level sets. *Fuzzy Sets and Systems*, 161:269–276.

Easton, R. J., Tucker, D. H., and Wayment, S. G. (1967). On the existence almost everywhere of the cross partial derivatives. *Math. Z.*, 102:171–176.

Edgar, G. A. (1990). *Measure, topology and fractal geometry.* Springer, New York.

Edgar, G. A. (1998). *Integral, probability and fractal measures.* Springer, New York.

Ekeland, I., Galichon, A., and Henry, M. (2012). Comonotonic measures of multivariate risks. *Math. Finance*, 22(1):109–132.

Elidan, G. (2013). Copulas in machine learning. In Jaworski, P., Durante, F., and Härdle, W. K., editors, *Copulae in mathematical and quantitative finance*, volume 213 of *Lecture Notes in Statistics*, pages 39–60. Springer, Berlin/Heidelberg.

Embrechts, P. (2006). Discussion of: "Copulas: tales and facts" [Mikosch, 2006a]. *Extremes*, 9:45–47.

Embrechts, P. (2009). Copulas: a personal view. *Journal of Risk and Insurance*, 76:639–650.

Embrechts, P. and Hofert, M. (2013). A note on generalized inverses. *Math. Methods Oper. Res.*, 77:423–432.

Embrechts, P., McNeil, A. J., and Straumann, D. (2002). Correlation and dependence in risk management: properties and pitfalls. In Dempster, M., editor, *Risk management: value at risk and beyond*, pages 176–223. Cambridge University Press, Cambridge.

Embrechts, P. and Puccetti, G. (2006). Bounds for functions of multivariate risks. *J. Multivariate Anal.*, 97:526–547.

Embrechts, P., Puccetti, G., and Rüschendorf, L. (2013). Model uncertainty and VaR aggregation. *J. Bank. Financ.*, 37(8):2750–2764.

Eyraud, H. (1936). Les principes de la mesure des corrélations. *Ann. Univ. Lyon, III. Ser., Sect. A*, 1:30–47.

Falconer, K. J. (2003). *Fractal geometry. Mathematical foundations and applications*. Wiley, Hoboken, NJ.

Fang, H. B., Fang, K. T., and Kotz, S. (2002). The meta-elliptical distributions with given marginals. *J. Multivariate Anal.*, 82:1–16.

Fang, K. T., Kotz, S., and Ng, K. W. (1990). *Symmetric multivariate and related distributions*, volume 36 of *Monographs on Statistics and Applied Probability*. Chapman and Hall, London.

Farlie, D. J. G. (1960). The performance of some correlation coefficients for a general bivariate distribution. *Biometrika*, 47:307–323.

Faugeras, O. P. (2013). Sklar's theorem derived using probabilistic continuation and two consistency results. *J. Multivariate. Anal.*, 122:271–277.

Fermanian, J.-D. (2013). An overview of the goodness-of-fit test problem for copulas. In Jaworski, P., Durante, F., and Härdle, W. K., editors, *Copulae in Mathematical and Quantitative Finance*, volume 213 of *Lecture Notes in Statistics*, pages 61–89. Springer, Berlin/Heidelberg.

Fermanian, J.-D. and Wegkamp, M. H. (2012). Time-dependent copulas. *J. Multivariate Anal.*, 110:19–29.

Fernández-Sánchez, J., Nelsen, R. B., and Úbeda Flores, M. (2011a). Multivariate copulas, quasi-copulas and lattices. *Stat. Probab. Lett.*, 81:1365–1369.

Fernández-Sánchez, J., Rodríguez-Lallena, J. A., and Úbeda Flores, M. (2011b). Bivariate quasi–copulas and doubly stochastic signed measures. *Fuzzy Sets and Systems*, 168:81–88.

Fernández-Sánchez, J. and Trutschnig, W. (2015). Conditioning-based metrics on the space of multivariate copulas and their interrelation with uniform and level-wise convergence and Iterated Function Systems. *J. Theoret. Probab.*, in press.

Fernández-Sánchez, J. and Trutschnig, W. (2015). Some members of the class of (quasi-)copulas with given diagonal from the Markov kernel perspective. *Comm. Statist. Theory Methods*, in press.

Fernández-Sánchez, J. and Úbeda Flores, M. (2014). A note on quasi-copulas and signed measures. *Fuzzy Sets and Systems*, 234:109–112.

Féron, R. (1956). Sur les tableaux de corrélation dont les marges sont données, cas de l'espace à trois dimensions. *Publ. Inst. Statist. Univ. Paris*, 5:3–12.

Fischer, M. and Klein, I. (2007). Constructing generalized FGM copulas by means of certain univariate distributions. *Metrika*, 65:243–260.

Fisher, N. I. and Sen, P. K., editors (1994). *The collected works of Wassily Hoeffding*. Springer, New York.

Flondor, P., Georgescu, G., and Iorgulescu, A. (2001). Pseudo-t-norms and pseudo-BL-algebras. *Soft Computing*, 5:355–371.

Fodor, J. C. and Keresztfalvi, T. (1995). Nonstandard conjunctions and implications in fuzzy logic. *Internat. J. Approx. Reason.*, 12(2):69–84.

Foschi, R. (2013). Hyper-dependence, hyper-ageing properties and analogies between them: a semigroup-based approach. *Kybernetika (Prague)*, 49:96–113.

Foschi, R. and Spizzichino, F. (2008). Semigroups of semicopulas and evolution of dependence at increase of age. *Mathware & Soft Computing*, 15:95–111.

Frahm, G., Junker, M., and Szimayer, A. (2003). Elliptical copulas: applicability and limitations. *Statist. Probab. Lett.*, 63:275–286.

Frank, M. J. (1979). On the simultaneous associativity of $f(x, y)$ and $x+y-f(x, y)$. *Aequationes Math.*, 19:194–226.

Frank, M. J., Nelsen, R. B., and Schweizer, B. (1987). Best-possible bounds for the distribution of a sum—a problem of Kolmogorov. *Probab. Theory Related Fields*, 74(2):199–211.

Fréchet, M. (1907). Sur les ensembles de fonctions et les opérateurs linéaires. *C.R. Acad. Sci. Paris*, 144:1414–1416.

Fréchet, M. (1951). Sur les tableaux de corrélation dont les marges sont données. *Ann. Univ. Lyon, Sect. A (3)*, 14:53–77.

Fréchet, M. (1956). Sur les tableaux de corrélation dont les marges sont données. *C. R. Acad. Sci. Paris*, 242:2426–2428.

Fréchet, M. (1958). Remarques au sujet de la note précédente. *C. R. Acad. Sci. Paris Sér. I Math.*, 246:2719–2720.

Fredricks, G. and Nelsen, R. B. (1997). Copulas constructed from diagonal sections. In Beneš, V. and Štěpán, J., editors, *Distributions with given marginal and moment problems*, pages 129–136. Kluwer, Dordrecht.

Fredricks, G. A. and Nelsen, R. B. (2003). The Bertino family of copulas. In Cuadras, C. M., Fortiana, J., and Rodríguez Lallena, J. A., editors, *Distributions with given marginals and statistical modelling*, pages 81–91. Kluwer, Dordrecht.

Fredricks, G. A., Nelsen, R. B., and Rodríguez-Lallena, J. A. (2005). Copulas with fractal support. *Insur. Math. Econom.*, 37:42–48.

Galambos, J. (1975). Order statistics of samples from multivariate distributions. *J. Amer. Statist. Assoc.*, 70:674–680.

Galambos, J. (1977). Bonferroni inequalities. *Ann. Probab.*, 5:577–581.

Galambos, J. (1978). *The asymptotic theory of extreme order statistics.* Wiley, New York/Chichester/Brisbane.

Garling, D. (2007). *Inequalities: a journey into linear analysis.* Cambridge University Press.

Genest, C. (1987). Frank's family of bivariate distributions. *Biometrika*, 74:549–555.

Genest, C. (2005a). Proceedings of the conference on dependence modeling: statistical theory and applications to finance and insurance. *Canadian J. Statist.*, 33 (3).

Genest, C. (2005b). Proceedings of the conference on dependence modeling: statistical theory and applications to finance and insurance. *Insur. Math. Econom.*, 37 (1).

Genest, C. and Favre, A. C. (2007). Everything you always wanted to know about copula modeling but were afraid to ask. *J. Hydrol. Eng.*, 12:347–368.

Genest, C., Favre, A. C., Béliveau, J., and Jacques, C. (2007). Metaelliptical copulas and their use in frequency analysis of multivariate hydrological data. *Water Resour. Res.*, 43:1–12.

Genest, C., Gendron, M., and Bourdeau-Brien, M. (2009a). The advent of copulas in finance. *Europ. J. Finance*, 53:609–618.

Genest, C., Ghoudi, K., and Rivest, L.-P. (1995). A semiparametric estimation procedure of dependence parameters in multivariate families of distributions. *Biometrika*, 82(3):543–552.

Genest, C., Ghoudi, K., and Rivest, L.-P. (1998). Discussion of "Understanding relationships using copulas". *N. Am. Actuar. J.*, 2:143–149.

Genest, C. and MacKay, J. (1986a). Copules archimédiennes et familles de lois bidimensionnelles dont les marges sont données. *Canad. J. Statist.*, 14:280–283.

Genest, C. and MacKay, J. (1986b). The joy of copulas: bivariate distributions with uniform marginals. *Amer. Statist.*, 40:280–283.

Genest, C., Nešlehová, J., and Ziegel, J. (2011). Inference in multivariate Archimedean copula models. *TEST*, 20(2):223–256.

Genest, C., Nešlehová, J. G., and Rémillard, B. (2014). On the empirical multilinear copula process for count data. *Bernoulli*, 20(3):1344–1371.

Genest, C. and Nešlehová, J. (2007). A primer on copulas for count data. *Astin Bull.*, 37:415–515.

Genest, C. and Nešlehová, J. (2013). Assessing and modeling asymmetry in bivariate continuous data. In Jaworski, P., Durante, F., and Härdle, W. K., editors, *Copulae in mathematical and quantitative finance*, volume 213 of *Lecture Notes in Statistics*, pages 91–114. Springer, Berlin/Heidelberg.

Genest, C., Quesada Molina, J. J., Rodríguez Lallena, J. A., and Sempi, C. (1999). A characterization of quasi-copulas. *J. Multivariate Anal.*, 69:193–205.

Genest, C. and Rémillard, B. (2006). Discussion of: "Copulas: tales and facts" by T. Mikosch, *Extremes* 9: 3–20 (2006). *Extremes*, 9:27–36.

Genest, C., Rémillard, B., and Beaudoin, D. (2009b). Goodness-of-fit tests for copulas: a review and a power study. *Insurance Math. Econom.*, 44:199–213.

Genest, C. and Rivest, L.-P. (1989). A characterization of Gumbel family of extreme value distributions. *Statist. Probab. Lett.*, 8:207–211.

Genest, C. and Rivest, L.-P. (1993). Statistical inference procedures for bivariate Archimedean copulas. *J. Amer. Statist. Assoc.*, 88:1034–1043.

Genest, C. and Rivest, L.-P. (2001). On the multivariate probability integral transform. *Statist. Probab. Lett.*, 53:391–399.

Ghiselli Ricci, R. (2013a). Exchangeable copulas. *Fuzzy Sets and Systems*, 220:88–98.

Ghiselli Ricci, R. (2013b). On differential properties of copulas. *Fuzzy Sets and Systems*, 220:78–87.

Ghosh, S. and Henderson, S. G. (2009). Patchwork distributions. In Alexopoulos, C., Goldsman, D., and Wilson, J., editors, *Advancing the frontiers of simulation: a Festschrift in honor of George Samuel Fishman*, pages 65–86. Springer.

Ghoudi, K., Khoudraji, A., and Rivest, L.-P. (1998). Propriétés statistiques des copules de valeurs extrêmes bidimensionnelles. *Canad. J. Statist.*, 26:187–197.

Gijbels, I., Omelka, M., and Sznajder, D. (2010). Positive quadrant dependence tests for copulas. *Canad. J. Statist.*, 38(4):555–581.

González-Barrios, J. M. and Hernández-Cedillo, M. (2013). Construction of multivariate copulas in n-boxes. *Kybernetika (Prague)*, 49:73–95.

Grabisch, M., Marichal, J.-L., Mesiar, R., and Pap, E. (2009). *Aggregation functions*, volume 127 of *Encyclopedia of Mathematics and its Applications*. Cambridge Univesity Press, New York.

Grimmett, G. and Stirzaker, D. (2001). *Probability and random processes*. Oxford University Press. 3rd ed.

Gruber, P. M. (2007). *Convex and discrete geometry*. Springer, Berlin/Heidelberg.

Gudendorf, G. and Segers, J. (2010). Extreme-value copulas. In Jaworski, P., Durante, F., Härdle, W. K., and Rychlik, T., editors, *Copula theory and its applications*, volume 198 of *Lecture Notes in Statistics – Proceedings*, pages 127–145. Springer, Berlin/Heidelberg.

Gumbel, E. J. (1958). Distributions à plusieurs variables dont les marges sont données. *C. R. Acad. Sci. Paris*, 246:2717–2719.

Gumbel, E. J. (1960a). Bivariate exponential distributions. *J. Amer. Statist. Assoc.*, 55:698–707.

Gumbel, E. J. (1960b). Distribution des valeurs extrêmes en plusieurs dimensions. *Publ. Inst. Stast. Univ. Paris*, 9:171–173.

Gumbel, E. J. (1961). Bivariate logistic distributions. *J. Amer. Statist. Assoc.*, 56:335–349.

Hájek, P. (1998). *Metamathematics of fuzzy logic*, volume 4 of *Trends in Logic— Studia Logica Library*. Kluwer, Dordrecht.

Hájek, P. and Mesiar, R. (2008). On copulas, quasicopulas and fuzzy logic. *Soft Comput.*, 12:123–1243.

Halmos, P. R. (1974). *Measure theory*. Springer-Verlag.

Hamacher, H. (1978). *Über logische Aggregationen nicht-binär explizierter Entscheidungskriterien*. Rita G. Fischer Verlag.

Hammersley, J. M. (1974). Postulates for subadditive processes. *Ann. Probability*, 2:652–680.

Härdle, W. K., Okhrin, O., and Wang, W. (2015). Hidden Markov structures for dynamic copulae. *Econometric Theory*, in press.

Hering, C., Hofert, M., Mai, J.-F., and Scherer, M. (2010). Constructing hierarchical Archimedean copulas with Lévy subordinators. *J. Multivariate Anal.*, 101(6):1428–1433.

Hoeffding, V. (1940). Masztabinvariante Korrelationstheorie. *Schriften des Mathrematisches Instituts und des Instituts für Angewandte Mathematik des Universität Berlin*, 5:181–233. English translation as "Scale invariant correlation theory" in Fisher and Sen [1994], pp. 57–107.

Hoeffding, V. (1941). Masztabinvariante Korrelationsmasse für diskontinuierliche Verteilungen. *Arch. Math. Wirtschafts- u. Sozialforschung*, 7:49–70. English translation as "Scale–invariant correlation for discontinuous disributions" in Fisher and Sen [1994], pp. 109–133.

Hofer, M. and Iacò, M. R. (2014). Optimal bounds for integrals with respect to copulas and applications. *J. Optim. Theory Appl.*, 161(3):999–1011.

Hofert, M. (2010). Construction and sampling of nested Archimedean copulas. In Jaworski, P., Durante, F., Härdle, W. K., and Rychlik, T., editors, *Copula theory and its applications*, volume 198 of *Lecture Notes in Statistics – Proceedings*, pages 3–31. Springer, Berlin/Heidelberg.

Hofert, M., Kojadinovic, I., Maechler, M., and Yan, J. (2014). *Copula: multivariate dependence with copulas*. R package version 0.999-12.

Hougaard, P. (1986). A class of multivariate failure time distributions. *Biometrika*, 73:671–678.

Hougaard, P. (2000). *Analysis of multivariate survival data*. Statistics for Biology and Health. Springer-Verlag, New York.

Hürlimann, W. (2003). Hutchinson–Lai's conjecture for bivariate extreme value copulas. *Stat. Probab. Lett.*, 61:191–198.

Hüsler, J. and Reiss, R.-D. (1989). Maxima of normal random vectors: between independence and complete dependence. *Statist. Probab. Lett.*, 7(4):283–286.

Hutchinson, T. P. and Lai, C. D. (1990). *Continuous bivariate distributions. Emphasising applications*. Rumsby Scientific Publishing, Adelaide.

Ibragimov, R. (2009). Copula-based characterizations for higher order Markov processes. *Econometric Theory*, 25:819–846.

Jágr, V., Komorníková, M., and Mesiar, R. (2010). Conditioning stable copulas. *Neural Netw. World*, 20(1):69–79.

Janssen, P., Swanepoel, P., and Veraverbeke, N. (2012). Large sample behavior of the Bernstein copula estimator. *J. Statist. Plann. Inference*, 142(5):1189–1197.

Jaworski, P. (2009). On copulas and their diagonals. *Inform. Sci.*, 179:2863–2871.

Jaworski, P. (2013a). Invariant dependence structure under univariate truncation: the high–dimensional case. *Statistics*, 47:1064–1074.

Jaworski, P. (2013b). The limit properties of copulas under univariate conditioning. In Jaworski, P., Durante, F., and Härdle, W. K., editors, *Copulæ in mathematical and quantitative finance*, volume 213 of *Lecture Notes in Statistics*, pages 129–163. Springer, Berlin/Heidelberg.

Jaworski, P., Durante, F., and Härdle, W. K., editors (2013). *Copulae in mathematical and quantitative finance*, volume 213 of *Lecture Notes in Statistics - Proceedings*. Springer, Berlin/Heidelberg.

Jaworski, P., Durante, F., Härdle, W. K., and Rychlik, T., editors (2010). *Copula Theory and its Applications*, volume 198 of *Lecture Notes in Statistics - Proceedings*. Springer, Berlin/Heidelberg.

Jaworski, P. and Rychlik, T. (2008). On distributions of order statistics for absolutely continuous copulas with applications to reliability. *Kybernetika (Prague)*, 44:757–776.

Joe, H. (1993). Parametric families of multivariate distributions with given marginals. *J. Multivariate Anal.*, 46:262–282.

Joe, H. (1997). *Multivariate models and dependence concepts*. Chapman & Hall, London.

Joe, H. (2005). Asymptotic efficiency of the two-stage estimation method for copula-based models. *J. Multivariate Anal.*, 94(2):401–419.

Joe, H. (2006). Discussion of: "Copulas: tales and facts" by T. Mikosch, *Extremes* 9: 3-20 (2006). *Extremes*, 9:37–41.

Joe, H. (2014). *Dependence modeling with copulas*. Chapman & Hall/CRC, London.

Johnson, N. L. and Kotz, S. (2007). Cloning of distributions. *Statistics*, 41(2):145–152.

Jones, S. (2009). Of couples and copulas. *Financial Times*. Published on April 24, 2009.

Jwaid, T., De Baets, B., and De Meyer, H. (2014). Ortholinear and paralinear semi-copulas. *Fuzzy Sets and Systems*, 252:76–98.

Kallenberg, O. (1997). *Foundations of modern probability*. Springer, New York.

Kallsen, J. and Tankov, P. (2006). Characterization of dependence of multidimensional Lévy processes using Lévy copulas. *J. Multivariate Anal.*, 97:1551–1572.

Kantorovich, L. and Vulich, B. (1937). Sur la représentation des opérations linéaires. *Compositio Math.*, 5:119–165.

Kelley, J. L. (1955). *General topology*. Van Nostrand, New York. Reprinted by Springer, New York/Berlin, 1975.

Khoudraji, A. (1995). *Contributions à l'étude des copules et à la modélisation des valeurs extrêmes bivariées*. PhD thesis, Université Laval, Québec (Canada).

Kim, G., Silvapulle, M. J., and Silvapulle, P. (2007). Comparison of semiparametric and parametric methods for estimating copulas. *Comput. Statist. Data Anal.*, 51(6):2836–2850.

Kimberling, C. H. (1974). A probabilistic interpretation of complete monotonicity. *Aequationes Math.*, 10:152–164.

Kimeldorf, G. and Sampson, A. R. (1975). Uniform representations of bivariate distributions. *Comm. Statist.*, 4:617–627.

Kimeldorf, G. and Sampson, A. R. (1978). Monotone dependence. *Ann. Stat.*, 6:895–903.

Kingman, J. F. C. and Taylor, S. J. (1966). *Introduction to measure and probability*. Cambridge University Press.

Klement, E. P. and Kolesárová, A. (2007). Intervals of 1-Lipschitz aggregation operators, quasi-copulas, and copulas with given affine section. *Monatsh. Math.*, 152(2):151–167.

Klement, E. P. and Mesiar, R., editors (2005). *Logical, algebraic, analytic, and probabilistic aspects of triangular norms*. Elsevier B.V., Amsterdam.

Klement, E. P. and Mesiar, R. (2006). How non-symmetric can a copula be? *Comment. Math. Univ. Carolinae*, 47:141–148.

Klement, E. P., Mesiar, R., and Pap, E. (1999). Quasi- and pseudo- inverses of monotone functions, and the construction of t–norms. *Fuzzy Sets and Systems*, 104:3–13.

Klement, E. P., Mesiar, R., and Pap, E. (2000). *Triangular norms*. Kluwer, Dordrecht.

Klement, E. P., Mesiar, R., and Pap, E. (2002). Invariant copulas. *Kybernetika (Prague)*, 38:275–285.

Klement, E. P., Mesiar, R., and Pap, E. (2005). Archimax copulas and invariance under transformations. *C. R. Math. Acad. Sci. Paris*, 340:755–758.

Klement, E. P., Mesiar, R., and Pap, E. (2010). A universal integral as common frame for Choquet and Sugeno integral. *IEEE Trans. Fuzzy Systems*, 18:178–187.

Klenke, A. (2008). *Probability theory. A comprehensive course.* Springer, Berlin/Heidelberg.

Klüppelberg, C. and Resnick, S. I. (2008). The Pareto copula, aggregation of risks, and the emperor's socks. *J. Appl. Probab.*, 45:67–84.

Kojadinovic, I. and Yan, J. (2010). Comparison of three semiparametric methods for estimating dependence parameters in copula models. *Insur. Math. Econ.*, 47:52–63.

Kolesárová, A., Mesiar, R., and Kalická, J. (2013). On a new construction of 1–Lipschitz aggregation functions, quasi–copulas and copulas. *Fuzzy Sets and Systems*, 226:19–31.

Kolesárová, A., Mesiar, R., Mordelová, J., and Sempi, C. (2006). Discrete copulas. *IEEE Trans. Fuzzy Syst.*, 14:698–705.

Kolesárová, A., Mesiar, R., and Sempi, C. (2008). Measure–preserving transformations, copulæ and compatibility. *Mediterr. J. Math.*, 5:325–339.

Kollo, T. (2009). Preface. *J. Statist. Plan. Infer.*, 139:3740.

Kolmogorov, A. N. (1933). *Grundbegriffe der Wahrscheinlichkeitsrechnung.* Springer, Berlin. English translation, *Foundations of probability*, Chelsea, New York, 1950.

Kortschak, D. and Albrecher, H. (2009). Asymptotic results for the sum of dependent non-identically distributed random variables. *Methodol. Comput. Appl. Probab.*, 11(3):279–306.

Kotz, S., Balakrishnan, N., and Johnson, N. L. (2000). *Continuous multivariate distributions.* Wiley, New York. 2nd ed.

Krause, G. (1981). *A strengthened form of Ling's theorem on associative functions.* PhD Dissertation, Illinois Institute of Technology.

Kruskal, W. H. (1958). Ordinal measures of association. *J. Amer. Statist. Soc.*, 53:814–861.

Kulpa, T. (1999). On approximation of copulas. *Int. J. Math. Math. Sci.*, 22:259–269.

Kurowicka, D. and Joe, H., editors (2011). *Dependence modeling. Vine copula handbook.* World Scientific, Singapore.

Lagerås, A. N. (2010). Copulas for Markovian dependence. *Bernoulli*, 16:331–342.

Lai, C. D. and Balakrishnan, N. (2009). *Continuous bivariate distributions.* Springer, New York.

Lancaster, H. O. (1963). Correlation and complete dependence of random variables. *Ann. Math. Statist.*, 34:1315–1321.

Larsson, M. and Nešlehová, J. (2011). Extremal behavior of Archimedean copulas. *Adv. in Appl. Probab.*, 43(1):195–216.

Lasota, A. and Mackey, M. C. (1994). *Chaos, fractals, and noise*, volume 97 of *Applied Mathematical Sciences*. Springer, New York. 2nd ed.

Li, D. (2001). On default correlation: a copula function approach. *Journal of Fixed Income*, 9:43–54.

Li, H. (2013). Toward a copula theory for multivariate regular variation. In Jaworski, P., Durante, F., and Härdle, W. K., editors, *Copulae in mathematical and quantitative finance*, volume 213 of *Lecture Notes in Statistics*, pages 177–199. Springer, Berlin/Heidelberg.

Li, X., Mikusiński, P., Sherwood, H., and Taylor, M. D. (1997). On approximation of copulas. In Beneš, V. and Štěpán, J., editors, *Distributions with given marginal and moment problems*, pages 107–116. Kluwer Dordrecht.

Li, X., Mikusiński, P., and Taylor, M. D. (1998). Strong approximation of copulas. *J. Math. Anal.Appl.*, 225:608–623.

Li, X., Mikusiński, P., and Taylor, M. D. (2003). Some integration-by-parts formulas involving 2-copulas. In Cuadras, C. M., Fortiana, J., and Rodríguez Lallena, J., editors, *Distributions with given marginals and statistical modelling*, pages 153–159. Kluwer, Dordrecht.

Liebscher, E. (2008). Construction of asymmetric multivariate copulas. *J. Multivariate Anal.*, 99:2234–2250. Erratum 102: 869–870, 2011.

Lindner, A. (2006). Discussion of: "Copulas: tales and facts" by T. Mikosch, *Extremes* 9: 3-20 (2006). *Extremes*, 9:43–44.

Ling, C. H. (1965). Representation of associative functions. *Publ. Math. Debrecen*, 12:189–212.

Loève, M. (1977). *Probability theory I*, volume 45 of *Graduate Texts in Mathematics*. Springer, New York. 4th ed.

Lojasiewicz, S. (1988). *An introduction to the theory of real functions*. Wiley–Interscience, Chichester. 3rd ed.

Longla, M. and Peligrad, M. (2012). Some aspects of modeling dependence in copula–based Markov chains. *J. Multivariate Anal.*, 111:234–240.

MacKenzie, D. and Spears, T. (2014a). "A device for being able to book P&L": the organizational embedding of the Gaussian copula. *Social Studies of Science*, 44(3):418–440.

MacKenzie, D. and Spears, T. (2014b). "The formula that killed Wall Street": the Gaussian copula and modelling practices in investment banking. *Social Studies of Science*, 44(3):393–417.

Mai, J.-F., Schenk, S., and Scherer, M. (2015). Exchangeable exogenous shock models. *Bernoulli*, in press.

Mai, J.-F. and Scherer, M. (2009). Lévy-frailty copulas. *J. Multivariate Anal.*, 100:1567–1585.

Mai, J.-F. and Scherer, M. (2012a). *H*-extendible copulas. *J. Multivariate Anal.*, 110:151–160.

Mai, J.-F. and Scherer, M. (2012b). *Simulating copulas. Stochastic models, sampling algorithms and applications.* World Scientific, Singapore.

Mai, J.-F. and Scherer, M. (2014). *Financial engineering with copulas explained.* Palgrave MacMillan, Hampshire, UK.

Makarov, G. D. (1981). Estimates for the distribution function of the sum of two random variables with given marginal distributions. *Teor. Veroyatnost. i Primenen.*, 26(4):815–817.

Malevergne, Y. and Sornette, D. (2006). *Extreme financial risks.* Springer, Berlin.

Malov, S. V. (2001). On finite-dimensional Archimedean copulas. In Balakrishnan, N., Ibragimov, I. A., and Nevzorov, V. B., editors, *Asymptotic methods in probability and statistics with applications*, pages 19–35. Birkhäuser, Boston.

Manner, H. and Segers, J. (2011). Tails of correlation mixtures of elliptical copulas. *Insurance Math. Econom.*, 48(1):153–160.

Mardia, K. V. (1970). A translation family of bivariate distributions and Fréchet's bounds. *Sankhyā Series A*, 32:119–122.

Mardia, K. V. (1972). A multi-sample uniform scores test on a circle and its parametric competitor. *J. Roy. Statist. Soc. Ser. B.*, 34:102–113.

Marshall, A. and Olkin, I. (1967a). A generalized bivariate exponential distribution. *J. Appl. Probability*, 4:291–302.

Marshall, A. and Olkin, I. (1967b). A multivariate exponential distribution. *J. Amer. Statist. Assoc.*, 62:30–44.

Marshall, A. W. (1996). Copulas, marginals, and joint distributions. In Rüschendorf, L., Schweizer, B., and Taylor, M. D., editors, *Distributions with fixed marginals and related topics*, volume 28 of *Lecture Notes—Monograph Series*, pages 213–222. Institute of Mathematical Statistics, Hayward, CA.

Marshall, A. W. and Olkin, I. (1979). *Inequalities: Theory of Majorization and its Applications.* Academic Press, New York. 2nd ed., with Arnold, B. C., Springer, New York, 2011.

Marshall, A. W. and Olkin, I. (1988). Families of multivariate distributions. *J. Amer. Statist. Assoc.*, 83(403):834–841.

Mayor, G., Suñer, J., and Torrens, J. (2005). Copula–like operations on finite settings. *IEEE Trans. Fuzzy Syst.*, 13:468–477.

McNeil, A. J., Frey, R., and Embrechts, P. (2005). *Quantitative risk management. Concepts, techniques and tools.* Princeton University Press.

McNeil, A. J. and Nešlehová, J. (2009). Multivariate Archimedean copulas, d–monotone functions and l_1–norm symmetric distributions. *Ann. Statist.*, 37:3059–3097.

McNeil, A. J. and Nešlehová, J. (2010). From Archimedean to Liouville copulas. *J. Multivariate Anal.*, 101:1772–1790.

Menger, K. (1942). Statistical metrics. *Proc. Nat. Acad. Sci. U.S.A.*, 8:535–537.

Mesiar, R., Jágr, V., Juráňová, M., and Komorníková, M. (2008). Univariate conditioning of copulas. *Kybernetika (Prague)*, 44(6):807–816.

Mesiar, R. and Sempi, C. (2010). Ordinal sums and idempotents of copulas. *Aequationes Math.*, 79:39–52.

Meyer, C. (2013). The bivariate normal copula. *Comm. Statist. Theory Methods*, 42:2402–2422.

Mikosch, T. (2006a). Copulas: tales and facts. *Extremes*, 9:3–20.

Mikosch, T. (2006b). Copulas: tales and facts—rejoinder. *Extremes*, 9:55–62.

Mikusiński, P., Sherwood, H., and Taylor, M. D. (1991–92). The Fréchet bounds revisited. *Real Anal. Exch.*, 17:759–764.

Mikusiński, P., Sherwood, H., and Taylor, M. D. (1992). Shuffles of min. *Stochastica*, 13:61–74.

Mikusiński, P. and Taylor, M. D. (2009). Markov operators and n–copulas. *Ann. Math. Polon.*, 96:75–95.

Mikusiński, P. and Taylor, M. D. (2010). Some approximation and n–copulas. *Metrika*, 72:385–414.

Moore, D. S. and Spruill, M. C. (1975). Unified large-sample theory of general chi-square statistics for tests of fit. *Ann. Statist.*, 3:599–616.

Morgenstern, D. (1956). Einfache Beispiele zweidimensionaler Verteilungen. *Mitteilungsbl. Math. Statist.*, 8:234–235.

Morillas, P. M. (2005). A method to obtain new copulas from a given one. *Metrika*, 61:169–184.

Moynihan, R. (1978). On τ_t-semigroups of probability distribution functions II. *Aequationes Math.*, 17:19–40.

Müller, A. and Scarsini, M. (2005). Archimedean copulæ and positive dependence. *J. Multivariate Anal.*, 93:434–445.

Nappo, G. and Spizzichino, F. (2009). Kendall distributions and level sets in bivariate exchangeable survival models. *Inf. Sci.*, 179:2878–2890.

Navarro, J. and Spizzichino, F. (2010). On the relationships between copulas of order statistics and marginal distributions. *Statist. Probab. Lett.*, 80(5-6):473–479.

Nelsen, R. B. (1986). Properties of a one-parameter family of bivariate distributions with specified marginals. *Comm. Statist. A—Theory Methods*, 15:3277–3285.

Nelsen, R. B. (1993). Some concepts of bivariate symmetry. *J. Nonparametr. Statist.*, 3:95–101.

Nelsen, R. B. (1999). *An introduction to copulas*, volume 139 of *Lecture Notes in Statistics*. Springer, New York.

Nelsen, R. B. (2005). Copulas and quasi-copulas: an introduction to their properties and applications. In Klement, E. P. and Mesiar, R., editors, *Logical, algebraic analytic, and probabilistic aspects of triangular norms*, pages 391–413. Elsevier, New York.

Nelsen, R. B. (2006). *An introduction to copulas*. Springer, New York.

Nelsen, R. B. (2007). Extremes of nonexchangeability. *Statist. Papers*, 48:329–336.

Nelsen, R. B. and Fredricks, G. A. (1997). Diagonal copulas. In Beneš, V. and Štěpán, J., editors, *Distributions with given marginal and moment problems*, pages 121–128. Kluwer, Dordrecht.

Nelsen, R. B., Quesada Molina, J. J., Rodríguez Lallena, J. A., and Úbeda Flores, M. (2001). Distribution functions of copulas: a class of bivariate probability integral transforms. *Statist. Probab. Lett.*, 54:263–268.

Nelsen, R. B., Quesada Molina, J. J., Rodríguez Lallena, J. A., and Úbeda Flores, M. (2002). Some new properties of quasi-copulas. In Cuadras, C. M., Fortiana, J., and Rodríguez Lallena, J. A., editors, *Distributions with given marginals and statistical modelling*, pages 187–194. Kluwer, Dordrecht.

Nelsen, R. B., Quesada Molina, J. J., Rodríguez Lallena, J. A., and Úbeda Flores, M. (2003). Kendall distribution functions. *Statist. Probab. Lett.*, 65:263–268.

Nelsen, R. B., Quesada Molina, J. J., Rodríguez Lallena, J. A., and Úbeda Flores, M. (2004). Best-possible bounds on sets of bivariate distribution functions. *J. Multivariate Anal.*, 90:348–358.

Nelsen, R. B., Quesada Molina, J. J., Rodríguez Lallena, J. A., and Úbeda Flores, M. (2009). Kendall distribution functions and associative copulas. *Fuzzy Sets and Systems*, 160:52–57.

Nelsen, R. B., Quesada Molina, J. J., Rodríguez Lallena, J. A., and Úbeda Flores, M. (2010). Quasi-copulas and signed measures. *Fuzzy Sets and Systems*, 161:2328–2336.

Nelsen, R. B., Quesada Molina, J. J., Schweizer, B., and Sempi, C. (1996). Derivability of some operations on distribution functions. In Rüschendorf, L., Schweizer, B., and Taylor, M. D., editors, *Distributions with fixed marginals and related topics*, volume 28 of *Lecture Notes–Monograph Series*, pages 233–243. Institute of Mathematical Statistics, Hayward, CA.

Nelsen, R. B. and Úbeda Flores, M. (2005). The lattice-theoretic structure of sets of bivariate copulas and quasi copulas. *C. R. Acad. Sci. Paris Sér. I*, 341:583–586.

Oakes, D. (1982). A model for association in bivariate survival data. *J. Roy. Statist. Soc. Ser. B*, 44:414–422.

Oakes, D. (2005). On the preservation of copula structure under truncation. *Canad. J. Statist.*, 33(3):465–468.

Okhrin, O., Okhrin, Y., and Schmid, W. (2013). On the structure and estimation of hierarchical Archimedean copulas. *J. Econometrics*, 173(2):189–204.

Olsen, E. T., Darsow, W., and Nguyen, B. (1996). Copulas and Markov operators. In Rüschendorf, L., Schweizer, B., and Taylor, M. D., editors, *Distributions with fixed marginals and related topics*, volume 28 of *Lecture Notes—Monograph Series*, pages 244–259. Institute of Mathematical Statistics, Hayward, CA.

Omelka, M., Gijbels, I., and Veraverbeke, N. (2009). Improved kernel estimation of copulas: weak convergence and goodness-of-fit testing. *Ann. Stat.*, 37(5B):3023–3058.

Pap, E., editor (2002). *Handbook of measure theory, Vol. I & II*. Elsevier North–Holland, Amsterdam.

Patton, A. J. (2012). A review of copula models for economic time series. *J. Multivariate Anal.*, 110:4–18.

Patton, A. J. (2013). Copula methods for forecasting multivariate time series. In Elliott, G. and Timmermann, A., editors, *Handbook of economic forecasting*, volume 2, pages 899–960. Elsevier, Oxford.

Peck, J. E. L. (1959). Doubly stochastic measures. *Michigan Math. J.*, 6:217–220.

Pellerey, F. (2008). On univariate and bivariate aging for dependent lifetimes with Archimedean survival copulas. *Kybernetika (Prague)*, 44:795–806.

Peng, L. (2006). Discussion of: "Copulas: tales and facts" by T. Mikosch, *Extremes* 9: 3-20 (2006). *Extremes*, 9:49–50.

Pfanzagl, J. (1967). Characterization of conditional expectations. *Ann. Math. Statist.*, 28:415–421.

Pickands, J. (1981). Multivariate extreme value distributions. *Bull. Internat. Statist. Inst.*, pages 859–878.

Puccetti, G. and Rüschendorf, L. (2012). Computation of sharp bounds on the distribution of a function of dependent risks. *J. Comput. Appl. Math.*, 236(7):1833–1840.

Puccetti, G. and Scarsini, M. (2010). Multivariate comonotonicity. *J. Multivariate Anal.*, 101:291–304.

Puccetti, G. and Wang, R. (2014). General extremal dependence concepts. Preprint, available at http://ssrn.com/abstract=2436392.

Quesada-Molina, J. J. and Rodríguez-Lallena, J. A. (1995). Bivariate copulas with quadratic sections. *J. Nonparametr. Statist.*, 5:323–337.

Quesada Molina, J. J. and Sempi, C. (2005). Discrete quasi–copulæ. *Insurance: Math. and Econom.*, 37:27–41.

R Core Team (2014). *R: A Language and Environment for Statistical Computing*. R Foundation for Statistical Computing, Vienna, Austria.

Rao, M. M. (1987). *Measure theory and integration*. Wiley, New York.

Rényi, A. (1959). On measures of dependence. *Acta Math. Acad. Sci. Hung.*, 10:441–451. Also in Turán, P., editor, *Selected papers of Alfréd Rényi*. Vol. 1, pages 402–412. Akadémiai Kiadó, Budapest, 1976.

Ressel, P. (2011). A revision of Kimberling's results—with an application to max-infinite divisibility of some Archimedean copulas. *Statist. Probab. Lett.*, 81:207–211.

Ressel, P. (2013). Homogeneous distributions–And a spectral representation of classical mean values and stable tail dependence functions. *J. Multivariate Anal.*, 117:246–256.

Riesz, F. (1907). Sur une espèce de géométrie analytique des fonctions sommables. *C. R. Acad. Sci. Paris*, 144:1409–1411.

Rodríguez-Lallena, J. A. and Úbeda Flores, M. (2004a). Best-possible bounds on sets of multivariate distribution functions. *Comm. Statist. Theory Methods*, 33:805–820.

Rodríguez-Lallena, J. A. and Úbeda Flores, M. (2004b). A new class of bivariate copulas. *Statist. Probab. Lett.*, 66:315–325.

Rodríguez-Lallena, J. A. and Úbeda Flores, M. (2009). Some new characterizations and properties of quasi-copulas. *Fuzzy Sets and Systems*, 160:717–725.

Rodríguez-Lallena, J. A. and Úbeda Flores, M. (2010). Multivariate copulas with quadratic sections in one variable. *Metrika*, 72:331–349.

Rogers, L. and Williams, D. (2000). *Diffusions, Markov processes and martingales. Volume 1: Foundations.* Cambridge University Press. 2nd ed.

Royden, H. (1988). *Real analysis.* Macmillan, New York. 3rd. ed.

Ruankong, P., Santiwipanont, T., and Sumetkijakan, S. (2013). Shuffles of copulas and a new measure of dependence. *J. Math. Anal. Appl.*, 398(1):392–402.

Ruankong, P. and Sumetkijakan, S. (2011). On a generalized ∗–product for copulas. In *Proceedings of the Annual Pure and Applied Mathematics Conference, Chulalongkorn University, Thailand*, pages 13–21. arXiv:1204.1627.

Rüschendorf, L. (1981a). Sharpness of the Fréchet bounds. *Z. Wahrscheinlichkeitstheorie verw. Gebiete*, 57:293–302.

Rüschendorf, L. (1981b). Stochastically ordered distributions and monotonicity of the OC-function of sequential probability ratio tests. *Mathematische Operationsforschung und Statistik Series Statistics*, 12:327–338.

Rüschendorf, L. (1982). Random variables with maximum sums. *Adv. in Appl. Probab.*, 14(3):623–632.

Rüschendorf, L. (1985). Construction of multivariate distributions with given marginals. *Ann. Inst. Statist. Math.*, 37:225–233.

Rüschendorf, L. (2005). Stochastic ordering of risks, influence of dependence and a.s. constructions. In N., B., Bairamov, I. G., and Gebizlioglu, O. L., editors, *Advances on models, characterizations and applications*, pages 19–56. Chapman & Hall/CRC, Boca Raton, FL.

Rüschendorf, L. (2009). On the distributional transform, Sklar's theorem, and the empirical copula process. *J. Statist. Plann. Inference*, 139:3921–3927.

Rüschendorf, L. (2013). *Mathematical risk analysis. Dependence, risk bounds, optimal allocations and portfolios.* Springer, Heidelberg.

Rüschendorf, L., Schweizer, B., and Taylor, M. D., editors (1996). *Distributions with fixed marginals and related topics*, volume 28 of *Lecture Notes— Monograph Series*. Institute of Mathematical Statistics, Hayward, CA.

Rychlik, T. (1993). Bounds for expectation of l-estimates for dependent samples. *Statistics*, 24:1–7.

Rychlik, T. (1994). Distributions and expectations of order statistics for possibly dependent random variables. *J. Multivariate. Anal.*, 48:31–42.

Ryff, J. V. (1963). On the representation of doubly stochastic operators. *Pacific J. Math*, 13:1379–1386.

Saks, S. (1937). *Theory of the integral.* G. E. Stechert & Co., New York. 2nd revised ed., Engl. translat. by L. C. Young. With two additional notes by Stefan Banach.

Salmon, F. (2009). Recipe for disaster: The formula that killed Wall Street. *Wired Magazine*. Published on February 23.

Salmon, F. (2012). The formula that killed Wall Street. *Significance*, 9(1):16–20.

Salvadori, G., De Michele, C., and Durante, F. (2011). On the return period and design in a multivariate framework. *Hydrol. Earth Syst. Sci.*, 15:3293–3305.

Salvadori, G., De Michele, C., Kottegoda, N. T., and Rosso, R. (2007). *Extremes in nature. An approach using copulas.* Springer, Dordrecht.

Salvadori, G., Durante, F., and De Michele, C. (2013). Multivariate return period calculation via survival functions. *Water Resour. Res.*, 49(4):2308–2311.

Saminger-Platz, S. and Sempi, C. (2008). A primer on triangle functions. I. *Aequationes Math.*, 76:201–240.

Saminger-Platz, S. and Sempi, C. (2010). A primer on triangle functions. II. *Aequationes Math.*, 80:239–268.

Sancetta, A. and Satchell, S. (2004). The Bernstein copula and its applications to modeling and approximations of multivariate distributions. *Econometric Theory*, 20:535–562.

Sander, W. (2005). Some aspects of functional equations. In Klement, E. P. and Mesiar, R., editors, *Logical, algebraic, analytic and probabilistic aspects of triangular norms*, pages 143–187. Elsevier Science B.V., Amsterdam.

Sarmanov, O. V. (1966). Generalized normal correlation and two-dimensional Fréchet classes. *Dokl. Akad. Nauk SSSR*, 168:32–35.

Scarsini, M. (1984a). On measures of concordance. *Stochastica*, 8:201–218.

Scarsini, M. (1984b). Strong measures of concordance and convergence in probability. *Riv. Mat. Sci. Econom. Social.*, 7:39–44.

Scarsini, M. (1989). Copulæ of probability measures on product spaces. *J. Multivariate Anal.*, 31:201–219.

Scarsini, M. (1996). Copulæ of capacities on product spaces. In Rüschendorf, L., Schweizer, B., and Taylor, M. D., editors, *Distributions with fixed marginals and related topics*, volume 28 of *Lecture Notes—Monograph Series*, pages 307–318. Institute of Mathematical Statistics, Hayward, CA.

Schmid, F., Schmidt, R., Blumentritt, T., Gaißer, S., and Ruppert, M. (2010). Copula-based measures of multivariate association. In Jaworski, P., Durante, F., Härdle, W. K., and Rychlik, T., editors, *Copula theory and its applications*, volume 198 of *Lecture Notes in Statistics – Proceedings*, pages 209–236. Springer, Berlin/Heidelberg.

Schmidt, R. (2002). Tail dependence for elliptically contoured distributions. *Math. Meth. Per. Res.*, 55:301–327.

Schmitz, V. (2004). Revealing the dependence structure between $X_{(1)}$ and $X_{(n)}$. *J. Statist. Plann. Inference*, 123:41–47.

Schönbucker, P. (2003). *Credit derivatives pricing models: models, pricing, implementation.* Wiley, Chichester.

Schur, I. (1923). Über eine Klasse von Mittelbildungen mit Anwendungen auf die Determinantentheorie. *Sitzungsber. Berl. Math. Ges.*, 22:9–20.

Schweizer, B. (1991). Thirty years of copulas. In Dall'Aglio, G., Kotz, S., and Salinetti, G., editors, *Probability distributions with given marginals*, pages 13–50. Kluwer, Dordrecht.

Schweizer, B. (2007). Introduction to copulas. *J. Hydrological Engineering*, 12:346.

Schweizer, B. and Sklar, A. (1958). Espaces métriques aléatoires. *C. R. Acad. Sci. Paris*, 247:2092–2094.

Schweizer, B. and Sklar, A. (1974). Operations on distribution functions not derivable from operations on random variables. *Studia Math.*, 52:43–52.

Schweizer, B. and Sklar, A. (1983). *Probabilistic metric spaces.* North-Holland, New York. Reprinted, Dover, Mineola, NY, 2005.

Schweizer, B. and Wolff, E. F. (1976). Sur une mesure de dépendance pour les variables aléatoires. *C. R. Acad. Sci. Paris Sér. A–B*, 283:A659–A661.

Schweizer, B. and Wolff, E. F. (1981). On nonparametric measures of dependence for random variables. *Ann. Statist.*, 9:879–885.

Seethoff, T. L. and Shiflett, R. C. (1978). Doubly stochastic measures with prescribed support. *Z. Wahrscheinlichkeitstheor. Verw. Geb.*, 41:283–288.

Segers, J. (2006). Efficient estimation of copula parameter. Discussion of: "Copulas: tales and facts". *Extremes*, 9:51–53.

Segers, J. (2012). Asymptotics of empirical copula processes under non-restrictive smoothness assumptions. *Bernoulli*, 18(3):764–782.

Sempi, C. (2002). Conditional expectations and idempotent copulæ. In Cuadras, C. M., Fortiana, J., and Rodríguez Lallena, J. A., editors, *Distributions with given marginals and statistical modelling*, pages 223–228. Kluwer, Dordrecht.

Sempi, C. (2004). Convergence of copulas: critical remarks. *Rad. Mat.*, 12:1–8.

Seneta, E. (1981). *Non-negative matrices and Markov chains*. Springer, New York.

Serfozo, R. (1982). Convergence of Lebesgue integrals with varying measures. *Sankhyā*, 44:380–402.

Shaked, M. and Shanthikumar, J. G. (2007). *Stochastic orders*. Springer Series in Statistics. Springer, New York.

Sherwood, H. and Taylor, M. D. (1988). Doubly stochastic measures with hairpin support. *Probab. Theory Related Fields*, 78:617–626.

Siburg, K. F. and Stoimenov, P. A. (2008a). Gluing copulas. *Commun. Statist.– Theory and Methods*, 37:3124–3134.

Siburg, K. F. and Stoimenov, P. A. (2008b). A scalar product for copulas. *J. Math. Anal. Appl.*, 34:429–439.

Sibuya, M. (1960). Bivariate extreme statistics. I. *Ann. Inst. Statist. Math. Tokyo*, 11:195–210.

Sklar, A. (1959). Fonctions de répartition à n dimensions et leurs marges. *Publ. Inst. Statist. Univ. Paris*, 8:229–231.

Sklar, A. (1973). Random variables, joint distribution functions and copulas. *Kybernetika (Prague)*, 9:449–460.

Sklar, A. (1996). Random variables, distribution functions, and copulas—a personal look backward and forward. In Rüschendorf, L., Schweizer, B., and Taylor, M. D., editors, *Distributions with fixed marginals and related topics*, volume 28 of *Lecture Notes—Monograph Series*, pages 1–14. Institute of Mathematical Statistics, Hayward, CA.

Sklar, A. (2010). Fonctions de répartition à n dimensions et leurs marges. *Publ. Inst. Statist. Univ. Paris*, 54:3–6. Reprint of the original paper with an introduction by Denis Bosq.

Song, P. X.-K. (2007). *Correlated data analysis: modeling, analytics, and applications*. Springer Series in Statistics. Springer, New York.

Spizzichino, F. (2001). *Subjective probability models for lifetimes*, volume 91 of *Monographs on Statistics and Applied Probability*. Chapman & Hall/CRC, Boca Raton, FL.

Suárez-García, F. and Gil-Álvarez, P. (1986). Two families of fuzzy integrals. *Fuzzy Sets and Systems*, 18:67–81.

Takahasi, K. (1965). Note on the multivariate Burr's distribution. *Ann. Inst. Statist. Math.*, 17:257–260.

Tao, T. (2011). *An introduction to measure theory*, volume 126 of *Graduate Studies in Mathematics*. American Mathematical Society, Providence, RI.

Tawn, J. A. (1988). Bivariate extreme value theory: models and estimation. *Biometrika*, 75:397–415.

Taylor, M. D. (2007). Multivariate measures of concordance. *Ann. Inst. Stat. Math.*, 59:789–806.

Tiit, E. M. (1996). Mixtures of multivariate quasi-extremal distributions having given marginals. In *Distributions with fixed marginals and related topics (Seattle, WA, 1993)*, volume 28 of *Lecture Notes—Monograph Series*, pages 337–357. Institute of Mathematical Statistics, Hayward, CA.

Trivedi, P. K. and Zimmer, D. M. (2007). *Copula modeling: an introduction for practitioners*. Now Publishers, Hanover, MA.

Trutschnig, W. (2011). On a strong metric on the space of copulas and its induced dependence measure. *J. Math. Anal. Appl.*, 348:690–705.

Trutschnig, W. (2012). Some results on the convergence of (quasi-)copulas. *Fuzzy Sets and Systems*, 191:113–121.

Trutschnig, W. (2013). On Cesáro convergence of iterates of the star product of copulas. *Statist. Probab. Lett.*, 83:357–365.

Trutschnig, W. and Fernández Sánchez, J. (2012). Idempotent and multivariate copulas with fractal support. *J. Statist. Plann. Inference*, 142:3086–3096.

Trutschnig, W. and Fernández Sánchez, J. (2013). Some results on shuffles of two-dimensional copulas. *J. Statist. Plann. Inference*, 143(2):251–260.

Trutschnig, W. and Fernández Sánchez, J. (2014). Copulas with continuous, strictly increasing singular conditional distribution functions. *J. Math. Anal. Appl.*, 410(2):1014–1027.

Úbeda Flores, M. (2008). On the best-possible upper bound on sets of copulas with given diagonal sections. *Soft Comput.*, 12:1019–1025.

Valdez, E. A. and Xiao, Y. (2011). On the distortion of a copula and its margins. *Scand. Actuar. J.*, 4:292–317.

Vitale, R. A. (1990). On stochastic dependence and a class of degenerate distributions. In *Topics in statistical dependence*, volume 16 of *Lecture Notes—Monograph Series*, pages 459–469. Institute of Mathematical Statistics, Hayward, CA.

Vitale, R. A. (1996). Parametrizing doubly stochastic measures. In Rüschendorf, L., Schweizer, B., and Taylor, M. D., editors, *Distributions with fixed marginals and related topics*, volume 28 of *Lecture Notes—Monograph Series*, pages 358–366. Institute of Mathematical Statistics, Hayward, CA.

Vitali, G. (1905). Sulle funzioni integrali. *Atti R. Accad. Sci. Torino Cl. Sci. Fis. Mat. Natur.*, 40:1021–1034. Reprinted in *G. Vitali, Opere sull'analisi reale e complessa. Carteggio*, pages 205–220. Cremonese, Roma, 1984.

von Neumann, J. (1953). A certain zero-sum two-person game equivalent to the optimal assignment problem. In *Contributions to the theory of games, vol. 2*, Annals of Mathematics Studies, no. 28, pages 5–12. Princeton University Press.

Walters, P. (1982). *An introduction to ergodic theory*, volume 79 of *Graduate Texts in Mathematics*. Springer, New York/Heidelberg/Berlin.

Whitehouse, M. (2005). How a formula ignited market that burned some big investors. *The Wall Street Journal*. Published on September 12.

Williams, D. (1991). *Probability with martingales*. Cambridge University Press, Cambridge/New York.

Williamson, R. E. (1956). Multiply monotone functions and their Laplace transforms. *Duke Math. J.*, 23:189–207.

Wolff, E. F. (1977). *Measures of dependence derived form copulas*. PhD thesis, University of Massachusetts.

Yang, J., Chen, Z., Wang, F., and Wang, R. (2015). Composite Bernstein copulas. *Astin Bull.*, in press.

Zadeh, L. A. (1965). Fuzzy sets. *Inform. and Control*, 8:338–353.

Index